Sustaining Cape Town

Imagining a Livable City

Edited by Mark Swilling

In association with

Amy Davison (Research Coordination)

Luke Metelerkamp (Photography Coordination)

Barbara Nussbaum (Content Editor)

Lisa Thompson-Smeddle (Research Coordination)

SUSTAINABILITY INSTITUTE

SUN PRESS

Sustaining Cape Town – Imagining a Livable City

Published by SUN MeDIA Stellenbosch and the Sustainability Institute

Coordinators: Mark Swilling (Editor), Lisa Thompson-Smeddle (Research Coordination), Amy Davison (Research Coordination), Barbara Nussbaum (Content Editor), Luke Metelerkamp (Photography Coordination)

Funding provided by the Sustainability Institute and the City of Cape Town
www.sustainabilityinstitute.net

First edition 2010
ISBN 978-1-920338-30-5

Cover artwork graffiti artists: mode2, falko, dreph, faith47 & phiks;
Cover photograph: Jim T'Water Naude
Cover design: Luke Metelerkamp and Nelia Landman

Photographic credits: Peter Hardcastle (p 276), Luke Metelerkamp (pp 24, 82, 136, 160, 184, 198, 212, 228, 244 & 270), Mandla Mnyakama (p 42), Jim T'Water Naude (p 60), Karen Schimmering (p 4) & Anna Telford (p 102)

Artistic design by Elbie Els
Typesetting by Nelia Landman
Publisher: Liezel Meintjes

Set in Amerigo BT 9/11

SUN PReSS is an imprint of SUN MeDIA Stellenbosch. Academic, professional and reference works are published under this imprint in print and electronic format. This publication may be ordered directly from www.sun-e-shop.co.za.

www.africansunmedia.co.za
www.sun-e-shop.co.za

Acknowledgements

I would like to acknowledge in the first instance the City of Cape Town and the Sustainability Institute for providing the funds for the commissioning of the papers that culminated in these chapters and the Sustainability Institute specifically for providing the funds for the editing and publishing of the book. In particular, the sustained effort of Lisa Thompson-Smeddle from the Sustainability Institute and Amy Davison from the City of Cape Town are acknowledged.

The authors of the chapters need to be thanked for their ongoing willingness to deal with four separate editorial rounds – initially in response to the blind reviews of each of the original papers prepared for the workshop with City of Cape Town officials, then a second round of detailed content edits, a third round of blind reviews of the entire manuscript, and a final round of technical edits. The diligent efforts of the reviewers has contributed significantly to the improvement of the chapters and the careful work of the content editor, Barbara Nussbaum, made it possible to link what were a diverse range of contributions written from the perspective of very different disciplines to the introductory chapter.

Luke Metelerkamp needs to be thanked for all the creative energy he put into the process of assembling the photographs that make this such a handsome book. Paul Weinberg and his photography students are also thanked for the access they gave to their photographs, and *mode2, falko, dreph, faith47* and *phiks* are thanked for allowing us to use their graffiti for the book cover.

Finally, I would like to thank Liezel Meintjes from Sun Media for her patience and consistency and Ruenda Odendaal from the Sustainability Institute for all the administrative support during the last stretch of the project. Last, but not least, I want to thank all the officials and experts who gave of their valuable time to talk to the researchers and to those Capetonians who talked to the researchers and who agreed to be photographed.

Mark Swilling
18 October 2010

Contents

Preface 1

1. Sustainability and a Sense of the City: Ways of Seeing Cape Town's Futures 3
Mark Swilling

1. Cape Town's Growth 5
2. An unsustainable mess 6
3. Second Urbanisation Wave – Global Context 8
4. The Emergence of Slum Cities 10
5. Sustainability and the Polycrisis 11
6. Towards Green Economies 12
7. Decoupling Resources and Environmental Impact 14
8. Cities and Sustainability 15
9. Rethinking Cape Town 17
10. Imagining a Sustainable Cape Town 18
11. The Focus of this Book 18
12. Concluding Thoughts 20
References 21

2. Municipal Finances, Service Delivery and Prospects for Sustainable Resource Use in
Cape Town 23
Mark Swilling & Martin de Wit

1. Core Challenges 25
2. Overview of Municipal Finances 28
3. Services for all: from Uniformity to Contextual Diversity 30
4. Sectoral Approaches 34
5. Conclusion: Financial Implications of Sustainability Interventions ... 36
References 38

3. Cape Town DenCity: Towards Sustainable Urban Form 41
Kathryn Ewing & Nisa Mammon

1. Introduction 43
2. Challenging Sustainable Urban Form in Cape Town 44
3. The Compact City – a sustainable response to urban form 48
4. Conclusion 57
References 57

4. Social Justice and Sustainable Use of Natural Resources in Cape Town 59
Mazibuko Jara

1. Introduction 61
2. Defining social justice and injustice 62
3. Social injustice, inequality and human development in Cape Town ... 63
4. The past and present causes and drivers of social injustice and inequality in Cape Town 67
5. Sustainable livelihoods, sustainable development and social justice in Cape Town 70
6. Cape Town: the link between social justice and the consumption of ecological resources 71
7. Applying the concepts of sustainable livelihoods and sustainable development to achieve social justice
in Cape Town 73
8. Social justice in Cape Town: Practical suggestions 75
9. Conclusion 78
References 78

5. Options facing Cape Town from a 'Sustainable City' Perspective: Economic and
Industrial Development 81
Mazibuko Jara

1. Introduction 83
2. Implications of the Global Economy for Cape Town 83

3. Cape Town recognises the challenges 83
4. The 'sustainable city' model 85
5. Cape Town economy: a continuing 'consumption city' 86
6. Size, sectors and growth 86
7. Towards a 'sustainable city' model for economic growth and development 92
8. Conclusion 99
References 99

6. Water and Sanitation in the City of Cape Town: Sustainability Uncertain 101
Kevin Winter

1. Introduction 103
2. Troubled waters: an overview of the City's water supply and demand 104
3. Imperatives of sustainability 106
4. Cape Town: the unsustainable city 108
5. Towards sustainability 110
6. Water and Sanitation options 111
7. Conclusion 113
Acknowledgements 114
References 114

7. In This Changing City: Exploring Urbanisation in Cape Town 115
Luke Metelerkamp

List of Images 119

8. Sustainable Energy 135
Frank Spencer

1. Introduction 137
2. Energy Policy and Cape Town 137
3. Current Cape Town Energy Scenario Baseline 139
4. Current Projections for a Future Scenario 145
5. Future Cape Town Energy Scenario 151
6. Technical Interventions 155
7. Key Goals for Sustainable Energy 156
8. Conclusion 157
References 158

9. Sustainable Integrated Solid Waste Management 159
Sally-Anne Engeldow

1. Introduction 161
2. Definitions of waste 162
3. Solid waste management in the City of Cape Town 163
4. Waste Streams 168
5. Integrated waste management hierarchy 170
6. Key challenges in terms of sustainability 174
7. Opportunities for Sustainable Waste Management 177
8. Future Plans 180
References 180

10. Sustainable Transport Options: Passenger Transport in Cape Town 183
Roger Behrens & Peter Wilkinson

1. Introduction 185
2. Cape Town's transport system: an overview 185
3. Cape Town's transport system and the environment 188
4. Implications for the sustainability of transport operations in Cape Town 191
5. Towards more sustainable resource use in the transport sector 192
6. Conclusion 194
References 194

11. Natural Space and City Growth 197
Matthew Cullinan

1. Introduction 199
2. Natural Space and Sustainable Urban Development 199
3. Definitions and concepts 200
4. Functions of natural and green open space 205
5. Quality and Accessibility – how much is enough? 206
6. Strategic Direction 207
7. Conclusion 210
References 210

12. Urban Agriculture in the City of Cape Town 211
Gareth Haysom

1. Introduction 213
2. Global context: the call of sustainable agriculture 213
3. Cape Town's Food Security Challenge 214
4. Why urban agriculture? 215
5. Planning for sustainable urban agriculture 216
6. The City of Cape Town's Urban Agriculture Policy 217
7. Urban agriculture and the City of Cape Town: where to from here? 218
8. The call for an integrated approach 219
9. Creating an enabling environment 220
10. Complementary strategies for integrated implementation 220
11. Conclusion 222
References 222
Annexure 1: Urban Agriculture Integrative Policy Interventions 224

13. Sustainable Quality: Architecture for a Sustainable Cape Town 227
Mokena Makeka & Peta Brom

1. Introduction 229
2. Challenges to Innovation for Green Architecture 230
3. Looking forward – Recommendations 237
4. Case Study 1: The Public Transport Shared Services Centre 238
5. Case Study 2: The Oude Molen Eco Village 240
6. Conclusion 242
References 242

14. Governance, Housing and Sustainable Neighbourhoods 243
Paul Hendler

1. Introduction 245
2. The role of the state vis-à-vis the people 246
3. Sustainability in an affordable housing context 247
4. Delivery mechanisms to generate sustainable affordable housing 249
5. Analysing Cape Town's housing situation from a sustainability perspective 253
6. Transforming the political economy of housing 261
7. Strategic funding options 263
8. Conclusion 264
Appendix: Housing Delivery Typologies 266
References 267

15. Limits, Flows and Networks: Some Reflective Conclusions 269
Mark Swilling

1. Resource limits and innovation 272
2. Flows and structures 273
3. Networks and innovation 273
References 275

Contributors 277

Preface

The publication of *Sustaining Cape Town* contributes to a growing body of writing about the City of Cape Town that contributes to – and reflects on – a remarkable set of interlinked dialogues that are shaping the way citizens, journalists, policy-makers, regulators, investors, educators, care givers and visitors are starting to re-imagine Cape Town's possible futures. Cities do not change overnight, nor do they change just because the City Council adopts a new Integrated Development Plan. Cities seem to change very quickly while at the same time they appear not to be changing at all. What really fundamentally affects the future of the city is the gradual emergence of a shared way of thinking about the future – or what could be called a shared imaginary. This is not easy to achieve because a share imaginary is not the equivalent of a formal agreement about a policy framework reached between leadership elites. Instead it is something that is far deeper and is not really all that dependent on everyone agreeing. A shared imaginary within a city is deeper because it tends to emerge from countless interactions, dialogues, engagements, conflicts and seemingly pointless convictions to improve things that no single major interest or power can really control. A shared imaginary tends to become a kind of share framework of reference for adjudicating debates and disagreements without always having to refer back to formal policy documents. Yes, there are major players, many of which are discussed in one way or another in this book. But these major players rarely have as much influence as they would like to have.

Sustaining Cape Town has been a long-time coming and should be read as a major contribution to a particular trajectory within Cape Town's multiple dialogues about the future of the city, namely the view that Cape Town should become a sustainable city. Significantly, the original set of papers that have now been published here as chapters of this book were jointly commissioned by the City of Cape Town (CCT) and the Sustainability Institute. The CCT made it clear that the researchers should be independent of the council which meant they were neither former employees of the CCT nor regularly appointed as consultants. The brief to the researchers was to reflect on what was unsustainable about Cape Town and what options the city faces in the search for ways to become more sustainable. The papers were presented to officials from the various CCT Departments and then rewritten and edited without any editorial interference by the CCT. The result is a set of chapters that ask some very tough questions and which make some inconvenient suggestions. This is, however, achieved in a way that accepts as points of departure the various stated intentions about sustainability to be found in many policy suggestions published by the CCT, the Western Cape Provincial Government, the Cape Town Partnership, many University research centres and various civil society groupings and many other key players.

Although very rare, surely this must be exactly the role that a City Council must play within a city faced with socio-economic and ecological challenges that it cannot ever hope to resolve acting alone. By working in partnership with independent Cape Town-based researchers (based within and outside the Universities), the CCT has effectively legitimised the rich and deep process of dialogue that is so crucial for building a share imaginary. As significant, it has effectively created a space for intellectuals to reflect on the city in an engaged way so that they can ask questions and make suggestions that reflect aspirations or technical challenges that have not been adequately addressed during the busy formal proceedings of City Council meetings or the clipped business meetings in the boardrooms of the major financial institutions.

What is interesting, though, is while many of the city's most influential political, business and social actors agree with the normative idea that Cape Town should be a more sustainable city, there is very little critical reflection on what this means in practice. Too often there are very formalised definitions that refer to the need to integrate economic, social and environmental issues, but what does this verbiage refer to in practice? For many developers it means doing the Environmental Impact Assessments (EIAs) properly, but there are very few examples where the development plan was changed after the EIA report was completed. For many environmental NGOs it means protecting the conservation areas from developers. For the CCT it means many things, including *inter alia* densification, more public transport, investments in renewable energy and more waste recycling.

Sustaining Cape Town does not fall into the trap of offering a neat definition of sustainability. Instead, the different chapters are explorations of what more sustainable use of the City's various natural resources and ecosystems could mean and what the consequences would be of a business-as-usual approach. Most significantly of all, many of the chapters systematically explore in substantial technical detail a range of issues that have never been addressed before in this way. Whereas most studies that are interested in sustainability tend to address environmental impacts and how these can be mitigated, many of the chapters in this book address the socio-technical infrastructures that conduct the flow of resources through the city. The chapters on energy, water and sanitation, solid waste and transport, for example, are primarily interested in how these infrastructures are configured and what practical options can be considered that could result in a very different – yes, more sustainable – flow of resources through the city and, where applicable, their re-use. Read together with the other chapters on open space, densification, social justice and finance, the lasting impression is that sustainability is really much more about recognising limits as opportunities for innovation that could stimulate growth, more jobs and greater levels of equity.

I therefore strongly recommend this book to all Capetonians who have an interest in the future of the city, and to all those who make decisions about the way we use resources to build and operate our city.

Professor Edgar Pieterse
Director – African Centre for Cities
University of Cape Town

Sustainability and a Sense of the City

Ways of Seeing Cape Town's Futures

Mark Swilling

"I have known all along that the further we move away from April 27th, 1994, we'll become increasingly normal and more ordinary, shedding our well-deserved sense of specialness which we earned from our 'miraculous' transition … I am disappointed that we do not appear to have succeeded in defining the terms of our ordinariness. Our own brand of ordinariness ought to work at a higher level."

Njabulo Ndebele, from Njabulo Ndebele: *An artist interrupted*,
Cape Argus, Thursday, April 3, 2003

As we live in, move through, see, smell and hear the city all around us, we experience an extraordinary tension between its seemingly fixed 'sunk-in-concrete' structure, and the incessant movement, noise, activity and flow. While it looks the same every day, it never stops moving. When we imagine our city, we think of its structure and physical place (and even beauty), but we live everyday in the nooks and crannies of our homes, workplaces, means of transport and our leisure-time places. City living is always about being somewhere else and yet in the same place all the time.

But what changes the city? How does it shift in character from being one thing to another? When do we say, "Our city has changed"? And how do we imagine this change coming about? Are we even aware that 'the-city-as-structure' is changing? If so, do we like what is happening? Should other things be happening, and if so, who should be making these things happen? Do we think our children and grandchildren will be able to live in our city in 50 or 100 years from now? What impact do our individual and collective choices bring to the city of Cape Town? How do our actions – conscious and unconscious, informed and ignorant – shape the future of our city?

These may seem like strange questions, but in many ways we live out these questions every moment of every day as we traverse the capillaries and interstices of the city.

1. Cape Town's Growth

Are we aware that between 1977 and 2006, the size of Cape Town (in land area) increased by 40%? In other words, nearly half of what is now Cape Town has been built in the last 25 years. This remarkably rapid rate of urban change is reflected in the names of new suburbs that every Capetonian now takes for granted as part of everyday urban life: Parklands and Plattekloof; Pinehurst and Philippi East; Sunningdale and Soneike; West Beach, Wesbank and Wimbledon; Blouberg Sands and Blommendal; Bonnie Brae and Bloekombos; Dunoon, Delft, Durbanvale and Durmonte, Sunset Beach, Sunset Links and Summer Green; Marconi Beam and Milnerton Ridge; Milnerton Race Course and Mfuleni Happy Valley; Welgedacht, Oude Westhof and Wallacedene; Vierlanden, Vredekloof, Van Riebeeckshof; Phoenix, Joe Slovo Park; Glenwood and Goedemoed, Sonstraal Heights, Langeberg Ridge, Kleinbosch and Kleinbron, Okavango Park, Fisantekraal, Protea Village, Zevenwacht, Uitzicht, Kalkfontein and Khayelitsha: Harare and Southern 2/3, Nomzano, Lwandle, Weltevreden Valley, Browns Farm, Vrygrond, Houtbay Suburbs and Imizamo Yethu; Sunnydale and Capri (Sun Valley area).

The icons of this suburban growth are not the streets or community character, the architectural inspiration of the buildings, or the green spaces between them. No, if we want to take visitors to see the 'streets of Cape Town' we go to the Bo-Kaap, or Sea Point, Kalk Bay or the V&A Waterfront – spaces that 'have character' because they reflect the diversity of the city. By contrast the icons of the 'new Cape Town' are the malls and industrial areas: Century City, N1 City, Cape Gate, Somerset Mall, Longbeach Mall, Westgate Mall, Access Park, Khayelitsha Mall, Killarney Gardens and Montague Gardens. Yes, there are also investments in spaces for more public engagement than malls – Convenco,

Newlands Rugby/Cricket, V&A Waterfront, Urban Park (Cape Point to Table Mountain), Greenpoint Stadium, and the revamping of the CBD – which together play a much bigger role in reshaping the symbolic significance of what could be called Cape Town's distinct urbanism.

2. An unsustainable mess

But as things change, so much just remains the same. The city consumes 900 hectares per annum of undeveloped land (equal to 3600 soccer fields), but still 20% of the richest residential property owners occupy 41% of the residential land area (City of Cape Town, 2006). For the estate agents who market the northward-spreading suburbia along the Atlantic seaboard to white families fleeing increasingly multi-racial Cape Town suburbs, the implicit message is always very clear: "Democracy means that blacks will take over the city, so be prepared and move here."

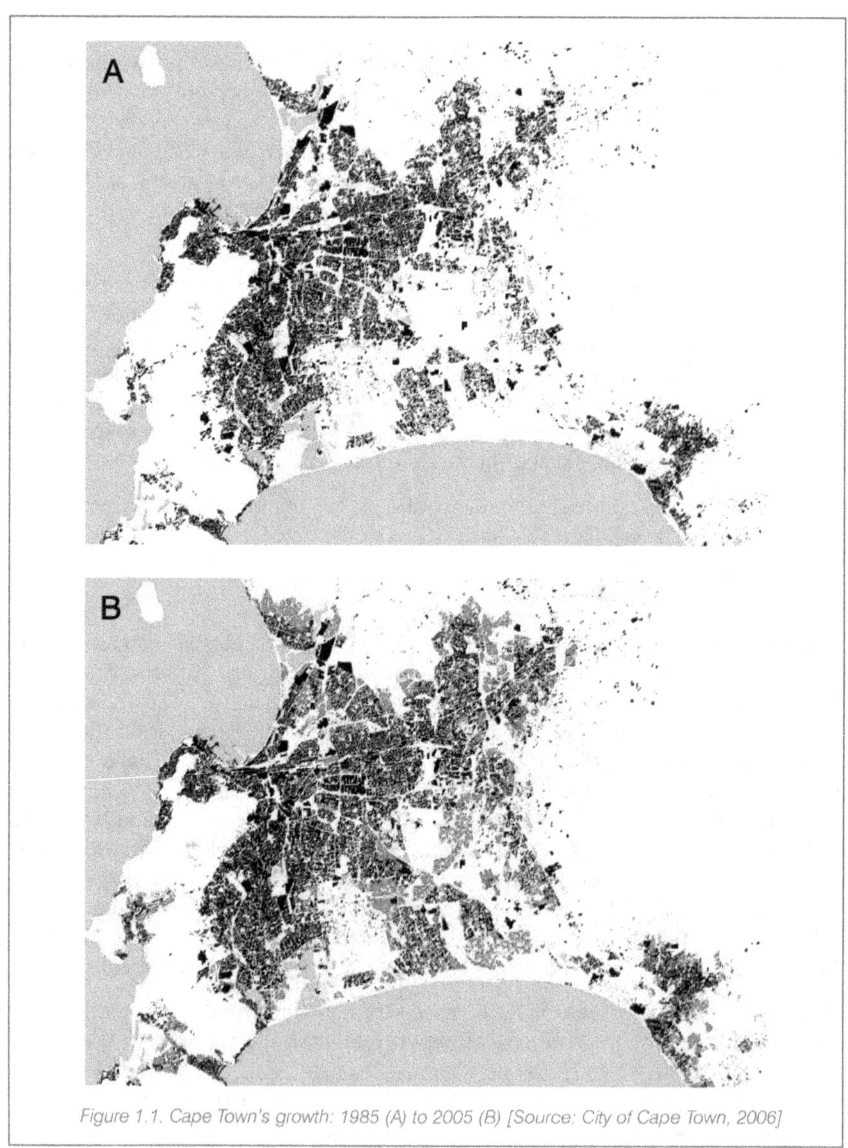

Figure 1.1. Cape Town's growth: 1985 (A) to 2005 (B) [Source: City of Cape Town, 2006]

In reality, the black majority are confined to increasingly densifying 'townships' that have effectively become the centre of the Cape Metropolitan Area – now stretching from Atlantis up the North Coast to Somerset West to the south. Landlocked by increasingly costly industrial land, it is not townships that are pushed onto the peripheries of the metropolitan area, but rather developments for the burgeoning lower middle classes. Though spatially dispersed, these groups generate the 20% return on investment for developers who have burnt their fingers in both luxurious developments for the rich and mass housing for the subsidised poor.

Of the 20 000 or so new residential units that are built in Cape Town each year, roughly two thirds are for the poor (mostly subsidised), while the remaining third is for the private market (mostly 'lower' and 'middle' middle classes). But this is woefully inadequate: the housing backlog for the urban poor is nearly 300 000 units, while every estate agent knows that there is an under-supply of housing for the 'lower' and 'middle' middle classes. The upper middle class and elites are over-supplied, hence the diminishing returns on investment at these levels since at least 2005. Basil Read, one of South Africa's largest construction companies, is putting together a development for 14 000 units near Durbanville on the N1 – unsurprisingly, this includes subsidised housing, 'gap housing' and lower-middle class housing, because this is where it is still possible to make a 20% or more return on investment.

The massive and rapid urban growth of Cape Town has not been coupled to investments in the kinds of urban infrastructure, like energy and transportation systems, that are appropriate for a world that is running out of atmosphere, water, oil, and sinks for liquid, solid and airborne waste. To make matters worse, low density urban sprawl has effectively required massive public investments and de facto subsidies to continuously extend the infrastructures that carry the resources required by contemporary city living and working – roads, electric cables, water and sewage pipes, telephone and fibre optic cables, storm water drains and, in a few places, railway lines. These are the infrastructures that conduct the resource flows through the city – with ecological sources and sinks on either end – and the connections to these metabolic flows is what we call 'service delivery'. When the poor don't get connected to these networks, we have what we call 'service protests'.

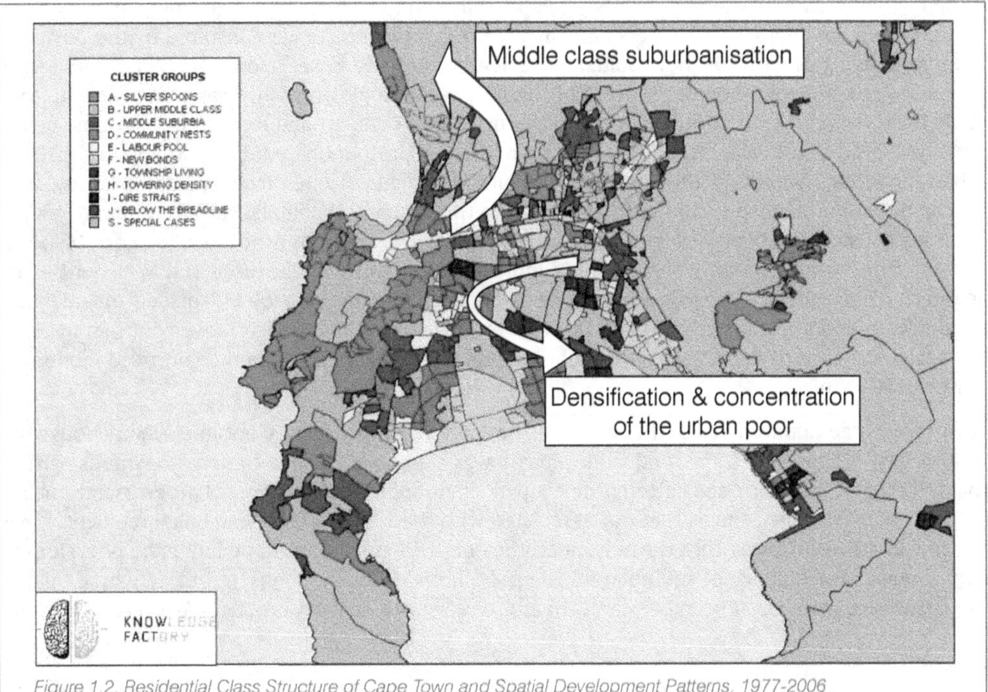

Figure 1.2. Residential Class Structure of Cape Town and Spatial Development Patterns, 1977-2006 [Source: Swilling, 2006]

But who designs these infrastructures? What technical criteria are used? And are these technical criteria ever questioned through social or political processes? Or is it assumed that engineers know better, so what they propose must be the best option? Maybe it is time to question whether the way the 900 hectares per annum of land is sliced and diced by these infrastructures is the best use of our so-called 'resources'? Are there perhaps other ways of managing the metabolic flows of sewage, water, electrons, storm water and data through our city that are cheaper, less resource intensive, more environmentally friendly, and easier to operate? Maybe we need to densify rather than sprawl? Where is the public debate about these technical criteria? How many researchers investigate the claims made about how best to design and construct infrastructures that lock us into particular modes of movement, electricity consumption, waste disposal and leisure?

These questions are not specific to Cape Town. They are the questions that confront anyone thinking about a longer-term future for the cities we inhabit in the world. They are questions we need to consider, in view of the fact that it will be an overwhelmingly urban world subverted by increasingly serious ecological challenges. Over time, all these changes will lead to fundamental shifts in our assumptions about what it will mean to live in the city. Those cities that choose to invest in a future that looks very different to the present will become the incubators of a more sustainable world.

3. Second Urbanisation Wave – Global Context

The first urbanisation wave took 200 years – 1750 to 1950 – and resulted in an increase in the number of urban dwellers in Europe and North America from 15 million to 423 million people (United Nations, 2006). It also resulted in the iconic images of 'the modern city' as architects, engineers and planners gave cultural form to this radical transformation of everyday life and work.

Marshall Berman's great classic text took its prescient title – All that is Solid Melts into Air – from a line in the Communist Manifesto. His work revealed the intimate connection between the rise of modernity – as the aesthetics of the new middle classes were shaped by the Northern industrial revolutions – and the drive to transform pre-industrial cities into paragons of urban modernity (Berman, 1988). This powerful movement – complete with Faust as its mythical hero – has its origins in Haussmann's Paris, Peter the Great's St. Petersburg, and the New York that Robert Moses built. All three were archetypal 'developers', and as extreme reductionists were inspired by the power of rational planning. How easily they could impose neatly measured boxes and straight lines onto messy complex spaces! Inspired by the powers of the new technologies of light, water and motion, they could build the perfect, frictionless, modern city using law, cash, art and force. It is a movement that gets mass-produced in the incarnations of Le Corbusier's profoundly reductionist, one-dimensional urban modernity, which, in turn, inspired the urban design of cities from Atlanta to Brasília. In Africa, modern urban expression either turned into the formalised colonial suburbs of many African cities, or became contrived new capitals, like Abuja, Nigeria. In South Africa, we had Lord Milner's[1] 'Kindergarten' who planned and built early Johannesburg, and we witnessed H.F. Verwoerd – the architect of apartheid – personally planning the apartheid township of Mdantsane outside East London as a model 'garden city' for blacks. Later on it was the 'Garden City' movement that inspired the design of much of Cape Town, from suburban models like Pinelands to the contrived modernity of Mitchells Plain.

For nearly three hundred years, the image of the modern industrial city became synonymous with what a 'city' is supposed to be. And what the city was supposed to be became synonymous with all that is 'civilized', 'western' and (latterly) 'developed'. It embodied the meaning of progress, rationality, secularism, universality, and all that was associated with the Western European Enlightenment. It has become in the new South Africa the only measurement of what it means to no longer be poor: for the rich, it means the defence of suburbia; for the poor, it means being connected to networked urban infrastructures (secured, if need be, via the ubiquitous 'service delivery protest'). All other cities were

---

1   This is what the group of young whizz kids were called that Lord Milner recruited to design and build post-Boer War Johannesburg as an icon of British colonial order.

either getting there, or – like in the case of the great pre-colonial African cities or historic core cities in many Islamic countries – were entirely forgotten or denied the status of being cities in their own right (Malik, 2001). The awesome power of this historically constructed lens is what makes it all the more difficult to 'see' the cities being created by the second urbanisation wave, as slum dwellers quietly make (Bayat, 2000) their way into ever-wider connections to the infrastructures and resource flows (especially water, energy and transport) of the city.

The second urbanisation wave will take less than 100 years – 1950 to 2030 – to be complete, and is taking place in developing countries where the urban population is projected to grow from 309 million to a staggering 3.9 billion (United Nations, 2006). In China alone, more people will become urbanised in 50 years than were urbanised in North America and Europe put together in over 200 years. As the global population increases from 6 to 9 billion, with the bulk of this increase located in Africa and Asia, this means that it is the urban centres of Africa and Asia (some of which don't exist yet) that will be home to the additional 3 billion people expected on the planet by 2050. Consequently, pressures will be greatest where the urban and institutional infrastructure is weakest, and where the cultural memories needed for urban living must still be created.

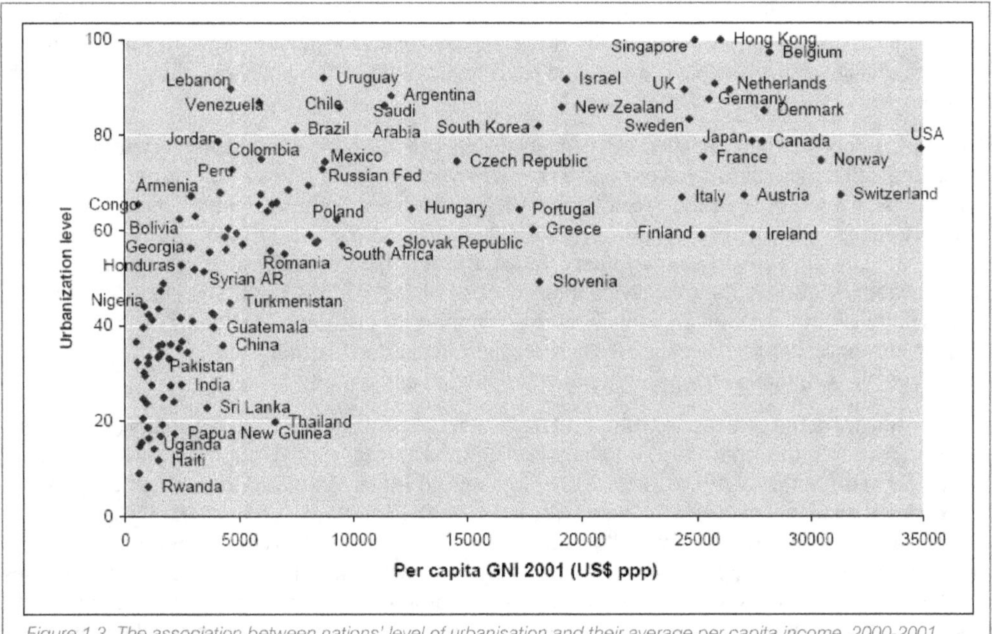

Figure 1.3. The association between nations' level of urbanisation and their average per capita income, 2000-2001 [Source: Satterthwaite, 2007]

For some, this confirms the sum of all fears: cities are the irredeemable sites of unsustainable consumption (Low *et al.*, 2000) and dystopic social marginalisation (Davis, 2005). There is undoubtedly some truth is this requiem for urban modernity, especially given the pattern demonstrated in Figure 1.3. Correlations between levels of urbanisation and growth in per capita GDP may be a proxy for development, but it is also a proxy for a process that has created the billion or so urban-based über-consumers that drive global ecological destruction. But we share the view of those who see cities as potential for something different. After all, by virtue of the shear concentration of intellectual resources for innovation created by an urban-centred science and education system, they provide the ideational, cultural and institutional context for fostering imaginaries about more sustainable futures.

Part of the reason for optimism is that contrary to popular belief, the urbanisation trends across all continents are not creating more mega-cities, but rather a rich global patchwork of smaller cities of around a million people or less. It is possible to estimate that, if current trends continue, by 2015 nearly 60% of the total urban population across high, low and middle income countries will live in cities of less than a million people. Approximately 25% will live in cities of 1-5 million people, and 5% will live in cities of 6-10 million people, while only 10% will live in cities of 10 million people or more (United Nations, 2006; calculated from National Research Council, 2003). Cape Town, in this context, is typical of the future: it is unlikely that it will become a megacity.

Although the bulk of future urban growth will be in the developing world, only 17% of these cities are growing at 4% per annum or more, while the majority (36%) is growing at a more moderate 2-4% per annum. What is staggering though, is that 40% of cities in the developed world actually shrunk during the 1990's (UN Habitat 2008b: 11), whereas 17 big cities (of 5 million people and larger) in the developing world also contracted during the decade leading up to 2000 (UN Habitat 2008b: 36). In contrast, between 1990 and 2000, 694 new cities were established in the developing world in locations where there were less than 100 000 people before 1990 – 510 of these became small cities (100 000-500 000 people), 132 became intermediate cities (500 000-1 million), and 52 grew rapidly into big cities (1-5 million). These trends, when read together, suggest that it is far too simplistic to depict our urban future as dominated by the (often dystopic) specter of megacities as the logical outcome of a linear, irreversible urbanization trajectory from 'small to big'. Something else is clearly going on.

Perhaps it is out of these smaller, more manageable cities – both old and new, contracting and expanding – that new cultural fusions will emerge. Maybe we can look to smaller cities to inspire the innovations that will be needed to realise what Peter Evans has called "liveability" – more sustainable and more equitable modes of urban living (Evans, 2002). After all, if it was the innovations for economic growth that resulted in the extraordinary concentrations of economic, political, communicative and knowledge power in the globally connected cities of the world (Borja & Castells, 1997), maybe sustainability innovations will generate a more geographically distributed hierarchy of smaller cities. Imagine the potential of these cities! Each of them, embedded within their respective, uniquely configured, bio-economic regional systems of production and reproduction.

A cursory glance at the cities across the world that are at the cutting edge of sustainability innovations, tends to reinforce this conjecture – names like Curitiba (Brazil) (Schwartz, 2004; Campbell, n.d.), Rizhao (China) (Worldwatch Institute, 2007), Melbourne (Australia) (Worldwatch Institute, 2007), Vancouver (Canada), Freiburg (Germany) (Beatley 2000), San José (USA) (Reed, 2007), Bogota (Columbia) (Newman & Kenworthy, 2007), and Portland (USA)(Taylor 2008) immediately come to mind. In all these cases, very specific local ecological drivers have connected with economic processes and knowledge networks, culminating in surprisingly profound bio-economic diversifications with positive developmental results. In almost every case, state institutions have played a key role – not only leading the way, but also creating space for other actors to innovate.

No cities in Africa have attempted to capture for themselves a similar identity, although the various sustainability initiatives in Cape Town may well position it as a leading city in the African context.

4.   The Emergence of Slum Cities

Although an emerging urban future of smaller rather than megacities is heartening from a sustainability perspective, the bad news is that the second urbanisation wave has created a global population of slums. According to the path-breaking UN Habitat report, The Challenge of Slums, there were a billion people living in slums by 2005 (United Nations Centre for Human Settlements 2003).[2] In other words, by 2005 one in three people living in the cities of the world lived in a slum.

---

2   This figure was revised downwards to 810 000 in the State of the World's Cities Report 2008/9 because of a decision to exclude those households that have access to an improved pit latrine (UN Habitat 2008b).

A slum is defined in the UN Habitat report as a settlement made up of households that lack one or more of the following features: access to improved water, access to improved sanitation facilities (minimally, a pit latrine with a slab), sufficient living area (not more than three people sharing the same room), structural quality and durability of dwellings, and security of tenure.

Significantly, although half of all slum dwellers are in Asian cities, it is only in sub-Saharan Africa that one finds cities where the majority of the population live in slums. No less than 62% of all urban dwellers in sub-Saharan Africa live in slums, compared to Asia – where it varies from 43% (Southern Asia) to 24% (Western Asia) – and Latin America and the Caribbean – where slums make up 27% of the urban population (UN Habitat 2008b). The large majority of cities in Sub-Saharan Africa are, therefore, slum cities. Given the fact that urbanisation rates in Africa are the highest in the world at 3.3% (UN Habitat 2008a), the slum cities of Sub-Saharan Africa will be with us for the foreseeable decades. Africa is now 40% urbanised and is projected to be 60% urbanised by 2050, which translates into an increase in the urban population from the current 373 million to 1,2 billion by 2050 (UN Habitat 2008a). If African governments continue to ignore this problem (by stubbornly insisting that slum dwellers are only a problem because they refuse to go back to the rural areas), the additional 800 million urban dwellers will land up in Africa's mushrooming slums.

Africa is becoming a continent of slum cities and, in so doing it is transforming entirely what we mean when we use the word 'city' to describe quite a unique set of urban dynamics and modalities (Pieterse, 2008; Swilling et al., 2003; Simone, 2001; Simone, 2004). Indeed, for many analysts and policy makers, African cities don't deserve to be called cities at all – a position that is only tenable if you assume that the 'Western City' is the only legitimate template for defining a city. Maybe it is time to realise that the iconic image of the 'Western City' that emerged from the specificities of the first urbanisation wave has become little more than a mirage. Maybe it's time to find non-western reference points for rethinking our deepest assumptions about the purpose, meaning and impact of the city (Malik, 2001; Swilling et al., 2003). And maybe it is also time to realise that industrialisation, modernisation, and (from the late 1980's onwards) high-tech informationalism – the traditional economic drivers of urbanisation – have not been the primary driving forces of African urbanisation and the emergent urbanisms we see across the diverse cities of the continent.

Something far more complex and difficult to grasp is going on when it comes to the making and shaping of slum urbanism (Pieterse, 2008; Swilling et al., 2003; Simone, 2004). Grasping this is key to understanding the ongoing struggles between poor communities – who forge their own means of urban survival – and the vast array of actors who share an interest in urban regulations that aim to limit and confine these encroachments to specific places, so that property values and middle class lifestyles can be preserved. Unlike most African cities, slum urbanism does not characterise how the majority live in Cape Town. The large majority of Cape Town's residents are connected to urban services, even though a third still live in shacks. This partial inclusion of the urban poor via connections to the urban infrastructure networks – without the simultaneous provision of housing – is a key aspect of Cape Town's urban development pattern.

5.   Sustainability and the Polycrisis

There is a broad global consensus that we face an unprecedented twin challenge created by inter-linked economic and ecological crises. As the economic and ecological crises mutually reinforce one another, decision makers across the public, private and non-profit sectors in both the developed and developing world intensify demands for practical solutions. A succession of global mainstream assessments over the past decade have together raised very serious questions about the sustainability of a global economic growth model that depends on material flows that have reached – or soon will reach – their natural limits (Intergovernmental Panel on Climate Change 2007; Watson et al., 2008; United Nations, 2005; Barbier, 2009; United Nations Environment Programme 2007; World Resources Institute, 2002; World Wildlife Fund, 2008; Gleick, 2006). The crises of resource depletion and the negative economic implications of climate change have even been recognized by mainstream reviews such as the Stern Report (Stern, 2007) and the International Energy Agency, which finally acknowledged in 2008 that the "era of cheap oil is over" (International Energy Agency, 2008).

Unfortunately, the scientific consensus reflected in the assessments cited above has had a very limited impact on mainstream economic policy makers connected to World Bank/IMF networks. This is most clearly evident in the institutional economics of the so-called 'post-Washington Consensus' as reflected in The Growth Report (published by the Growth and Development Commission), which brought together the most eminent actors in these networks (including Trevor Manuel, then Minister of Finance of South Africa). For this group of economists, the only constraints to accelerating traditional modes of economic growth are institutional (Commission on Growth and Development, 2008). Resource limits are ignored, except for a distracted reference to the IPCC (International Panel on Climate Change).

It is only in Europe and China that there is evidence that economic policy making is responding to the severity of the global ecological crisis. Nobel prizewinners Joseph Stiglitz and Amartya Sen headed up a commission initiated by French President Sarkozy, which recommended in 2009 that GDP is no longer an adequate measure of progress, suggesting that it be replaced by a Happiness Index (Stiglitz *et al.*, 2009) instead. This, combined with the EU's 'Green Growth' approach, is resulting in significant shifts. In October 2008, the 17th National Congress of the Communist Party of China endorsed President Hu Jintao's call for the building of a new "ecological civilization", which the Party has defined as the "summary of physical, spiritual and institutional achievements of human beings when they develop and utilize natural resources while taking initiative to protect nature." (Shengxian, 2009) This has become the cornerstone of China's 'green development' strategy, which is the strategic focus of China's massive economic recovery package.

Unfortunately, in his Medium Term Budget Policy Statement in October 2009, South Africa's new Finance Minister, Pravin Gordhan, ignored the global scientific consensus about the ecological limits to growth and explicitly endorsed The Growth Report. This may change, especially given the fact that South Africa tabled mitigation targets at the Copenhagen Climate Change negotiations in December 2009 that do recognise these limits. There is a major tension within the South African policy-making system between a 'jobs through growth' perspective that ignores the opportunities created by the 'Green Growth' perspective, and a traditional 'environmental impact' perspective that assumes that all development is bad for the environment. What is needed is a paradigm shift that makes it possible for South Africa to catch up to the rest of the world. That, interestingly enough, is most likely to happen at the city level where – like in Cape Town – real resource constraints are forcing local decision makers and citizens to rethink the old paradigms. Sustainability from this perspective is not a constraint, but a new way of seeing constraints as opportunities.

6.   Towards Green Economies

As global discussion of a 'Green New Deal' gathers momentum following its adoption by the G20 at the Pittsburg meeting 2009, the key dimensions of the 'polycrisis' (Morin, 1999) are finally being acknowledged.

In 2008, the International Energy Association predicted that demand for oil will increase by 45% by 2030, without any evidence that it will be possible to find this amount of oil at affordable prices. Increasingly, the costs of extraction across the world's oil fields will rise as the quality and quantity of reserves decline, thus further undermining traditional drivers of economic recovery (International Energy Agency, 2008).

The Millennium Ecosystem Assessment (MEA) found that 15 out of 24 key ecosystem services are degraded or used unsustainably, often with negative consequences for the poor – 1.3 billion people live in ecologically fragile environments located mainly in developing countries, half of whom are the rural poor (United Nations, 2005).

It has been estimated that the combined value of the fiscal stimulus packages assembled by the G20 is between US$2 and 3 trillion, or 3% of global GDP (Barbier, 2009). If these stimulus packages focus exclusively on economic recovery and ignore global warming, ecosystem breakdown, the end of cheap oil, and global poverty, the outcomes will contradict the original recovery intentions. Of the US$827 billion to be spent by the US Government, US$100 billion has been allocated to green investments. South Korea has an ambitious Green New Deal investment package worth US$36 billion which comprises the bulk of its recovery package, and China has earmarked 38% of its recovery package (equal to 4 trillion yuan) for green technology investments.

These examples testify that the causes of the current crisis are being recognised as far more complex than merely short-term economic factors. They are also investments that will drive unprecedented rates of innovation as governments and private sector players strive to convert these investments into competitive advantages within the global economy.

This logic was reflected in a shift in global ideological discourse expressed most clearly by UK Prime Minster Gordon Brown when he wrote in Newsweek:

"There can be little doubt that the economy of the 21st century will be low-carbon. What has become clear is that the push toward decarbonisation will be one of the major drivers of global and national economic growth over the next decade. And the economies that embrace the green revolution earliest will reap the greatest economic rewards. ... Just as the revolution in information and communication technologies provided a major motor of growth over the past 30 years, the transformation to low-carbon technologies will do so over the next. It is unsurprising, therefore, that over the past year governments across the world have made green investment a major part of their economic stimulus packages. They have recognised the vital role that spending on energy efficiency and infrastructure can have on demand and employment in the short term, while also laying the foundations for future growth."

Newsweek, 28 September 2009

In his 2008 Budget Speech, the then South African Minister of Finance, Trevor Manuel, expressed a remarkably similar and far-sighted analysis when he said:

"We have an opportunity over the decade ahead to shift the structure of our economy towards greater energy efficiency, and more responsible use of our natural resources and relevant resource-based knowledge and expertise. Our economic growth over the next decade and beyond cannot be built on the same principles and technologies, the same energy systems and the same transport modes, that we are familiar with today."

Minister of Finance, Parliament, 20 February, 2008

The assessment reports of the IPCC and the MEA succeeded respectively in placing carbon and ecosystem services at the centre of the global discussion about more sustainable futures. However, in the wake of the 4th Assessment Report of the IPCC, the United Nations Environment Programme set up the International Panel for Sustainable Resource Management (IPSRM), which has highlighted a third factor – material resource flows and their associated impacts. By resource flows we mean primarily the metabolic flow of fossil fuels, biomass, minerals and metals through the global economy. The focus is on extraction and domestic use of materials quantified in tons with the aim of analysing the relationship between economic growth and resource use. Material Flow Analysis (MFA) is the methodological tool that is used to conduct this analysis. MFA has matured over the past five years and has become an established method for assessing the sustainability of local, national and global economies (see Krausmann *et al.*, forthcoming; Krausman *et al.*, 2008; Haberl *et al.*, 2004; Fischer-Kowalski 1999; Fischer-Kowalski 1998; Bringezu & Bleischwitz 2009; Bringezu *et al.*, 2004; Bringezu & Schutz 2001).

Using MFA, it has been possible to compute global material flows for the first time. These flows include fossil fuels, biomass, minerals and metals. Over the period of accelerated globalisation and financialisation since 1980, total extraction of all these resources has increased by 36%, from 40 billion tons in 1980 to 55 billion tons in 2002 (see Figure 1.4).

Increased resource extraction did not, however, result in greater equity. The 1998 United Nations Human Development Report demonstrated that by the end of the 1990's, the richest 20% of the world's population was responsible for 86% of consumption expenditure, whereas the poorest 20% were responsible for only 1.3% of consumption expenditure (United Nations Development Programme, 1998). In other words, as total domestic consumption shot over the 50 billion ton/annum mark in the late 1990's, 20% of the world's population had the means to purchase 86% of this harvest. This, in turn, was made possible by declining real prices of primary resources, with negative consequences for Africa's resource-rich resource-exporting countries. Indeed, a World Bank report estimated that these prices were lower than the real value of these resources, which means that many African countries (including South Africa) have experienced a decline in their Gross National Incomes (which includes the value of financial, human and natural capital) even as exports of primary resources have increased (World Bank, 2005).

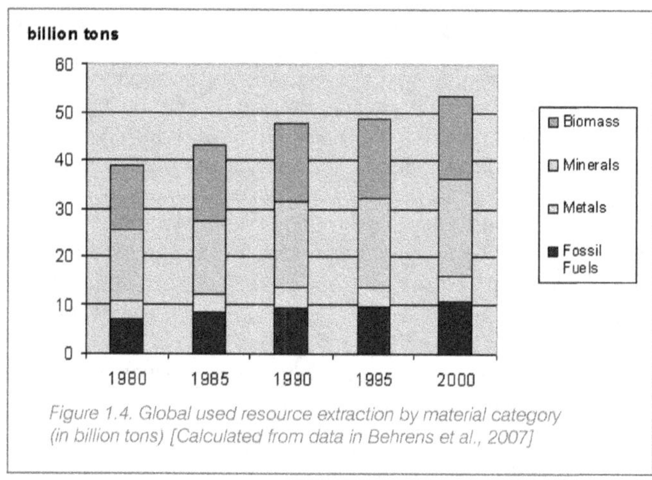

Figure 1.4. Global used resource extraction by material category (in billion tons) [Calculated from data in Behrens et al., 2007]

7.   Decoupling Resources and Environmental Impact

The IPSRM brings together two well consolidated intellectual traditions, namely industrial ecology and ecological economics. These traditions share the view that a new economic theory, which makes it possible to conceptualize the economic modalities of a "Green Economy" – i.e. an economy that grows by reducing rather than increasing resource consumption, or what in the literature is referred to as 'dematerialisation' – is required. The core concept at the centre of this work is 'decoupling'.

The notion of decoupling opens up a new way of thinking about the relationship between the rate of economic growth and the rate of resource consumption and its associated impacts. This implies two types of decoupling: 'resource decoupling', which refers to decoupling growth rates from resource extraction, and 'impact decoupling', which refers to decoupling growth rates from environmental impacts. Furthermore, it is possible to make a distinction between 'relative decoupling' and 'absolute decoupling': the former implies slower rates of resource use and impacts relative to economic growth rates, while the latter refers to negative rates of growth in resource consumption and related impacts. Policy and strategic decisions can be made that foster the relative/absolute decoupling of both resource use and impacts in contextually specific ways that reduce the significance of resource limits as a constraint to growth (Swilling *et al.*, 2009). The inevitable result will be the redefinition of growth, or what the Latin American ecological economist Gilberto Gallopin has called "non-material growth", which he distinguishes from the European notion of "zero growth" (Gallopin, 2003).

Decoupling makes it possible to think about the connections between economic growth, improved wellbeing through development, and sustainable resource use in ways that open up more creative options for innovation than has hitherto been the case.

After the Club of Rome report in the early 1970's, there was a pervasive assumption that growth was the cause of environmental destruction and therefore sustainable development would only be achieved through zero growth strategies. This conception was only tenable because the relationship between economic growth and resource use was seen to be directly proportional to one another.

For the Brundtland Commission Report, the focus was on how best to 'sustain' development (conventionally defined). In terms of this report, the only limits that mattered were institutional and technological factors, not natural resource constraints. The inter-substitutability of different forms of capital that Brundtland legitimised effectively reduced the conceptual space for decoupling. The result is that both developed and developing countries have been allowed to validate resource intensive growth paths to eliminate poverty, with the result that they are now coming up against the natural resource limits that have hitherto not received significant attention.

Both resource and impact decoupling are necessary – one without the other will prove ineffective when it comes to realising sustainability goals. In so doing, our understanding of economic growth also changes. Instead of seeing economic growth as inevitably driven only by value derived from an increasing quantity of material goods and assets, so-called 'non-material growth' (or prosperity or wellbeing) is driven by knowledge intensity and associated investments in both tangible and intangible assets:

- culture
- liveability
- education
- improved health
- environmental quality
- safety
- a sense of place, and
- individual and collective capabilities for enhancing wellbeing and freedom.

Developing countries should take very seriously the call for sustainable resource management, because it opens up new opportunities for strategic approaches to development and poverty eradication that may not have existed before. The current context actually highlights the possibilities for a wider range of strategies which now become more obvious. Sustainable resource management offers new opportunities for investments in innovations that could stimulate endogenous growth strategies in developing economies. These could be more effective in eradicating poverty than traditional strategies that depend on primary exports or exports of cheap manufactured goods, underpinned in both cases by resource depletion and/or environmental degradation. As oil prices rise, long distance trading regimes will be forced to restructure – the sooner developing countries prepare for this eventuality, the better off they will be when it happens.

Unfortunately, most developing countries think that sustainable resource management only refers to costly impact decoupling. Resource decoupling is where the economic opportunities for developing countries lie. In fact, impact decoupling reinforces adaptation with major benefits for millions of poor people who depend on the ecological sustainability of ecosystem goods and services (such as fishing, indigenous forests, good soils, and stable climates and their associated predictable rains).

8. Cities and Sustainability

The strategic centrality of cities in the global polycrisis is reflected in the importance given to cities in the 'solutions' embedded in the so-called 'rescue packages'. The evidence suggests that the publicly financed investments to stimulate the global recovery will be targeted primarily at investments to refurbish the ageing urban infrastructures of cities in the developed world, and the under-serviced over-burdened urban infrastructures of the burgeoning cities in the developing world (many of whom still only have the original colonial enclaves that were built for the settler elites during the first urbanisation wave).

The first global estimates of what it will take have already started to be published. The US-based global consulting firm, Booz Allen Hamilton, has compiled a detailed estimate of the investment required to meet demand for urban infrastructure over the next 25 years across all the cities of the world. (To give some context, Booz Allen Hamilton depends heavily for its $4.5 billion turnover on world-wide contracts to build infrastructures that are allocated to a handful of the world's largest engineering contractors.) Significantly called Lights! Water! Motion!, and published in the influential business journal, Strategy and Business, the report estimates that a total of $41 trillion is required to refurbish the old (in mainly developed country cities) and build new (mainly in the developing country cities) urban infrastructures over the period 2005-2030 – an amount that was greater than the value of all the stocks on the world's stock exchanges in 2007.

Of this, over 50% ($22.6 trillion) would be required for water systems,[3] $9 trillion for energy, $7.8 trillion for road and rail infrastructure, and $1.6 trillion for air- and sea-ports (Doshi *et al.*, 2007).

Echoing the Stern Report's new dictum that it is 'cheaper to fix things earlier rather than later', a revealing statement demonstrates the sales pitch that these powerful global firms use to land lucrative contracts. The authors of the Booz Allen Hamilton report write:

"Sooner or later, the money needed to modernize and expand the world's urban infrastructure will have to be spent. The demand and need are too great to ignore. The solutions may be applied in a reactive, ad hoc, and ineffective fashion, as they have been in the past, and in that case the price tag will probably be higher than $40 trillion. After all, infrastructure projects are notorious for cost overruns. But perhaps the money can be spent proactively and innovatively, with a pragmatic hand, a responsive ear, and a visionary eye. The potential payoff is not simply the survival of urban populations, but the next generation of great cities." (Emphasis added.)

Doshi et al. 2007: 4

To their credit, Booz Allen Hamilton acknowledges (albeit only in a side box) that the grand retooling of the world's urban infrastructures will mean finding new designs and technologies that will make it possible to use natural resources more sustainably:

"...[C]ities that ignore environmental impact will find themselves facing another collapse of infra-structure 30 or 40 years from now, and our children and grandchildren will bear a much higher price tag."

Doshi et al. 2007: 13

Just as economists look back on the investments in automobiles, roads, petro-chemicals and mass production systems made from the 1930's through to the post-WWII period (including the Marshal Plan in Europe) as the investments that 'resolved' the 1929-33 economic crisis and paved the way for the rollout of the age of oil and mass production (Perez, 2007), we predict that in ten to fifteen years' time researchers will look back and realise that the investments that helped 'resolve' the crisis of 2007-2010/11 were, in fact, investments in networked urban infrastructures – with, of course, 'Web 2.0'-type ICTs as their operating systems (e.g. 'smart grids', 'telecommuting', 'virtual shopping', remotely controlled 'intelligence systems', 'digitalization', etc). Retooling the world's cities for (hopefully more equitable) global re-industrialisation is what is at stake here; or, to use the language that the global consulting and development finance community might use: How do we create 'the next generation of great cities'? How can we transform cities in such a way that they may become the innovative geographical nodes of the post-crisis regimes of accumulation and development?

---

3   Although not specifically defined, we presume this means 'water and sanitation' infrastructure, because the one without the other does not make much technical sense.

The question is, of course, what kind of networked urban infrastructures will be built? Will these systems set cities up for sustainable socio-ecological metabolisms? Or will decision-makers set cities up against natures' resource bases and ecosystems?

Will these systems reinforce the stark techno-apartheid that is splintering cities around the world, or will the basis for greater equity, reduced levels of poverty, and greater opportunities to build a sense of community be created? To what extent will the multiple stakeholders who shape the lives of cities choose more sustainable modes of resource-use? Will the search for greater equity and a sense of place be reinforced or undermined?

9. Rethinking Cape Town

We can return now to a consideration of the challenges facing Cape Town as articulated in the chapters of this book. Whether we are fully aware of it or not, the global trends addressed in the foregoing sections will not only shape the material context of future economic and socio-demographic trajectories, but also the ideas that actors will draw on and use to make sense of the changing development dynamics of Cape Town.

The second urbanisation wave will undoubtedly undermine the persistent failure right across Africa (including South Africa) to fully comprehend the unique dynamics of the city. Hopefully this will result in critical choices that will harness the potential created by increasing concentrations of economic activities, communications, and people within cities. Quite simply, as the scale of urbanisation in Africa (and the rest of the world) results in the realisation that cities are here to stay, it is highly likely that this will lead to a search for policy solutions that draw more heavily on the growing – but to date largely ignored – body of inter- and trans-disciplinary research that aims to grasp the complexities of cities in Africa and the developing world (See Pieterse, 2008, also the work of the Africa Centre for Cities at University of Cape Town headed by Edgar Pieterse).

The best outcome we can hope for is a sense of the city – a pervasive realisation that social, economic, cultural, and environmental activities and strategies are highly dependent on the contextual specificities of the unique spatial relations of each and every city. Knowing these contextual specificities holds the key to successful applications of generally accepted prescriptions. Better still, we have an opportunity to generate unique 'home grown' solutions via increasingly productive learning networks of actors, researchers and storytellers (which include journalists, marketers, designers, cultural workers, etc.). Increasingly, thought leaders, in acknowledging the marked complexity of the challenges facing cities, believe that the vast potential for creativity and innovation capital lies in the intellectual capital between and among the wide and diverse range of stakeholders – citizens, policy makers, professionals, communities, storytellers and artists – who could shape and re-shape each and every city.

Similarly, the global polycrisis and the sustainability challenge have already started to undermine the validity of the assumptions that inform mainstream economic development paradigms across the ideological spectrum. Neo-liberal, welfarist, and developmental state perspectives have all tended to assume that the natural resource base is intact and that resource depletion and ecosystem degradation are externalities with little significance for economic and urban development policy (Swilling, forthcoming). A new language for thinking about the future in a new light has started to emerge, opening up the potential to generate a conglomeration of innovative technologies, investment opportunities, skills sets, services and value chains that have hitherto been literally unthinkable. When these now 'thinkable' alternatives are contextualised in space in particular cities, a new panorama of innovation is imaginable. There are many examples of cities around the world that have integrated sustainability thinking into their sense of the city. The result is a rich mix of often surprising outcomes.

10. Imagining a Sustainable Cape Town

The contributions to this book are informed in one way or another by the growing congruence of 'sustainability thinking' and a 'sense of the city'. The emergent outcome is that cities provide a unique context for realising that resource limits and degrading ecosystem services are not simply constraints to development, but opportunities for redefining what development means.

Gordon Brown (as cited earlier) is only half right when he observes that a growing proportion of the investments in economic recovery are investments in what he called the "green revolution". As argued, these investments are heavily concentrated in networked urban infrastructures, thus signalling the central importance of the city in both economic recovery and the transition to a more sustainable socio-ecological regime of accumulation and wellbeing. This is the reason why a number of chapters in this book are devoted to issues that are often ignored by publications that address the environmental challenges of cities, namely Cape Town's waste, energy, transportation, and water and sanitation infrastructures.

It is commonly accepted that Cape Town has a well-developed community of urban development professionals employed in the private sector (in particular in consulting companies), government, NGOs, and universities, who together are able to sustain a reasonably well-developed 'sense of the city'. Although this 'sense of the city' is highly contested, because professionals serve competing interests (from developers/financial institutions, to the homeless, to government departments), this contributes to the richness of the discussion.

Cape Town is also well-known for having a critical mass of professionals who are actively engaged in one way or another in applying 'sustainability thinking' to Cape Town's problems, and beyond. Spread out across university departments, NGOs, consulting firms, government departments (especially at provincial and local government levels), companies and NGOs are increasingly involved in sectors like waste, energy, transport, tourism, and organic food. The proliferation of open-air markets and new courses on some aspect of sustainability are just two obvious examples of this trend. Even within traditional development sectors, like low-income housing or food security, solutions are being generated from within these sectors that emphasise higher densities, more sustainable use of resources, and retention of urban agricultural land for food production. It is no coincidence that the vast majority of the most significant policy documents published in recent years by the Western Cape Provincial Government and the City of Cape Town identify sustainability as a core policy objective.

11. The Focus of this Book

The aim of this book is to connect these two conversations: Cape Town's 'sense of the city' needs to include the realisation that it is a city facing serious natural resource restrictions to the current patterns of urban development and the design and management of networked infrastructures. Without this, urban development will become increasingly expensive and thus inexorably exclude the urban poor.

Similarly, 'sustainability thinking' in Cape Town needs to move way beyond the traditional preoccupations with the negative environmental impacts of development in order to embrace the notion that the city can develop in a completely different way. This will mean going beyond the 'minimising damage' approach of the Cape Town-based Green Building Council that is sponsored by the property development industry. Sustainability will not be achieved by slowing down the destruction of our natural resource base. As Janice Birkeland has argued, while minimising damage is important, it is becoming more important to focus on the 'restoration of life' – designing urban developments that restore the ecosystem services that our society and economy depends on (Birkeland, 2008).

For Birkeland and many others, sustainability is not attained by slowing down the destruction of our natural resources, or by thinking that the amelioration of poverty is a substitute for eradicating it. What is needed is a purposive commitment to restoring what has been destroyed and building the real capabilities within poor communities to ensure that developmental initiatives are geared to meet the real needs of these communities, rather than preconceived technocratic prescriptions.

****

The chapters in this book were written by researchers from a wide range of disciplines. The research was jointly funded by the City of Cape Town and the Sustainability Institute, and presented at a Workshop on 13 November 2008 attended by officials from most of the City Departments. This generated important feedback and enriched the trans-disciplinary, inter-departmental discussion about the challenge of building a more sustainable city, which informed the formulation of the Introduction and the editing of the final chapters of this book.

The first four essays discuss key dimensions of Cape Town's urban system:

- *Chapter 2: Municipal finance* – Mark Swilling and Martin de Wit argue that Cape Town's impressive commitment since 1994 to the provision of universal services to all has come up against two constraints: firstly, there are serious limits to the financial resources needed to sustain an infrastructure investment programme that has an appropriate balance between new capital projects and adequate levels of maintenance and repair of existing infrastructure, and secondly, it has been assumed until recently that there are no significant natural resource limits to further expansive urban development with infrastructure services delivered via the long-established technologies.
- *Chapter 3: Urban form* – Kathryn Ewing and Nisa Mammon focus on the dynamics of Cape Town's challenging space-economy. They argue that the discussion of urban sustainability needs to go beyond the traditional environmental impact paradigm by focussing on the relationship between densification and design. Unless it is recognised that design plays a key role in shaping workable alternatives, many of the sustainability interventions will not be realised. Design is, in reality, where a 'sense of the city' fuses with sustainability thinking. It is not, however, an autonomous driver.
- *Chapter 4:* Social justice and sustainable resource use – Mazibuko Jara demonstrates that there are deep structural determinants of poverty, and that there are resource limits to developmental strategies that aim to deal with these causes of poverty. Creative alternatives, he argues, are needed that emphasize the importance of cooperative organisation and collective action. Poverty will not be eradicated by relying purely on state-managed top-down financial flows.
- *Chapter 5:* Economic and industrial development – Mazibuko Jara asserts that Cape Town's economy is in trouble – to grow in order to maintain a tax base that can finance a more sustainable urban infrastructure, it is clear that Cape Town's economy cannot afford the continued decline of manufacturing, or ongoing dependence on debt-financed consumerism. This raises the important challenge of redirecting investments into the kinds of activities that will enhance the emergence of a more sustainable city.

The next four essays focus on a key aspect of the city's infrastructure that the City of Cape Town (CCT) directly manages. These chapters are significant because they highlight the resource constraints that are forcing Cape Town to rethink its development paradigm.

- *Chapter 6: Water and sanitation* – Kevin Winter demonstrates that there are absolute limits to Cape Town's bulk water supply. He argues that this will force immediate innovations with respect to more efficient use of water, but in the longer run it will also compel Cape Town to find alternatives, such as desalination or sustainable harvesting of groundwater. At the moment, however, Cape Town's water management administration continues to focus on the deployment of traditional water management technologies, with relatively low prices for water. A key theme in the chapter is the idea that the integration of economic, social and institutional imperatives remains one of the key challenges in the management of environmental resources such as water.
- *Chapter 8:* Energy – Cape Town is supplied by the CCT and ESKOM, with little cooperation between the two. This makes city-wide energy planning impossible. At the same time, Cape Town's economic growth is constrained by supply limits. This explains why many key actors are seriously considering investments in renewable energy and energy efficiency. Frank Spencer reviews the options in this chapter.

- *Chapter 9:* Solid waste management – Sally-Anne Engledow shows that, although there is clear evidence of decoupling in the energy and water sectors, until recently solid waste volumes have grown much faster than economic and population growth rates. While the average Capetonian throws away more unrecycled waste every year, landfills have been filling up without any agreement on where to build the next landfill site. This has forced Cape Town's waste management administration to introduce waste recycling.
- *Chapter 10:* Transport – The key to securing a sustainable future for Cape Town will be the design and establishment of a public transport system for the city that links the rail, bus and taxi systems. Low-density sprawl – with investments in roads to cater for private cars – will simply mean committing public funds to subsidising the middle class. The majority of poor people use poor quality public transport. Roger Behrens and Peter Wilkinson consider the implications of moving the majority of all commuters into high quality public transport systems. Without this, both economic growth and quality of life in Cape Town will suffer.

The final set of chapters address arenas of engagement between government, business and NGO actors that intimately shape the quality of every life.

- *Chapter 11: Natural space* – Cape Town is blessed, because at its centre is the largest natural urban park in the world. This, plus Cape Town's natural beauty, makes the management of natural space a key issue. Matthew Cullinan reviews the various ways in which this issue is understood and addressed by the policy stakeholders.
- *Chapter 12: Urban agriculture* – As long distance transportation of food becomes increasingly expensive, cities around the world will have to source local food supplies. In this chapter, Gareth Haysom reviews the opportunities in Cape Town for increasing supply from local urban and peri-urban sources, and the kinds of interventions that will be required to support this over the long run.
- *Chapter 13: Sustainable architecture* – Buildings may be physical structures that consume vast resources (during construction and over their life cycle), but how much they consume and how much waste they produce is totally dependent on how they are designed. It is not buildings themselves that are unsustainable, but rather how they are designed and managed. In this chapter, Mokena Makeka and Peta Brom consider what it will take to change the design paradigm so that future buildings in Cape Town will not only require far fewer energy-intensive materials, but can even become positive producers of strategic resources (such as energy from solar power, or separated wastes that can be productively re-used) over their life cycle.
- *Chapter 14: Housing* – To date, housing delivery systems have focussed on the numbers of units produced, mainly on the peripheries of the cities. National housing policy has shifted, and recognises the limits of a 'one-size-fits-all' approach that ignores quality. Paul Hendler reviews the implications of housing policy reforms that have made possible a range of housing delivery systems. This increased flexibility makes it possible for cities to include sustainability into the housing delivery system.

12. Concluding Thoughts

Cape Town is going through an extraordinarily rapid urban growth and transformation process. This is not driven from the centre, but is an emergent outcome of a vast array of deals, as all sorts of interests and stakeholders engage each other through conflict, negotiation or collaborative action. Although frequent party-political changeovers have made it impossible to formulate and adopt a clear long-term strategic framework for the city, this has not prevented developers and communities from securing approvals for projects that have made things happen.

Not all that happened as a result contributed to greater equity, less poverty and more sustainable resource use, but it did unleash enormous developmental energy across the city. Since 2005, however, there has been a process within the CCT of slowly consolidating a long-term vision for Cape Town as a more equitable and sustainable city.

The chapters in this book are aimed at enriching the discussion about this longer-term vision, focussing in particular on the nexus between what has been called a 'sense of the city' and 'sustainability thinking'. A lot more research, however, is needed. In this volume we have only just scratched the surface, and left out important aspects of the discussion (such as, for example, marine resources, biodiversity, air quality, and the economic potential of employment-creating activities, such as recycling, biodiversity restoration, low-carbon buildings, and the manufacture of new kinds of goods, such as electric vehicles).

The challenge facing Cape Town is, therefore, quite clear: if it wants to position itself in a rapidly changing world in a way that generates the greatest benefits for all its citizens, this city will have to recognise that the challenges of the second urbanisation wave, coupled with mounting 'green economy' investments as resource constraints open up new opportunities, are redefining future possibilities. We can no longer imagine this future using concepts that emerged from what is now a rapidly disappearing era. We need to mobilise the courage to imagine futures that take full advantage of the new world that is opening up. To do this, we need to understand how innovations for sustainability actually work, and how they can be supported and accelerated. Only then will we be able to take up the challenge laid down by Njabulo Ndebele, when he wrote that "[our] own brand of ordinariness ought to work at a higher level."

References

Barbier, E.B. 2009. *A Global Green New Deal:* Report prepared for the Economics and Trade Branch, Division of Technology, Industry and Economics, United Nations Environment Programme.

Bayat, A. 2000. From 'Dangerous Classes' to 'Quiet Rebels': Politics of the Urban Subaltern in the Global South. *International Sociology*, 15(3):533-557.

Beatley, T. 2000. *Green Urbanism*. Washington D.C.: Island Press.

Behrens, A., Giljum, S., Kovanda, J. & Niza, S. 2007. The Material Basis of the Global Economy: Worldwide Patterns of Natural Resource Extraction and their Implications for Sustainable Resource use Policies. *Ecological Economics*, 64:444-453.

Berman, M. 1988. *All that is Solid Melts into Air*. New York: Penguin.

Birkeland, J. 2008. *Positive Development: From Viscious Circles to Virtuous Cycles through Built Environment Design*. London: Earthscan.

Borja, J. & Castells, M. 1997. *Local and Global: Management of Cities in the Information Age*. London: Earthscan.

Bringezu, S. & Bleischwitz, R. 2009. *Sustainable Resource Management: Global Trends, Visions and Policies*. Germany: Wuppertal Institute.

Bringezu, S. & Schutz, H. 2001. *Material use Indicators for the European Union, 1980-1997. Economy-Wide Material Flow Accounts and Balances and Derived Indicators of Resource use*. Luxembourg: Eurostat.

Bringezu, S., Schutz, H., Steger, S. & Baudisch, J. 2004. International Comparison of Resource Use and its Relation to Economic Growth: The Development of Total Material Requirement, Direct Material Inputs and Hidden Flows and the Structure of TMR. *Ecological Economics*, 51(1/2):97-124.

Campbell, T. n.d. IPPUC: *The Untold Secret of Curitiba – In-House Technical Capacity for Sustainable Environmental Planning*. Urban Age Magazine. San Rafael, Ca.: Urban Age Institute.

City of Cape Town. 2006. *Planning for Future Cape Town: An Argument for the Long-term Spatial Development of Cape Town*, Draft Document for Discussion, August 2006. Cape Town: City of Cape Town.

Commission on Growth and Development. 2008. *The Growth Report: Strategies for Sustained Growth and Inclusive Development*. Washington, D.C.: World Bank.

Davis, M. 2005. *Planet of Slums*. London: Verso.

Doshi, V., Schulam, G. & Gabaldon, D. 2007. Lights! Water! Motion! *Strategy and Business*, 47. Resilience Report - Reprint Number 07104:1-16.

Evans, P. 2002. *Liveable Cities: Urban Struggles for Livelihood and Sustainability*. Berkeley: University of California Press.

Fischer-Kowalski, M. 1999. Society's Metabolism: The Intellectual History of Materials Flow Analysis, Part II, 1970-1998. *Journal of Industrial Ecology*, 2(4):107-136.

Fischer-Kowalski, M. 1998. Society's Metabolism: The Intellectual History of Materials Flow Analysis, Part I, 1860-1970. *Journal of Industrial Ecology*, 2(1):61-78.

Gallopin, G. 2003. *A Systems Approach to Sustainability and Sustainable Development:* Project NET/00/063. Santiago: Economic Commission for Latin America.

Gleick, P. 2006. *The World's Water* (2006-2007): *The Biennial Report on Freshwater Resources*. Washington, D.C.: Island Press.

Haberl, H., Fischer-Kowalski, M., Krausman, F., Weisz, H. & Winiwarter, V. 2004. Progress Towards Sustainability? What the Conceptual Framework of Material and Energy Flow Accounting (MEFA) can offer. *Land use Policy*, 21:199-213.

Intergovernmental Panel on Climate Change. 2007. *Climate Change 2007*. Geneva: United Nations Environment Programme.

International Energy Agency. 2008. *World Energy Outlook*. Paris: International Energy Agency.

Krausmann, F., Schandl, H., Fischer-Kowalski, M. & Eisenmenger, N. 2008. The Global Socio-Metabolic Transition: Past and Present Metabolic Profiles and their Future Trajectories. *Journal of Industrial Ecology*, 12:637-656.

Krausmann, F., Gingrich, S., Eisenmenger, K.H., Haberl, H. & Fischer-Kowalski, M. (forthcoming). Growth in Global Materials use, GDP and Population during the 20th Century. *Ecological Economics*.

Low, N., Gleeson, B., Elander, I. & Lidskog, R. 2000. *Consuming Cities: The Urban Environment in the Global Economy After the Rio Declaration*. New York: Routledge.

Lundvall, B.A. 2007. *National Innovation System: Analytical Focusing Device and Policy Learning Tool*. Working Paper. Ostersund, Sweden: Swedish Institute for Growth Policy Studies.

Malik, A. 2001. After Modernity: Contemporary Non-Western Cities and Architecture. *Futures*, 33:873-882.

Morin, E. 1999. *Homeland Earth*. Cresskill, NJ: Hampton Press.

National Research Council. 2003. *Cities Transformed*. Washington, D.C.: National Academies Press.

Newman, P. & Kenworthy, J. 2007. Greening Urban Transportation. In: Worldwatch Institute. (Ed.). *State of the World: Our Urban Future*. New York & London: W.W. Norton & Company.

Perez, C. 2007. *Great Surges of Development and Alternative Forms of Globalization*. Working Papers in Technology Governance and Economic Dynamics. Norway and Estonia: The Other Canon Foundation (Norway) and Tallinin University of Technology (Tallinin).

Pieterse, E. 2008. *City Futures*. Cape Town: Juta.

Reed, C. 2007. *Mayor Reed's Green Vision for San Jose*. Available from http://www.sanjoseca.gov/mayor/goals/environment/GreenVision/GreenVision.asp. (Accessed 6 August 2009).

Satterthwaite, D. 2007. *The Transition to a Predominantly Urban World and its Underpinnings*. Human Settlements Discussion Paper. London: International Institute for Environment and Development.

Schwartz, H. 2004. *Urban Renewal, Municipal Revitalisation: The Case of Curitiba, Brazil*. Alexandra, VA: Hugh Schwartz.

Shengxian, Z. 2009. Keynote Address to the Annual General Meeting of the China Council for International Cooperation on Environment and Development. Beijing: 11-13 November.

Simone, A. 2004. *For the City Yet to Come: Changing African Life in Four Cities*. Durham NC and London: Duke University Press.

Simone, A. 2001. *Between Ghetto and Globe: Remaking Urban Life in Africa*. In: A. Tostensen, I. Tvedten and M. Vaa. (Ed.). Associational Life in African Cities: Popular Responses to the Urban Crisis. Stockholm: Elanders Gotab.

Stern, N. 2007. *Stern Review: Economics of Climate Change*. Cambridge: Cambridge University Press.

Stiglitz, J., Sen, A. & Fitoussi, J.P. 2009. *Report by the Commission on the Measurment of Economic Performance and Social Progress*. Report Commissioned by President Sarkozy, French Government.

Swilling, M. forthcoming. Greening Public Value: The Sustainability Challenge. In: J. Bennington and M. Moore. (Ed.). *In Search of Public Value: Beyond Private Choice*. London: Palgrave.

Swilling, M., van Weiszaecker, E., Fischer-Kowalski, M., Manalang, A., Yong, R., Moriguchi, Y. & Crane, W. 2009. *Decoupling and Sustainable Resource Management: A Review - Draft 1.6*. Paris: International Panel for Sustainable Resource Management, United Nations Environment Programme.

Swilling, M. 2006. Sustainability and Infrastructure Planning in South Africa: A Cape Town Case Study. *Environment and Urbanization*, 18(1):23-50.

Swilling, M., Khan, F. & Simone, A. 2003. 'My Soul I can See': The Limits of Governing African Cities in a Context of Globalisation and Complexity. In: McCarney, P. & Stren, R. *Governance on the Ground: Innovations and Discontinuities in Cities of the Developing World*. Baltimore and Washington D.C.: Woodrow Wilson Centre Press and Johns Hopkins University Press.

Taylor, A. 2008. Sustainable Cities and Local Food Systems: A Partnership between Restaurants and Farmers in Portland, Oregon. Stellenbosch: Stellenbosch University, School of Public Management and Planning. Masters Thesis. [Available at www.sustainabilityinstitute.net]

UN Habitat. 2008a. *The State of African Cities 2008: A Framework for Addressing Urban Challenges in Africa*. Nairobi: United Nations Human Settlements Programme (UN-Habitat).

UN Habitat. 2008b. *State of the World's Cities 2008/2009: Harmonious Cities*. London: Earthscan.

United Nations. 2006. *State of the World's Cities 2006/7*. London: Earthscan & UN Habitat.

United Nations. 2005. Millenium Ecosystem Assessment. New York: United Nations.

United Nations Centre for Human Settlements. 2003. *The Challenge of Slums: Global Report on Human Settlements*. London: Earthscan.

United Nations Development Programme. 1998. *Human Development Report 1998*. Available from http://hdr.undp.org/reports/global/1998/en/. (Accessed 22 October 2006).

United Nations Environment Programme. 2007. *Global Environment Outlook GEO 4: Environment for Development*. Nairobi: United Nations Environment Programme.

Watson, R.T., Wakhungu, J. & Herren, H.R. 2008. *International Assessment of Agricultural Science and Technology for Development (IAASTD)*. Available from www.agassessment.org. (Accessed 21 April 2008).

World Bank. 2005. *Where is the Wealth of Nations? Measuring Capital for the Twenty-First Century*. Washington, D.C.: World Bank.

World Resources Institute. 2002. *Decisions for the Earth: Balance, Voice and Power*. Washington D.C.: World Resources Institute.

World Wildlife Fund. 2008. *Living Planet Report 2008*. Gland, Switzerland: World Wildlife Fund.

Worldwatch Institute. 2007. *State of the World 2007: Our Urban Future*. Washington, D.C.: Worldwatch Institute.

Municipal Finances, Service Delivery and Prospects for Sustainable Resource Use in Cape Town

Mark Swilling & Martin de Wit

This chapter gives a broad overview of the complex changes reflected in the City of Cape Town through the lens of municipal finance. We begin by introducing a broad landscape, highlighting the contours and interplay between a network of stakeholders and the new realities of change. Serious limits to the available financial resources undermine the capacity required to deal with the major challenges. Calls for sustained investment in infrastructure jockey for attention for new capital projects and adequate levels of maintenance. The complex picture includes the transformation of political governance, the attempts of the municipality to provide more equitable service delivery among the diverse class groups in the City, rising poverty, the call for a sustainable approach, and the unstable supply of natural resources such as water and electricity.

Cape Town's municipal finance story has been one of extremely complex manipulations and trade-offs. A key thread has been the need to address poverty via redistributive measures to finance extended electricity, water, waste and sanitation (EWWS) services without undermining economic growth. Discussion shifts to a series of observations about municipal finance, and we outline various challenges: increased pressure on EWWS services, under- spending of budget allocations, and the structure and impact of intergovernmental grants and subsidies. Responses to this broad range of challenges are outlined, highlighting both opportunities and constraints. A variety of complexities and delicate balances underscore the need for creative approaches to fiscal viability in a dynamic city like Cape Town.

The chapter ends with both optimistic and sobering conclusions. What gives hope, is that the rate of consumption of materials and energy in Cape Town's metropolitan economy is beginning to decouple from economic growth rates. A case is also made for the possibility that municipal interventions themselves could reinforce, rather than reverse, this decoupling trend. Practical suggestions are identified for implementing budgetary planning and financial management inspired by the vision and benefits of a sustainable resource perspective.

1.   Core Challenges

Like all post-Apartheid cities, after 1994 Cape Town has faced the twin challenges of overcoming the spatial divisions created during the colonial/apartheid era and addressing the endemic poverty that these divisions reproduced for over three centuries. In 1993, the City of Cape Town was governed by 61 municipalities and managed by 17 separate administrations. In 1995-1996, the first democratic local government elections took place in integrated municipal areas. Initially, Cape Town established 7 local government authorities from the former 61 municipalities in order to strive for more autonomy. The Municipal Structures Act (Act no. 117 of 1998 as amended Act no. 33 of 2000) (Republic of South Africa a) outlined new systems of metropolitan government leading to the establishment of the Cape Town Unicity – a single-tiered (sphere) form of metropolitan government. The so-called 'Unicity' structure secured a single metropolitan tax base in an attempt to address development disparities. This process of municipal restructuring and its impact on service provision has been well documented (Van Donk *et al.*, 2008; Parnell & Pieterse, 2007; Jaglan, 2004; McDonald & Smith, 2002; Khosa, 2000). Formerly segregated white local authorities (WLA's) and black local authorities (BLA's) were amalgamated in an attempt to ensure greater equity between rich and poor communities, and to help standardise service delivery. This was achieved in particularly interesting ways in Cape Town, especially with respect to particularly progressive municipal finance policies that remained consistent despite regular party political changeovers.

This highly complex process of transforming and deracialising political governance was underpinned by the 'new public management' approach to institutional reform and the decentralisation of some national government functions to local government level. Privatisation/corporatisation of state-owned enterprises, the formulation of public-private partnerships, greater emphasis on service delivery on a cost recovery basis, and the implementation of performance management systems were also part of the process of institutional reform, albeit only partially implemented in the Cape Town context (Wilkinson, 2004; Watson, 2002). Despite references to neo-liberal approaches to service delivery in the Cape Town context (Miraftab, 2004; Smith, 2004; Smith & Hanson, 2003), the fact of the matter was that by 2008, all municipal services in Cape Town were still delivered either directly or through sub-contractors via the City of Cape Town's (CCT) integrated public service.

The restructuring process was far more arduous than initially anticipated. Local government authorities struggled to implement complex administrative changes and to deliver promised services in an unpredictable political environment. Former WLA's and BLA's, with different organisational structures, battled to implement complex new systems post-integration in financially viable ways.

There are nearly 800,000 households in Cape Town, with a rapidly expanding population of around 3,5 million people. The table below represents the class structure of these households – just over 50% are classifiable as poor and working households; 16% comprise the wealthy elite; and a relatively small 31% of households comprise the middle class.

Despite the many political changeovers in Cape Town's municipal government since 1994, a constant theme of successive administrations has been the need to address the service backlogs in the poorer areas of the city. This has had major implications for capital and operating expenditures in the EWWS sectors, which together account for nearly half of expenditure by the CCT. As will be demonstrated below, R9.3 billion (47%) of its budget for 2007/8 was spent on capital and operational expenditures for EWWS services. This equates to 8% of the Gross Geographic Product (GGP) of the Cape Town metropolitan economy. Given the magnitude of this expenditure and related incomes, it is imperative that researchers pay more attention to how this money is spent; who benefits; what the long-term impacts will be on the space economy, social structure and eco-systems; and what the alternative approaches may be.

Table 1: Household class structure in Cape Town

Cluster Group	% of suburbs	No of households	% of total households
Elite suburbs	14	54,630	7
Upper middle class	19	68,129	9
Sub-total	*33*	*122,759*	*16*
Middle suburbia	20	77,380	10
Inner city	1.5	17,564	2
Semi/skilled labour pool	9.5	42,404	6
New bonded areas	13.5	101,638	13
Sub-total	*44.5*	*238,986*	*31*
Traditional townships	4.5	80,980	11
Dense run-down high-rise	13	170,752	22
Urban and working poor	2	26,108	3
Below the poverty line (mainly informal)	3	111,770	15
Sub-total	*22.5*	*389 610*	*51*
Total	*100*	*751,355*	*98*

[Source: Swilling 2006]

The core challenge that has faced officials since 1994, has been to find fiscally viable ways to expand EWWS services into poorer areas while maintaining and operating EWWS services for the City as a whole. At the same time, since 2006 there has been a growing realisation that development strategies need to address the question of sustainable resource use. This has been recognised in the CCT's policy documents (City of Cape Town, 2008), and in the policy documents of the Western Cape Provincial Government (Western Cape Provincial Government, 2007). The academic literature has also started to reflect similar arguments (Crane and Swilling, 2008; Clark *et al.*, 2007; Petrie & Ocran, 2007; Sustainability Institute, 2007; Swilling, 2006, Swilling & Annecke, 2006). The core argument in these emerging policy documents and the academic literature, is that service delivery will not be able to address the needs of the poor if these services depend on traditional technologies and systems that are seen to be highly inefficient and ecologically unsustainable. This has resulted in repeated calls for greater integration of service delivery, and integrated planning to support an approach informed by a sustainable resource use perspective.

Sustainable resource use simply refers to living in a manner which is intrinsically compatible with the natural resource base and available eco-systems. More specifically, it means ensuring that resource consumption will not jeopardise the earth's life-support systems now and in the future. Over the last two centuries, although many nations have experienced unprecedented increases in living standards, the global stock of natural resources has dramatically declined, with increasingly negative consequences for the approximately two billion people who live in poverty (World Wildlife Fund *et al.*, 2006; United Nations, 2005; Hurrell & Woods, 1995). Global warming and its negative consequences for the poorest people in the world currently tops the agendas of the most important global meetings (Intergovernmental Panel on Climate Change, 2007; Stern, 2006) – witness the unprecedented attention given to global climate change in Copenhagen in December 2009.

At present, 'sustainable growth' and 'sustainable resource use' are concepts widely accepted in theoretical and political discourses worldwide (Behrens *et al.*, 2007; Fischer-Kowalski & Haberl, 2007; Mebratu, 1998; Pezzoli, 1997). However, a general consensus on the practical implications of sustainable resource use for municipal finance and service delivery is still required (Swilling, 2004).

Following recent work that applies material flow analysis to urban systems (Hodson & Marvin, 2009; Heynen *et al.*, 2006; Guy *et al.*, 2001), it has been possible to quantify the resource flows that the City of Cape Town depends on. It has been estimated that for the period 1996-98, Cape Town consumed 365 million tons of 'raw materials' per annum, and disposed of 208 million tons of waste into local water, air, and land sinks (Gasson, 2002). The most significant annual resource flows through Cape Town's urban systems are as follows:[1]

- **Electricity**: approximately 10 billion KWh of grid-supplied electricity for the year 2006/7;
- **Oil**: over 2 billion litres of crude oil for the year 2006/7;
- **CO2**: nearly 20 million tons of CO2 emitted for the year 2006/7 from all energy users, including transport, which is nearly 7 tons per capita (which is lower than the national average of 9.8, but equivalent to per capita emissions in Italy, France, Spain, Poland, and Malaysia);
- **Water**: 247 million kilolitres, or 82 kilolitres per person, of water per annum in 2006, which is higher than the global average of 57 kilolitres per person (and also higher than the European average);
- **Sewage**: 200 million kilolitres of sewage for the year 2005/6, or 67 kilolitres per person per annum;
- **Solid waste**: 2,9 million tons of solid waste in 2007/8 (which is over 2 kgs per person per day, higher than the European average), of which about 0.4 million tons was recycled;
- **Building materials**: by the early 2000's, 6 million tons of building materials entered Cape Town per annum for conversion into buildings and infrastructure, with an output of about 1 million tons of builders' rubble – much of which is recycled and re-used again as inputs (estimates range from 35% to 75%).

Progress towards a more sustainable city means working out ways to initially do more with a bit less (relative decoupling), but eventually to find ways of making sure that more is done by reducing the absolute quantity of resource inputs (as has happened with water use – excluding borehole resources – since 2001). For a growing city, the waste outputs will always be less than the required material inputs, but over time this ratio changes. Capturing nutrients from sewage to meet the

---

1   Note: This summary data is derived from the results of a three-year research project managed by Mark Swilling (MCA, 2006; see Sustainability Institute 2007a; Sustainability Institute 2007b; Sustainability Institute 2007c). Supplementary data has been drawn from Gasson (2002) and Hansen (2010).

nutrient requirements of locally produced food is probably already possible, with major job creation benefits. Using solar power to desalinate water is another closed loop with major benefits which Cape Town will need.

It will be argued that in response to the severe fiscal constraints within which officials have to work to extend EWWS services to areas with limited cost recovery capacity, while still maintaining and operating these services on a city-wide basis, there has been a move away from a uniform approach ('one-size-fits-all') to a recognition that unique responses are required for specific contexts ('horses for courses'). There is, therefore, growing acceptance of the need for innovation and experimentation. While for some this is evidence of an anti-poor 'neo-liberal' response, there is no necessary reason why a break from standardised uniform service via a centralised public service will automatically result in worse or more expensive services for the poor. From a sustainable resource use perspective, much about service delivery and the financing thereof will need to change.

A sustainable resource use approach with pro-poor results will need to make provision for new technologies and systems that could substantially decouple rising consumption from resource use on a city-wide basis. It is unlikely that this will happen if the traditional conception of public sector service delivery is maintained. For example, a 10% saving on a R9.3 billion expenditure on EWWS services is R930 million, which is more than twice the size of the housing subsidy grant that Cape Town received from National/Provincial Government in 2007/2008. Imagine if this saving could be captured and redirected into pro-poor development! The stakes, therefore, are high. Although this shift can happen on a piece-meal basis in pioneer ('EcoVillage') projects, it will also need to be developed at a city-wide level. For this to happen, integrated planning approaches will be required that go beyond the current multi-disciplinary approach to a trans-disciplinary approach. In the former, different sectors are simply added together to compile the Integrated Development Plan, whereas the latter approach embraces complexity modelling in a way that empowers decision makers to think and plan in ways that take into account a much wider set of feedback loops than is possible at the sectoral level (De Wit and Swilling, 2008).

2. Overview of Municipal Finances

Cape Town's municipal finance system is complex and has only recently emerged from a process of fundamental restructuring. The general observations that follow have emerged after three years of intensive research on Cape Town's EWWS services and related financing strategies (Sustainability Institute 2007b, 2007c, 2007d).

1. Despite the backlogs and increased pressures on EWWS services, the City's budgeted expenditure on EWWS services is fairly constant at around half the total budget. The City budgeted around R9.3 billion (or 47%) of its budget on capital and operational expenditures on EWWS services for 2007/8. This compares to R8.5 billion (or 50%) in 2006/7, and R7.8 billion (or 45%) in 2005/6. However, these numbers do mask a large variation in budgeted capital and operational expenditures, and in expenditure on various services. Contrary to what one would expect, budgeted operational expenditure on EWWS services is budgeted to drop from around 50% of total operational expenditure in 2005/6, to 45% in 2007/8. The overall budgeted capital expenditure is also budgeted to drop substantially from R4.1 billion in 2005/6 to R2.8 billion in 2008/9. These trends are reflected in Table 2 below.

Table 2: Budget allocations for EWWS, 2005-2009

Year	Total budget allocated to EWWS	% of budget allocated to EWWS	% spent on operational budget
2005/6	7.8 billion	45%	50%
2006/7	8.5 billion	50%	45%
2007/8	9.3 billion	47%	45%

2. In 2005/6, Cape Town only spent 37% (or R1.5 billion) of its R4.1 billion capital budget – a situation mainly attributed to the slow delivery on the N2 Gateway low-cost housing project (South African Cities Network, 2007). However, this trend of under-spending was persistent in the City over the last decade, with actual expenditure within a range of roughly 60-70% of the capital budget (City of Cape Town, 2007a). Despite this trend, capital expenditure on EWWS services is still budgeted to rise from R1 billion in 2005/6 to R1.8 billion in 2008/9. This amounts to 25% of total budgeted capital expenditure in 2005/6, and 64% in 2008/9.

3. Are these budgeted capital expenditures sufficient to deal with the EWWS backlogs that have to be made up in order to ensure bulk infrastructure and sustainable provision of services? According to South African Cities Network (2007), housing alone in the City will need an estimated 'top-up' of R550m pa (at a subsidy of R25,000 per dwelling unit),[2] with an additional operational cost of R250m pa for the provision of free basic services (with a backlog of 265,000 units at the time). When adjusted in a linear way to the current backlog of 300,100, this figure is likely to be closer to R750m pa over 15 years. According to Cape Town's Integrated Development Plan (CCT, 2006), solid waste would need capital expenditure of between R130m and R230m pa in the period 2007/8 to 2010/11. Water and sanitation services would need R1.1bn in 2007/8 and 2008/9, R840m in 2009/10, and R690m in 2010/11 – with large increases mainly driven by the need for new bulk water infrastructure, wastewater treatment extensions, and sewer reticulation systems. The capital budget for electricity services is also in the order of R500m pa in the period 2007/8-2009/10. Given that these capital budgets were designed to take account of the backlogs and future growth, and are a good reflection of actual costs – which is a strong and, as yet, untested assumption to make – rough estimated capital requirements for EWWS services are in the order of R2-R2.3bn per annum over the next three to four years.

4. Can the City generate enough revenue to finance the required operational and capital expenditure? It is apparent that the city has become increasingly dependent on government grants in recent years. Grants have increased from below 10% of total revenue before 2005/6, to 22% (R3.8bn) in 2007/8. In addition, income from EWWS service charges in 2007/8 are budgeted at R5.2bn, with 80% of service charges from electricity (R3bn) and water (R1.1bn) alone. Sanitation service charges are responsible for around R640m, and refuse service charges for R480m. Property rates bring in an additional R3.5bn. Municipalities, in general, generate surpluses on the sale of electricity to their customers, which in many instances are used to cross-subsidise other costs. National Treasury estimated that the City of Cape Town generated a surplus of 18.2% on the sale of electricity in 2000 (Intergovernmental Fiscal Review 2004 cited in de Wit and Swilling 2008).

5. The nature of inter-governmental grants is such that not all can be attributed to revenue. Government grants consist of an equitable share or an unconditional grant, and a conditional grant, where the municipality has to satisfy the condition of the grant before it can be recognised as revenue (SACN, 2007). Equitable share provisions are unconditional, but mostly associated with free basic services to the indigent. These grants have risen sharply from R105m in 2003 to R464m in 2008/9, and, according to the Division of Revenue Act (DORA 2 2008) allocations, to a provision of R831m in 2010/11 (CCT, 2008). The same rapid rise can be seen in total government grants, from R630m in 2004/5 to R1bn in 2005/6, and to a budgeted R3.8bn in 2008/9. Grants are dependent on the performance of the national economy, and over-dependence creates significant risks to the City in case of a national economic downturn.

6. Tariff increases are limited by several factors, most notably by the Total Municipal Account (TMA) payable by households. Tariff increases are usually aimed at increases not much higher than CPIX (consumer price index), but in case of exceptional external circumstances, such as ESKOM's tariff increase to municipalities, steeper tariff increases were implemented (see Table 3 for proposed

---

2   Needless to say R25,000 is far too low to produce a decent product – most housing practitioners in Cape Town currently assume that nothing less than R75,000 per month can produce a decent product.

tariff increases for 2008/9).[3] ESKOM's tariff increases, in turn, are approved by the National Energy Regulator of South Africa (NERSA). Tariffs are further influenced by the projections for operational expenditure, consumer behaviour and affordability, legal and political balancing of the budgets, and the 5-year cycle effects when new census results become available. All combined, this has the effect that the longer-term affordability of offering EWWS services are not well-configured into the current process, and continues to fuel the need for cross-subsidies and dependency on government grants.

Table 3: Proposed tariff increases 2008/9

	Increase
Rates	7.3%
Refuse	7.5%
Electricity	15.0%
Water	9.2%
Sanitation	6.0%

[Source: IDP Review 2008/9]

In summary, the combined forces of unstable supply of natural resources such as water and mainly fossil-fuel based electricity, environmental pollution from inadequate sanitation and storm water, the rapid filling of landfill airspace, net migration, rising poverty, the commitment to free basic service provision, EWWS services backlogs, housing and infrastructure maintenance backlogs, and high levels of unemployment create massive and complex pressures on the City of Cape Town and the metropolitan economy as a whole. This is happening within the context of a relatively constant proportional budget allocation to the operational costs of EWWS services and an under-spending of capital expenditure on such services. This is the situation that is giving rise to longer-term, integrated reflection on the contextual diversity facing the City with a view to improving longer-term, strategic decision making in such a complex environment.

3. Services for all: from Uniformity to Contextual Diversity[4]

The great majority of Cape Town's households have access to basic services. However, although 98% of households have access to electricity, about 30,000 still lack access to water, and around 40,000 lack access to sanitation and adequate solid waste services. Given that well over 100,000 households live in shacks (most of whom have services), it follows that housing provision has lagged behind service provision. This may have something to do with the fact that housing expenditure requires cooperation across all three spheres of government in order to effectively access and discharge housing subsidies allocated by the National Department of Housing to the Western Cape Provincial Government.

The tension between equity and cost recovery has bedevilled service provision policy in Cape Town since at least 1994. Whereas equity has been the focus of tariff policy, the different EWWS departments have aspired to retain financial surpluses and achieve cost recovery.

Since 2000, tariff policy has significantly benefitted the poor. Water tariffs have included a progressive 5-step structure that resulted in large consumers paying more per litre than small consumers, and a 7% surtax was levied on businesses to cross-subsidise the 6,000 litres of free water that had to be provided to all households in line with national government policy. As for sanitation, the first 4,200 litres were provided free of charge. Solid waste was standardised via a 240 litre bin, and the service was provided free of charge for properties valued at less than R88,000 and heavily subsidised for

---

3   Note: Electricity price increases do not include additional increases by Eskom.

4   This section draws on useful research by Jaglan (2008), enriched by the results of various workshops and discussions with officials conducted during 2007/2008.

houses worth between R88,000 and R160,000. A progressive approach to electricity tariffs has proved much more difficult, because ESKOM supplies approximately one third of Cape Town's households directly. Nevertheless, all consumers on the Domestic 2 Tariff, who on average consume less than 450 kWh per month, get 50kWh free electricity per month. After ESKOM refused to apply this to the areas they serve, the City had to step in to pay ESKOM for this benefit to the poorer households.

As far as rates are concerned, the 2007/8 budget provides for properties valued at less than R88,000 (about 85,000 property owners) to pay zero rates, get free refuse collection, and a basket of other free services. Properties valued at R199,000 receive a R20 per month discount on their rates, and households that earn less than R1,740 per month and are listed on the City's Indigent Register, receive a 100% rates rebate and a R20 per month subsidy on their services account. Finally, since 2003/4, tariff increases have taken into account the affordability levels of poor households – the result being average increases at above inflation, but with much lower increases for poor households (and in some cases even decreases) compensated for by much higher increases for richer households.

Table 4: Key elements of the Progressive Equity Model in Cape Town

Water	First 6000 litres free
Sanitation	First 4800 litres free
Electricity	First 50 Kwh free
Solid waste	Free service on properties valued <R88,000 and heavily subsidised services on properties of R88,000-R160,000
Rates	Properties <R88,000: zero rates
	Properties R88,000> <R199,000: 20% discount
	Households earning < R1740/month receive 100% rebate and R20 subsidy on services account
Tariff increases	Since 2003/4, average increases at above inflation, but with much lower increases for poor households (and in some cases even decreases), compensated for by much higher increases for richer households.

To complement the progressive aims of rates and tariff policies, the general approach to services from the mid-1990's onwards was that the levels and standards applied in the former white areas must be applied to all areas. This had major implications for capital budgets, reinforced by increasingly large intergovernmental transfers. However, it is one thing to extend infrastructure using capital budgets and transfers; it is a completely different matter to make sure that operating budgets are expanded accordingly in order to maintain and repair these infrastructures into the future, and that provision is made in capital budgets for refurbishment and upgrades.

While tariff/rates policy and capital expenditure policy aimed to achieve equity via cross-subsidies, since 2004 the energy and water/sanitation departments have become increasingly strident about the need for so-called 'economic tariffs' for each service. By this they mean a 'corporatised' cost recovery model that would allow each sector to define its own costs of operations (maintenance and repairs) and capital expenditure, so that revenues from the sale of their services could be ploughed back into their sectors, rather than used to cross-subsidise the rates account and corporate services. The underlying reason for this response from these two departments is that the progressive equity model pursued via tariff/rates policy and capital expenditures came to be financed by surpluses creamed off the sale of electricity and water services. Whereas 10%-11% of expenditure by the electricity department in the 1980's and early 1990's was transferred to the rates account, this had increased to 15% by 2004/6. Similarly, the water sector has contributed significantly: the total contribution as a percentage of expenditure increased from less than 11% in 2001, to nearly 19% by 2004/6. The water and electricity departments have argued since at least 2002 that these contributions to the rates account and to corporate services undermine their capacity to finance essential upgrades and repairs.

From 2005 onwards, municipal engineers and the consulting industry started issuing increasingly strong warnings that cross-subsidisation coupled to funding of service expansion to achieve uniform levels and standards of service were undermining the operating budgets. By 2006, major infrastructure projects had to be postponed and serious disruptions, due to under-maintained infrastructures, began to emerge. In 2007, the municipality declared restrictions on new developments in numerous suburbs because of overloading of existing bulk infrastructure, in particular sanitation. These, coupled to rising levels of bad debt, reinforced calls to move towards a sectoral cost recovery model. By the start of 2007, there was a general consensus that the roll-out of basic services to meet the needs of the poor could no longer be at the expense of essential maintenance and refurbishment of existing city-wide infrastructures. Once this principle had been accepted, the choices were to either cut back on capital investments and/or to reduce transfers to the rates account and corporate services.

The N2 Gateway experience during 2006/7 added to the fiscal pressures on the EWWS Departments. This development brought home the realisation that massive increases in housing subsidies, and subsidies for related infrastructure, from National Government would translate into long-term pressures on the municipal operating and expenditure budgets to maintain, repair, and refurbish these infrastructures. Although transfers from the national fiscus via the Equitable Share and related mechanisms are designed to assist in this regard, there have been and will continue to be shortfalls.

Five largely disconnected responses to these complex pressures were discernible during the course of 2006/7:

1. A political decision was made to increase rates and tariffs in the 2007/8 budget by an average of 15%, with higher than average increases for wealthier areas to cross-subsidise the poorer households. The trebling of property values between the General Valuation done in 2000 and the one completed in 2007 certainly helped to support these increases, especially since the latest valuation comprised the actual market value of the total property and not just the estimated land value.

2. There has been increased discussion amongst officials and politicians about the idea of focusing basic service provision and 'free services' on the 'poorest of the poor' (in particular those living in expanding informal settlements). This suggests a move away from the cherished goal of achieving uniform standards and levels of service for all areas financed via a progressive rates/tariffs policy. There is a chance that 'free services' might steadily be narrowed down to benefit only the poorest of the poor (people in informal settlements and registered indigents in formal houses). In turn, this would increase revenue generation from households that previously paid very little because of their property valuations. There are even suggestions that the 6,000 litre free water for all policy could be replaced by free water for the poorest of the poor. This shift in thinking was driven largely by severe fiscal constraints, even though it could have negative political implications in communities that might lose out on subsidies. In Cape Town, this kind of political fallout can also have racial implications, especially if constituencies who have lived in Cape Town for decades waiting for houses perceive such a move as favouring 'new arrivals' (people who have migrated in from other regions in South Africa and Africa).

3. There has been a move towards a reduction in the cross-subsidy of the rates account and Corporate Services from electricity and water services in order to finance a wide range of new infrastructure, upgrades, and refurbishment projects within these respective sectors.

4. A wide range of private businesses, NGOs, CBOs (Community-based organisations), entrepreneurs, and informal sector operators have been included into service delivery systems. Typically, these stakeholders are more responsive to the unique conditions of each specific context. This is particularly evident in the solid waste sector, but increasingly so in the electricity and water sectors, and cleansing as well. Examples include the mushrooming of community-based waste collectors in the poorest areas, the business-subsidised City Improvement Districts in the CBD, independent energy services companies (ESCOs) in the electricity sector, totally self-managed large-scale property developments like Century City, sub-contracted water piping repair services, contracted-out meter reading, and a new generation of black-empowered waste removal and recycling operators sub-contracted by the Waste Department. Recognising contextual specificity and including non-state operators into the value chain will inevitably lead to a move away from

uniformity ('one-size-fits-all') with results that could – but not necessarily – have beneficial consequences for the urban poor. Because these kinds of diversified arrangements entail the restructuring of labour contracts, they rapidly become the focus of conflict between the South African Municipal Workers Union (SAWMU) – who opposes job losses/casualisation – and the City that defends its mission to be more efficient.

5. The termination of the proposed Regional Electricity Distributor (RED1) arrangement, which anticipated the establishment of a fully corporatised electricity distributor for metropolitan Cape Town. This would have been the precursor to full cost recovery and the (gradual) curtailment of the transfers of profits from electricity distribution to the rates account. An aggressive intervention to defend the City's dependence on the approximately R300 million it gets from electricity surpluses provided officials with a clear signal that the City Council was going to be very pragmatic when it comes to the management of municipal finances.

Whether or not these five responses could combine to generate solutions to the complex mix of pressures and problems confronting decision makers remains to be seen. Much, however, will depend on whether the economy can grow in a way that –

– reduces the levels of unemployment,
– improves the incomes of the working poor, and
– increases the size of the middle class.

If these three conditions are not met, moving away from the current 'free services' could exacerbate poverty, and increasing redistributive demands on businesses and richer households could stimulate relocations to other localities and discourage locational decisions in favour of Cape Town. Both trends could undermine job-creating economic growth, especially in light of rising fuel costs, food prices, and the onset of a national economic downturn that was exacerbated by the global economic crisis since 2007/8.

Unfortunately, the complex relationships between service provision, municipal revenues, and growth and poverty are characterised by numerous negative feedback loops.[5] If the water and energy sectors found a way to claw back more of their surpluses, this would increase pressure on the rates account and corporate services budget, ultimately resulting in expenditure cutbacks – and possibly even retrenchments. The huge jump in electricity tariffs necessitated by ESKOM's programme to increase primary generation will not only increase the living costs of Capetonian households, it will also crowd out any attempts by the City to extract higher incomes from rates and non-electricity services. To add to the pressures, new rates policies point towards increased revenues from residential areas (albeit more limited than in the past), but not from commercial and industrial areas, especially with respect to publicly-owned properties (Republic of South Africa, 2006). Intergovernmental transfers, discounted loan finance, and income from municipal bonds will all contribute to increase revenues for capital investments in infrastructure, but only limited funds for ongoing maintenance and repairs.

This, in turn, could either constrain new investments or increase pressures to recover costs, again with negative impacts on the poor. While the CCT experiences demands for extending infrastructure and improving operations and maintenance in response to population and economic growth, the National Treasury has adopted a policy of capping the growth in what is referred to as the Total Municipal Account (TMA), in accordance with national inflation targets. However, Cape Town's future growth rate (including the now increasingly unrealistic 6% target) is surprisingly dependent on land (re-)development and the associated extension and upgrading of infrastructure. The CBD, for example, is attracting billions of Rands of investment, which is constrained by severely limited EWWS capacity in the CBD. All this suggests that nationally imposed constraints on expanding municipal expenditure (no matter how justifiable) could, if applied mechanistically, clearly retard Cape Town's economic growth, thus exacerbating its problems over the long term. Equally, if the economy grows, this will

---

5   The analysis in this paragraph was generated from the results of a workshop with existing and former officials from the Municipality's Finance Department.

exacerbate inward migration by job seekers and increase demands for improved infrastructure from businesses wanting to expand, thus increasing service delivery pressures, which, in turn, will require increases in the size of the TMA that will fall foul of National Treasury's inflation targets.

To make matters worse, if the wage levels of the lowest paid do not keep up with CPIX (as has been the trend), then investments in extending basic services to the poorest of the poor have no chance of ever being recovered. In other words, if accelerated economic growth depends on ultra-cheap labour (in, for example, the construction sector which is booming), this will further undermine the municipality's access to households who can pay for their services. Alternatively this could result in a growing number of unserviced informal settlements as EWWS Departments simply steer clear of investments in locales where costs cannot be recovered. Finally, as long as capex budgets are used to extend basic services to the poor (as they should), without making provision for investments in refurbishment, upgrading and maintenance of existing infrastructure, pressures on operating budgets to keep collapsing infrastructure going will relentlessly continue with inevitably diminishing returns. However, 2007 did see major decisions being taken to upgrade and refurbish key infrastructures in the EWWS sectors, which is why revenue increases of 15% were necessary and a quiet shift away from costly 'free services for all' began to be considered.

The story line thus far has been about extremely complex manipulations and trade-offs driven by the need to address poverty via redistributive measures to finance extended EWWS services without undermining economic growth. The dominant casualty has been the capacity to maintain, repair, upgrade, and refurbish existing and newly created bulk infrastructure. All this was happening within a nationally set economic policy framework that until 2004 did little to significantly reduce unemployment, or increase the incomes of poor and working class households. The decline of Cape Town's manufacturing sector was simply part of a national trend dominated by the rapid expansion of the services sector and resource-based industries (Republic of South Africa, 2007; Hirsch, 2006; Bhorat *et al.*, 2005, King & Levine, 1994). Ideally, the municipality should be extending basic services to poor households that benefit from significant increases in the number of jobs available and wage increases that keep up with inflation. Failing that, their inability to pay for these services will of necessity need to be financed via horizontal and vertical redistributive mechanisms that could fall foul of nationally imposed inflation-linked caps. The good news is that, since 2004, formal employment levels in Cape Town did start to increase, a trend that ended by 2008.

The question now is: To what extent could a sustainable resource use approach contribute to a solution? The answer is probably 'not much' if a short time frame is adopted, but perhaps more optimistic from a medium- to long-term perspective. The potential of a sustainable resource use approach to infrastructure is that it could result in system-wide efficiencies ('doing more with less') and cost reductions at strategic points in the system (in particular capital cost and certain operating costs). The ability to achieve sustainable resource use first warrants an analysis of the natural resource system and its relation to the economy and population. This will be explored in the next section.

4. Sectoral Approaches

The City of Cape is the economic powerhouse of the Western Cape, accounting for more than three-quarters of the province's economic activity (Western Cape Provincial Government, 2007). The City's economy grew at an average annual rate of more than 4% between 2000 and 2006, reaching a high of 5.4% in 2004. The economy is increasingly orientated towards financial and business services, as well as wholesale and retail trade (which includes tourism). From 2000 to 2004, the economy was mostly driven by consumption, but a shift towards fixed investment spending in areas such as construction, electricity, gas and water, finance and business services, and trade stimulated growth during the period 2004-2008. Although public sector-led infrastructure investment remains the primary strategy for economic growth, the global economic crisis placed severe constraints on what is possible and affordable.

As already suggested, the good news is that, although employment and unemployment statistics are not considered to be very reliable, there is some optimism that overall unemployment in the City may be on the retreat. According to the annual Labour Force Survey, unemployment (in the age group 15-64, excluding discouraged job seekers) was measured at 21% in September 2005, compared to 15% in September 2006 (Republic of South Africa, 2007). It is interesting to note that this estimated drop in unemployment took place at the same time that the economy was not significantly increasing its growth effort, possibly signalling that underlying structural factors have started to be addressed, rather than growth itself creating new employment. The key point for this analysis is that absolute figures on unemployment are still substantial though, amounting to almost 225,000 unemployed people in the City, compared to 212,000 in 2000 and 175,000 in 1995 (City of Cape Town, 2006).

If we assume that this will continue resulting in declining unemployment levels as fixed investment accelerates, then very close attention needs to be paid to the budget planning frameworks that are used in the EWWS Departments. It would be a great tragedy if these planning frameworks resulted in expenditures that effectively killed off this growth period before it can gain real and meaningful momentum.

Most EWWS sectors do their forward planning by correlating growth in demand for their services to economic and population growth projections. The evidence, however, suggests that there is a decoupling of resource consumption from economic and population growth in certain EWWS sectors. This has major policy implications for capital budgets and operational expenditure.

The City experienced large year-on-year changes in the use of most services, most notably waste and energy use, and most notably from 2003/4 onwards. This does not mean, however, that economic and population growth necessarily translates into higher pressure on all EWWS services. When measured over the period 2000 to 2006, stronger sensitivities to growth are apparent in energy use and solid waste generation, but lower sensitivities are apparent for electricity use and influent received at wastewater treatment plants. The percentage change in water used per one percentage change in economic and population growth is the most volatile, and even negative for five out of the ten years. An interesting disparity is that, although growth in energy use has strongly increased from 2004 to 2006, growth in electricity use has declined from a high year-on-year growth of 4.8% in 2001/2 to 2.1% in 2006/7.

This analysis suggests that future projections of the demand for EWWS services cannot simply be generated via the linear extrapolation of economic growth or a population coefficient. The growth factors that are used for projections by the City are as follows:

- **For electricity use:** 2.7% in 2007/8 and 3.5% thereafter (City of Cape Town, 2007a). The 2008 IDP review factored a 0% increase of electricity growth, mainly due to the implementation of an energy savings plan, but this was only for one year (City of Cape Town, 2008).
- **For water use:** 3% pa unconstrained demand (Department of Water Affairs and Forestry cited in de Wit *et al.*, 2008)
- **For solid waste:** 7% (without waste minimisation) up to 2007 (City of Cape Town, 2007b), but since then declining rapidly.

The growth factors referred to above are important projections that provide strategic estimates of required capital and operating expenditure on EWWS services. However, they are out of line with actual rate of growth of these resource-intensive services. This is especially the case for projections on water use. The average growth in water use from 2001 to 2006 was -2%, and for the ten years from 1996 to 2006 only 0.04%, compared to an average economic growth of 3.9% pa over the same period. It is the volatility in the growth of water use that complicates planning for future water use. Water use growth peaked at 4.9% in 2003/4, compared to -12.4% year-on-year growth in 2000/01. The obvious driver here was the 2001/02 drought, coupled to municipal awareness campaigns and enforced restrictions. This proves the point: consumption behaviour can be changed.

Electricity is a similar story. The growth in electricity use over the last few years is also lower than the current estimates, which are tied to economic growth projections (excluding the single reference to 0% growth in the IDP). Growth in electricity use is expected to be even lower as the effects of rising electricity tariffs from 2007 onwards translate into consumption behaviours.

Solid waste grew by an average of 8.9% over the period 2001 to 2006, with a high of 20.4% in 2003/4, and a low of 2.6% in 2001/2. A projected growth factor of 7% triggered the introduction of the City's waste recycling experiment with 60,000 households in 2007/8, resulting in significant declines. This, plus other waste characterisation studies, could generate the knowledge basis for turning solid waste management into a sustainability leader. The problem, however, is that solid waste recycling might save on landfill costs, but these savings might be required to fund the recycling system. The advantage, of course, is that waste recycling translates immediately into job-creating growth along some long value chains, from collection right through to high-tech plastics manufacturing.

In short, there is a disjuncture between the actual year-on-year rates of growth and the growth rates used by budget planners, which are derived in large part from economic and population growth coefficients. By contrast, planning for sustainable resource use needs to focus how to decouple growth rates from rates of resource use.

## 5.	Conclusion: Financial Implications of Sustainability Interventions

We conclude on a hopeful note. There is evidence that without much collective or concerted effort, the rate of consumption of materials and energy (natural resources) in the Cape Town metropolitan economy is starting to decouple from economic growth rates, especially with respect to the energy and water sectors. However, even in the waste sector, the shock effect of extraordinarily large increases in the quantities of solid waste in recent years, coupled to extremely rapid increases in transportation and landfill costs, has already triggered significant recycling interventions by the City and private sector players. At the same time, despite a slow-down in economic growth rates since 2006, unemployment levels (as a percentage of the labour force) seem to be dropping as fixed investment – rather than purely consumption – kicks in as an economic driver.

Are these two trends coincidental, or are there some causal linkages? We think that there are causal linkages. Our conclusion is that job-creating growth – that is growth that can benefit the poor and not just the consumption-addicted middle class – can actually be stimulated by municipal interventions. Ideally, these could reinforce, rather than reverse, the beginnings of the decoupling trend that we have identified. These qualitative interventions, however, will depend to a large extent on a new approach to budgetary planning and financial management that is inspired by the potential returns of a sustainable resource use approach.

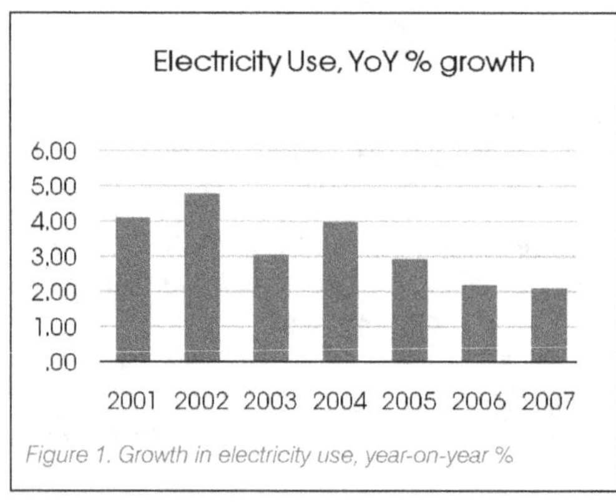

Figure 1. Growth in electricity use, year-on-year %

Our key conclusions are, therefore, as follows:

1. Budgetary planning in the EWWS sectors can no longer rely entirely on economic and population growth coefficients as a basis for forward projections for capital and operating expenditure. This is because of the decoupling trends that have been identified (water, electricity, solid waste). A more complex planning matrix is required, supported by decision-support systems that can cope with complexity (De Wit and Swilling, 2008).

2. Decoupling in the water sector was caused by the City's response to drought, and by the impact of the energy crisis that started with the 2006 blackouts. Decoupling in the waste sector, however, is potentially more promising, because it is a coordinated and concerted effort. These intended and unintended causes of decoupling suggest that much more could be achieved if there was a concentrated multi-stakeholder drive, coupled to a reduction of (often implicit) subsidies for wasteful behaviour and prices that reflect the true cost of scarcity, to significantly reduce resource consumption in the EWWS sectors and beyond. Good examples include the energy efficiency drive by the Cape Town Partnership – working in collaboration with the City and building owners in the public and private sectors – and the adjustment of water tariffs in response to recent droughts.

3. It needs to be recognised, however, that sustainable resource use approaches will negatively affect municipal revenues if existing tariff structures and rates policies remain unchanged. To cite a few obvious examples: selling less electricity and water means declining revenues from these trading services; less waste transported to the landfills means declining revenues from landfills; more waste depots to recycle waste means higher operating costs; and the creation of markets for recycled waste might mean that higher value waste items are bought up by private sector players. Green building bylaws could result in commercial and residential buildings that use 40% less water and 60% less energy per square meter of floor space, including low-income housing; they might even capture all their solid waste and sell it to private sector operators. These and many other examples point to the need for a fundamental rethink of tariff and rates policies that anticipate, stimulate and promote resource efficiencies and more sustainable resource use. The bottom line is that the cost per unit will need to go up to incentivise efficiencies, while reinforcing the progressive tariff structure that Cape Town has pioneered.

4. The diversification of service delivery agents should be welcomed rather than opposed, because it creates openings for innovation. Innovation, supported by public sector investments in social learning and learning networks, will be the key enabler of new approaches utilising new technologies and institutional arrangements. However, trade unions should be incorporated into – rather than excluded from – these restructuring processes, and pro-poor goals should at all times be the primary focus.

5. Sustainable resource use approaches can have numerous positive job-creating economic spin-offs. Solid waste recycling has already been referred to. Giving waste items that litter communities a financial value will generate livelihood opportunities for thousands of informal sector operators. At the same time, manufacturers in the Cape Town region will have a distinct advantage because they will be able to purchase materials at reduced cost. For example, plastic manufacturers who are obliged to buy virgin polymer from Sasol (at prices linked to the global oil price) are desperate to purchase recycled plastics – in fact, the fastest growing firms in the plastics sector in Cape Town are now recyclers. Other examples include many opportunities for researchers to work with innovative entrepreneurs and community-based organisations to develop new construction and solar energy systems. And the rapid rise in food prices has encouraged the creation of new short-run agricultural value chains that generate food supplies grown by urban and peri-urban farmers, using tried and tested organic farming technologies, for city dwellers.

6. By enhancing the City's capacity to fully spend the capital budgets of, in particular, the EWWS Departments, substantial capital injections will be made into the Cape Town metropolitan economy, which will reinforce the kind of growth that has pro-poor effects and benefits. Consistent underspending of capital budgets means that unspent funds banked via national banking systems are in all likelihood being loaned out to businesses operating in other localities. This does very little to complement the increasingly significant quantities of fixed capital investment that currently drive growth.

References

Behrens, A., Giljum, S., Kovanda, J. & Niza, S. 2007. The material basis of the global economy: worldwide patterns of natural resource extraction and their implications for sustainable resource use policies. *Ecological Economics*, 64:444-453.

Bhorat, H., Oosthuizen, M. & Poswell, L. 2005. *The Post-Apartheid South African Economy in Perspective: Growth, Poverty and Economic Policy*. Cape Town: Development Policy Research Unit, University of Cape Town.

City of Cape Town. 2006. *Integrated Development Plan*. Cape Town: City of Cape Town. Available from www.capetown.gov.za. (Accessed: 4 February 2007)

City of Cape Town. 2007a. *Annexure A. Budget 2007/2008 to 2009/2010*. Cape Town: City of Cape Town.

City of Cape Town. 2007b. *Report to the Chairperson: Utility Services Portfolio Committee Solid Waste Management. Department Sector Plan for Integrated Waste Management and Service Delivery in Cape Town*. Cape Town: City of Cape Town. Unpublished document.

City of Cape Town. 2008. *5 Year Plan for Cape Town: Integrated Development Plan (IDP) 2007/8-2010/12*. Cape Town: City of Cape Town.

City of Cape Town. 2008. Integrated Development Plan Review. Cape Town: City of Cape Town.

Clark, G., Dexter, P. & Parnell, S. 2007. *Rethinking Regional Development in the Western Cape*, Cape Town: University of Cape Town.

Crane, W. & Swilling, M. 2008. Environment, Sustainable Resource Use and the Cape Town Functional Region – An Overview. *Urban Forum*, 19:263-287.

De Wit, M. & Swilling, M. 2008. *Sustainable Urban Development in Cape Town: Planning for Natural Resource-Based Service Provision with a Systems Dynamics Model*. Winelands Conference on Sustainable Futures, April 2008, Spier Wine Estate, Stellenbosch: Stellenbosch University.

Fischer-Kowalski, M. & Haberl, H. 2007. *Socioecological Transitions and Global Change: Trajectories of Social Metabolism and Land Use*. Cheltenham, U.K.: Edward Elgar.

Gasson, B. 2002. *The Ecological Footprint of Cape Town: Unsustainable Resource Use and Planning Implications*. National Conference of the South African Planning Institution, 18-20 September 2002, Durban.

Guy, S., Marvin, S. & Moss, T. 2001. *Urban Infrastructure in Transition*. London: Earthscan.

Heynen, N., Kaika, M. & Swyngedouw, E. 2006. *In the Nature of Cities: Urban Political Ecology and the Politics of Urban Metabolism*. London and New York: Routledge.

Hirsch, A. 2006. *South Africa's Development Path: Government's Programme of Action*. Pretoria: Office of the President. (Unpublished document.)

Hodson, M. & Marvin, S. 2009. Urban Ecological Security: A New Urban Paradigm? *International Journal of Urban and Regional Research*, 33(1):193-215.

Hurrell, A. & Woods, N. 1995. Globalisation and inequality. *Millenium*, 24(3):447-470.

Intergovernmental Panel on Climate Change. 2007. *Climate Change 2007*. Geneva: United Nations Environment Programme.

Jaglan, E. 2008. Differing networked services in Cape Town: echoes of splintering urbanism? *GeoForum*, 39(6):1897-1906.

Jaglan, S. 2004. Water Delivery and Institution Building in Cape Town: The Problems of Urban Integration. *Urban Forum*, 15(3):231-253.

Khosa, M. M. 2000. *Empowerment Through Service Delivery*. Pretoria: Human Sciences Research Council.

King, L. & Levine, R. 1994. Capital Fundamentalism, Economic Development and Economic Growth. Washington, DC: *Carnegie-Rochester Conference Series on Public Policy*, 40:259-292.

MCA. 2006. *Capital Investment Patterns in Cape Town: 2001-2005*. Cape Town: Unpublished Research Report Commissioned by the Sustainability Institute.

McDonald, D. & Smith, L. 2002. *Privatizing Cape Town: Service Delivery and Policy Reforms Since 1996*. Johannesburg: Municipal Service Project, Occasional Paper Series No. 7.

Mebratu, D. 1998. Sustainability and Sustainable Development: Historical and Conceptual Review. *Environment Impact Assessment Review*, 18:493-510.

Miraftab, F. 2004. Neoliberalism and Casualization of Public Sector Services: The Case of Waste Collection Services in Cape Town, South Africa. *International Journal of Urban and Regional Research*, 28(4):847-892.

Parnell, S. & Pieterse, E. 2007. Developmental Local Government. In: Parnell, S., Pieterse, E., Swilling, M. & Wooldridge, D. (Eds.). *Democratising Local Government: The South African Experiment*. Cape Town: University of Cape Town Press.

Petrie, B. & Ocran, M. 2007. *Socio-Economic Considerations in Responding to Climate Change in the Western Cape*. Cape Town: OneWorld Sustainable Investments prepared for Department of Environmental Affairs and Planning, Western Cape Government. Supplementary Report No. 7.

Pezzoli, K. 1997. Sustainable Development: A Transdisciplinary Overview of the Literature. *Journal of Environmental Planning and Management*, 40(5):549-574.

Republic of South Africa a. Municipal Structures Act (Act no. 117 of 1998 as ammended Act no. 33 of 2000). *Government Gazette*, No. 19614: Volume 402. 1998.

Republic of South Africa b. Department of Trade and Industry. 2007. *A National Industrial Policy Framework*. Pretoria: Department of Trade and Industry.

Republic of South Africa c. Statistics South Africa. 2007 *Labour Force Survey 2006*. Pretoria: Statistics South Africa.

Republic of South Africa d. 2006. Local Government: Municipal Property Act (Act 6 of 2004): Amendment of the Municipal Property Rates Regulations, Act 2006. *Government Gazette* Number 32187 – Regulation 468, 2006.

Smith, L. 2004. The murky waters of the second wave of neoliberalism: corporatisation as a service delivery model in Cape Town. *GeoForum*, 35:375-393.

Smith, L. & Hanson, S. 2003. Access to water for the urban poor in Cape Town: where equity meets cost recovery. *Urban Studies*, 40(8):1517-1548.

South African Cities Network. 2007. *State of the Cities Report*. Johannesburg: South African Cities Network.

Stern, N. 2006. *Stern Review: Economics of Climate Change*. London: Chancellor of the Exchequer, Government of the United Kingdom.

Sustainability Institute. 2007a. *Oude Molen Sustainable Neighbourhood*. Report Commissioned by Department of Public Works, Western Cape Provincial Government, pp. 1-82. Cape Town: Sustainability Institute.

Sustainability Institute. 2007b. *Energy: City of Cape Town Integrated Analysis Baseline Report*. Integrated Resources Management for Urban Development. Stellenbosch: Sustainability Institute.

Sustainability Institute. 2007c. *Integrated Analysis Solid Waste Baseline Report*. Stellenbosch: Sustainability Institute.

Sustainability Institute. 2007d. *Water and Sanitation in the City of Cape Town*. Stellenbosch: Sustainability Institute.

Swilling, M. 2004. Rethinking the Sustainability of the South African City. *Development Update*, 5.

Swilling, M. 2006. Sustainability and Infrastructure Planning in South Africa. *Environment and Urbanization*, 18(1):23-51.

Swilling, M. & Annecke, E. 2006. Building Sustainable Communities in South Africa: the Lynedoch Case Study. *Environment and Urbanization*, 18(2):23-50.

United Nations. 2005. *Millenium Ecosystem Assessment*. New York: United Nations. Available from http://www.maweb.org/en. (Accessed on 12 August 2006.)

Van Donk, M., Swilling, M., Pieterse, E. & Parnell, S. (Eds.). 2008. *Consolidating Developmental Local Government: Lessons from the South African Experience*. Cape Town: Isandla Institute and University of Cape Town Press.

Watson, V. 2002. *Change and Continuity in Spatial Planning: Metropolitan Planning in Cape Town under Political Transition*. London and New York: Routledge.

Western Cape Provincial Government. 2007. *iKapa Provincial Growth and Develoment Strategy*. Cape Town: Department of the Premier, Western Cape Provincial Government.

Wilkinson, P. 2004. Renegotiating Local Governance in a Post-Apartheid City: The Case of Cape Town. *Urban Forum*, 15(3):213-230.

World Wildlife Fund, Zoological Society of London & Global Footprint Network. 2006. *Living Planet Report 2006*. Gland, Switzerland: WWF, ZSL and Global Footprint Network.

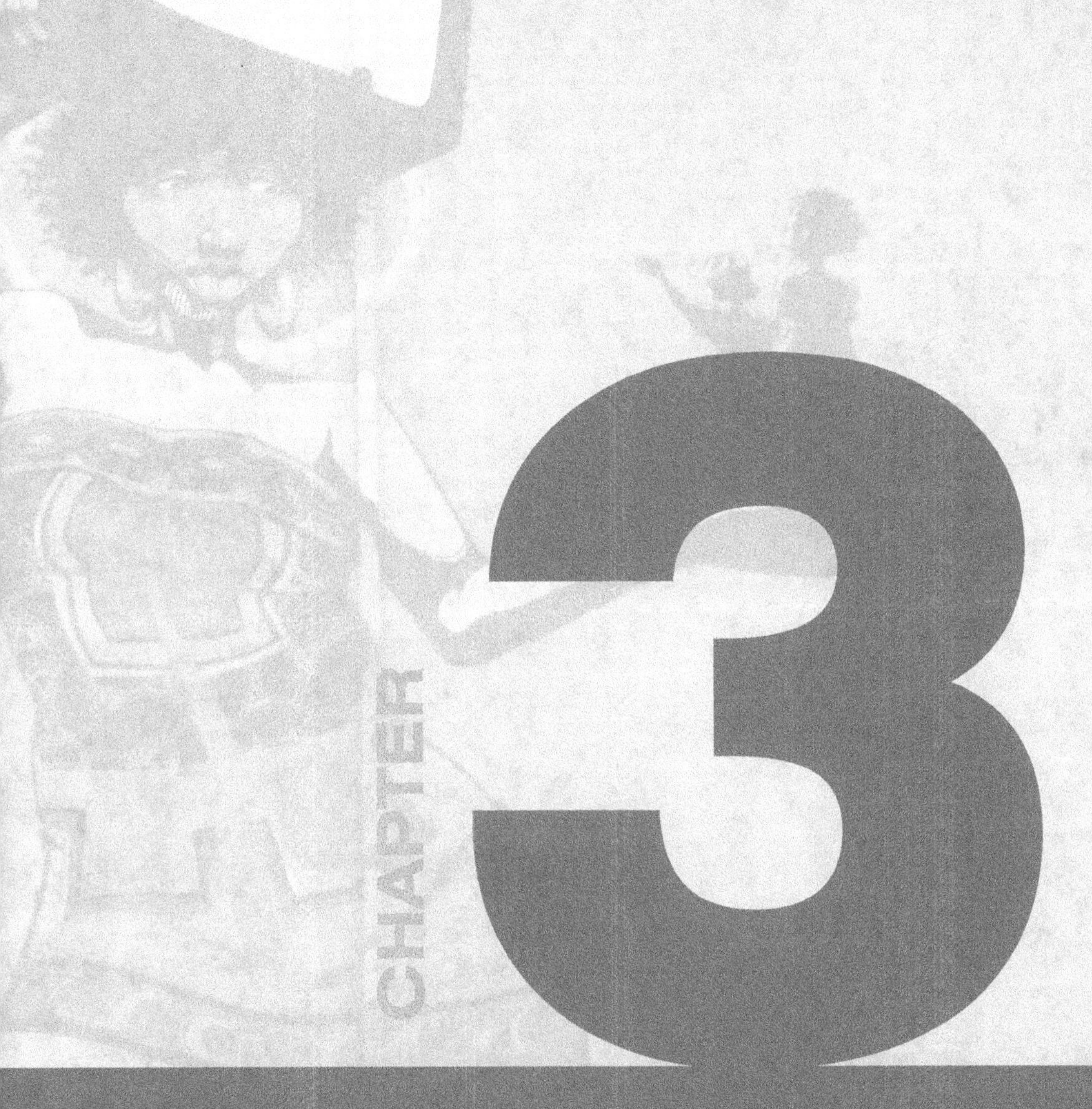

3

Cape Town DenCity

Towards Sustainable Urban Form

Kathryn Ewing & Nisa Mammon

"We are far from having a shared vision of an ideal human environment. It is said that world sustainability will depend to a large degree on what will happen in cities, particularly in fast-growing cities in developing countries. Yet, what is a sustainable city?"

Enrique Peñalosa, 2007: 319

1. Introduction

Most discourse on sustainability places emphasis on the environmental agenda ('green and brown' issues). As such, it includes pollution, energy and water consumption, urban waste, ecological footprinting, and so on, as well as macro-scale processes, issues of governance and institutional capacity (Keiner *et al.*, 2005; Burgess *et al.*, 1997). Although vital for dialogue, most of this theory largely sidelines the challenge of designing cities as liveable, adaptable environments; abundant in culture, heritage and identity (Zetter and Watson, 2006). This chapter not only intends to address achievable goals towards sustainable urban form through the concept of densification, but also promotes sustainability from a design perspective,[1] highlighting the importance of the relationship between design and densification.

Challenges to urban form in Cape Town are explored, outlining the ways in which separation, rather than inclusion, has characterised the development of urban form. We also examine and expose the various reasons why the range of urban forms and public space fall short of its potential as agents of urban development.

We ask the following key questions: Does densification improve or decrease sustainability? Conversely, how does urban sprawl promote or prevent sustainability, and why? The chapter documents the fragmentation and the nature of urban sprawl in the city, and makes a strong case that sustainability can only be achieved by promoting a compact city and through a general spatial strategy of densification of the built area. Crucial goals required for a more holistic and systems approach are identified. These include goals about how Cape Town can shift towards a more sustainable approach to urban form by better understanding the role of land and landscape; how mobility and land use can be integrated over time; how multiple elements of the urban environment can be integrated into an urban sustainability agenda, such as transport, which connects different public spaces and integrated infrastructure. We examine how urban development processes can provide sustainable approaches to designing cities, neighbourhoods, places, and spaces in Cape Town. The chapter concludes with suggestions for a range of targets and practical recommendations aimed at more sustainable strategies.

---

1   Urban planning and design concerns the arrangement, appearance, and functionality of cities, and in particular the management and shaping of urban public space (i.e. the 'public environment', 'public realm' or 'public domain'), and the way public places are experienced. Public space includes the totality of spaces used freely on a day-to-day basis by the general public, such as streets, parks, bicycle ways, markets, and public infrastructure (i.e. PT facilities, bicycle storage, trading, ablutions, town halls, markets, etc.).

2.   Challenging Sustainable Urban Form in Cape Town

This chapter makes the case that there is a direct relationship between **urban form**[2] and **density**[3] in an urban context. Cities are composed of multiple urban forms, including buildings, roads, parks, and infrastructure. Buildings and infrastructure are among the largest culprits in contributing to global carbon emissions.[4] The urban systems[5] supporting the buildings and infrastructure rely on extensive energy and water consumption. They also generate waste, air and water pollution, and produce harmful by-products (Power, 2007). On the other hand, buildings and infrastructure also provide venues for learning, shelter for families, spaces for recreation, and places for income generation, all of which are vital for vibrant urban environments. These two contradictory aspects of urban form therefore need to find a balance between the urban environment and its capacity to absorb built form. This involves taking account of efficiency, cost effectiveness, limited resources, fair distribution, and equitable access.

One of the major problems in South African cities is the type, nature, and character of urban form. Sennett (2007: 292) asserts that "Today's ways of building cities – segregating functions, homogenizing populations, pre-empting through zoning and regulation the meaning of place – fail to provide communities the time and space needed for growth." This relates directly to Cape Town, where current planning patterns do little to inspire sustainability or resolve the relentless pressures of urban growth, given the familiarity of the settlement pattern and form of Cape Town. Further challenges to crafting sustainable urban form are social perceptions and individual aspirations towards suburbanisation.

2.1   Fragmented cityscapes and urban sprawl[6]

As a result of modernist (apartheid) 'reactive' planning and design practice, the current form of urban development in Cape Town is grossly unsustainable. Although National, provincial and municipal policies support and encourage sustainable human settlements,[7] both the public and private sector continue to roll out inappropriate urban form. Between the years 1985 to 2005,

---

2   Urban planning and design concerns the arrangement, appearance, and functionality of cities, and in particular the management and shaping of urban public space (i.e. the 'public environment', 'public realm' or 'public domain'), and the way public places are experienced. Public space includes the totality of spaces used freely on a day-to-day basis by the general public, such as streets, parks, bicycle ways, markets, and public infrastructure (i.e. PT facilities, bicycle storage, trading, ablutions, town halls, markets, etc.).

3   Density simply refers to "the amount of available space per person" (Maas *et al.*, 1998: 1). Densification (as defined by the City of Cape Town, 2008: 4) is "the increased use of space both horizontally and vertically within existing areas/ properties and new developments accompanied by an increased number of units and/or population thresholds."

4   Cities consume about 75% (50% buildings – construction, lighting, heating, air-conditioning, electricity-based appliances; 25% transport and 25% industry) of world energy, of which 79% is from fossil fuels (Ward, 2008: 4), and contribute an equal amount to global pollution (Battle, 2007: 391). In the United Kingdom, for example, buildings and infrastructure produce 50% of carbon emissions (Power, 2007: 364). One ton of CO_2 emissions occupy 556m3 of space at 25ºC at standard pressure. An Olympic size swimming pool is 2500m3. The average South African household consumes 12.81 tons of CO_2 per year. This can fill almost three Olympic size swimming pools (http://www.capetown.gov.za).

5   Urban systems include activities involved in trade, commerce, services, manufacturing, transport, and so on.

6   Urban sprawl generally has negative connotations due to extreme health and environmental concerns, but it is defined in this paper as described in a CCT Report (2006: 78) as the "gradual and uncontrolled spread of urban areas into the surrounding natural areas."

7   Sustainability principles are also encapsulated in other policy, such as municipal level Integrated Development Plans (IDP's); at provincial level in the form of the Western Cape Sustainable Human Settlement Strategy (WCSHSS) (DLGH, undated); and breaking New Ground (BNG) (Republic of South Africa, 2004) at national level. The BNG approach to housing delivery seeks to address the fundamental mismatch between delivery and development in the urban environment. This policy is a comprehensive and approved housing delivery policy programme that acknowledges the need to see housing as an instrument to spatial restructuring by creating sustainable human settlements. However, as stated by Mchunu (2006), South Africa is producing new forms of fragmentation, exclusion, and conflict based on wealth, rather than along racial lines (e.g. gated communities, road closures, business parks, shopping malls), with little regard for culture and public space, which is far from developing sustainable human settlements.

Cape Town has increased in area by 40%.[8] What is alarming is that growth has been characterised neither by coordinated direction and management, nor by alignment with infrastructure provision, capacity, and appropriate spatial planning (CCT, 2006). As mentioned in Chapter 10, transport is still dominated by private vehicles, rather than by efforts to promote clean, efficient, affordable, and safe public transport (PT) and non-motorised transport (NMT) that are inherently associated with more appropriate land uses. This has resulted in significant impacts on the natural and spatial environment of Cape Town, producing negative social and economic conditions.

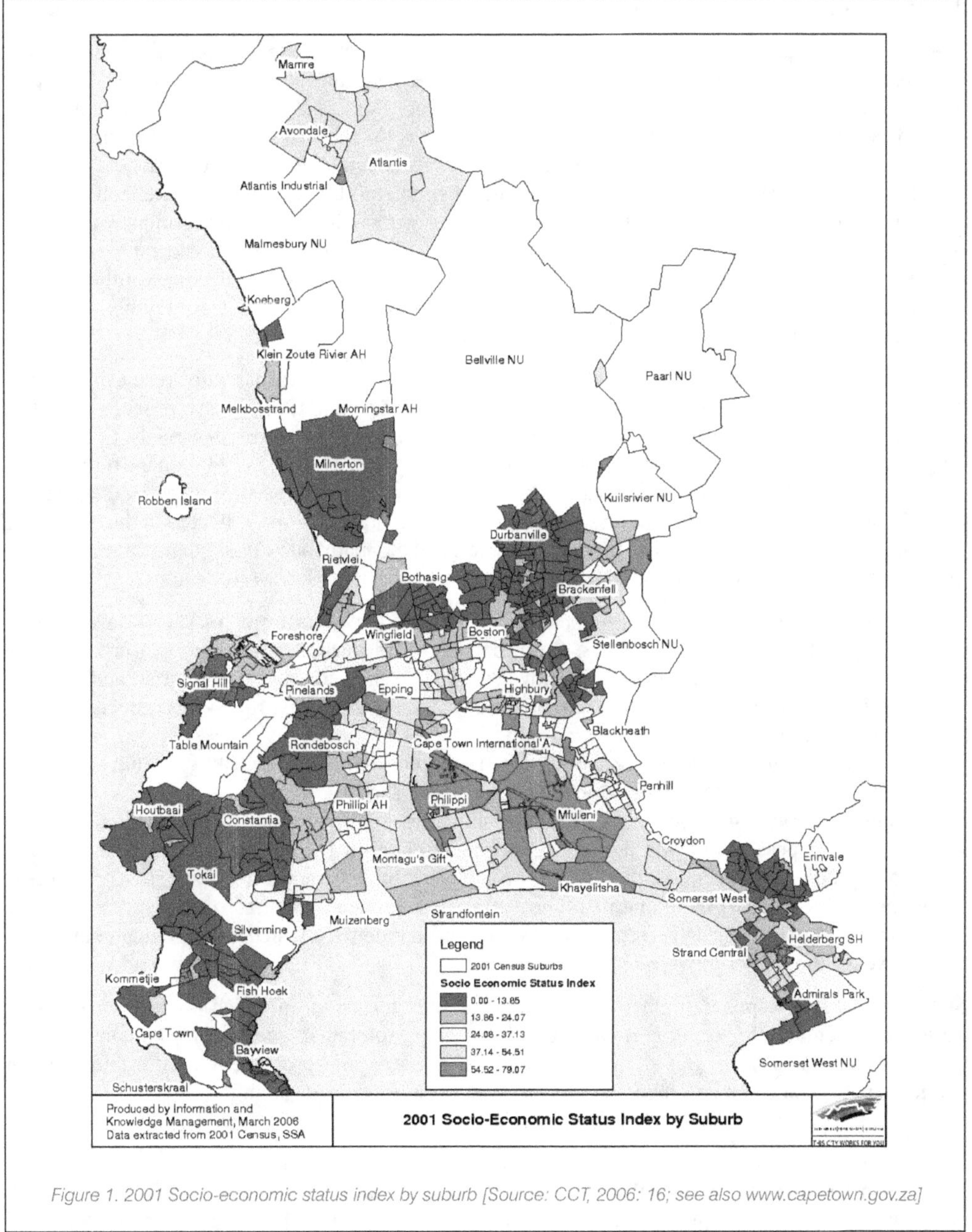

Figure 1. 2001 Socio-economic status index by suburb [Source: CCT, 2006: 16; see also www.capetown.gov.za]

---

8 From 1977 to 1988, Cape Town developed by an average of 701ha/year. In 2005, the city was developing at a rate of 1,232ha/year (CCT, 2006: 25).

Figure 1, which illustrates a socio-economic status index by suburb for Cape Town in 2001, reveals a city of immense inequity. Low-income settlements are generally inappropriately located, poorly adapted to local needs, and badly planned. The majority of households living in absolute poverty are clustered in the Metro South-East, which experiences the highest (and growing) density. Formal housing in these areas is largely built according to standardised formulae and concepts obsessed more with quantity than the quality of the home or the public realm. Informal housing is relatively dense, and these areas are typically deprived of adequate water supply, sanitation, solid waste management, education, and healthcare. Middle-income suburbia circles this low-income cluster to the north, west, and east in a horse-shoe type band.

The wealthy in Cape Town, however, continue to build unsuitable, dispersed developments on 'greenfields' that should ideally remain productive land in the urban economy. Examples of this tendency can be seen in the subdivision of large plots in Constantia and Durbanville, or the development of residential 'green' estates on or beyond the urban edge, for example in the Winelands. These trends exemplify responses to a market-led 'need' for security, privacy, and private investment. These wealthy estates – such as the Boschendal development north of Stellenbosch - are often justified on the grounds that 'development' that attracts large-scale investment will benefit the rural poor and farm workers. However, the land that gets developed should be reserved for food security and the rural economy. As a result, we are clearly "trying to solve new problems with outdated perceptions and planning" (Geis & Kutzmark, 2006). In Cape Town, 20% of the housing value in the city takes up 40% of developed land (CCT, 2006: 26).

Extreme differences in housing patterns between the wealthy and the urban poor result in spatially, financially, and ecologically flawed processes. These include limitations in the process of public participation right through to construction and occupation on the ground. Collectively, this inhibits long-term sustainability. Associated with these residential patterns, is the predominant tendency to develop separate, internalised shopping malls, accessible mainly by expensive motorised transport. There is also the decentralisation of retail and business to separate entities, reflected in the increasing movement of businesses out of the inner city to places like Canal Walk. Consequently, urban sprawl in Cape Town has resulted in a spatial pattern and urban form that is characterised by:

- *A population that is widely dispersed and spatially disproportionate* in terms of social development planning and economic possibilities: the urban poor are pushed away from the urban centre and employment areas; the wealthy occupy key valuable sites adjacent to mountain and sea; and suburbia 'in-between' has a low-density, single-dwelling residential pattern – this reflects a critical lack of integration of different income-groups;
- *A clear separation of spaces* for living, working, and shopping, with limited facilities within acceptable walking distance in both the wealthy and poor areas of the city;
- *Discontinuous roads structure* (built for predominantly motorised transport), marked by very large super blocks and giving limited access, particularly in the lowest income areas. A large majority of the population are then reliant on travelling long distances at great expense to access opportunities – these environments require ongoing investment in infrastructure; and
- *A move away from key activity centres*, such as the central inner city, which is becoming an exclusive enclave for the higher-income sector.

Such physical landscapes monumentalise separation over inclusion, in which public space fails to perform its democratic potential as a place of exchange, tolerance, and healing (Mammon *et al.*, 2008). We need to reverse this trend of fragmented enclaves that promote a discontinuous urban spatial structure. Such environments are unsustainable and socially disabling.

2.2   Densification and urban form

Cape Town has a relatively low urban density compared to other world cities (refer to Figure 2). Densities measured in 'people per km2' recorded for world cities include New York (9,610), Shanghai (2,590), London (4,800), Mexico City (3,700), Johannesburg (1,960) and Berlin (3,810) (Burdett &

Sudjic, 2007: 246-248). In 2005, Cape Town's density was 1,252 people per km2 on the basis of 3.1 million people, measured against built and non-built areas (Robertson & Dalvie, NM & Associates Planners and Designers, 2008). While significant residential and commercial/retail development has occurred, Cape Town's inner city area, urban and residential densities in particular are still relatively low. Furthermore, this residential market has targeted middle to upper income professionals, yuppies, and foreign owners while excluding local families and middle to lower income households. This constrains the capacity to build urban thresholds for sustainable urban development in the city.

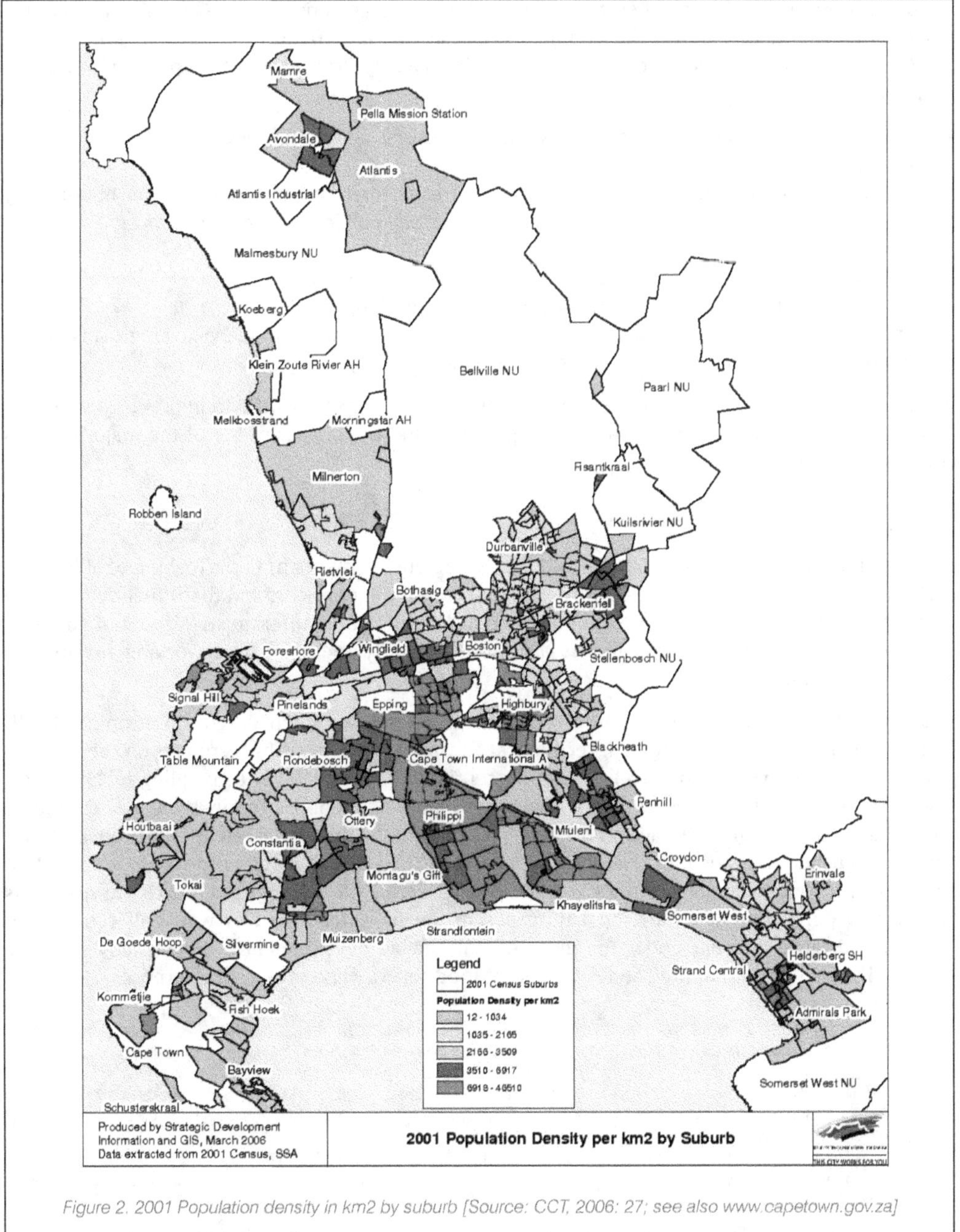

Figure 2. 2001 Population density in km2 by suburb [Source: CCT, 2006: 27; see also www.capetown.gov.za]

Densification is happening in reverse mode in Cape Town. Instead of densifying areas where urban resources are abundant and easily accessible – e.g. around key urban nodes and along major corridors or axes – densification currently happens mainly on the outer edges of the city. Areas such as Kensington and Rosebank, which are well-located and resourced local areas in Cape Town, have gross dwelling unit densities (du's) of approximately 11 and 44 per hectare respectively (CCT, 2008: 10), whereas areas such as Mfuleni (on the periphery of the city) has a gross density of approximately 55 du's per hectare. The recently completed housing in Joe Slovo,[9] which has limited proximity to urban opportunities and is not as well resourced as a local area when compared with Rosebank, has a gross density of 120du's per hectare. Average dwelling unit densities for Cape Town's informal settlements, largely located on the peripheries of the city, are 100du's per hectare, according to the CCT (2008: 10). This reverse mode densification has resulted in a spatial pattern and urban form that –

– does not acknowledge that land is a finite resource and non-renewable by nature in an urban context;
– requires commuting on transport that is unaffordable for those living away from opportunity and does not directly support investment in a viable public transport system as a result of dispersed thresholds;
– lacks equity for a diverse cross-section of Cape Town's population;
– is expensive in terms of service and infrastructure provision; and
– lacks a spatial logic that would facilitate economic opportunity and place-making using public structure and built form as a key ordering elements.

Against this background, it can be argued that densification does very little to improve Cape Town's sustainability in terms of urban form, spatial patterns, and the quality of life of the majority of its citizens as seen today.

3. The Compact City – a sustainable response to urban form

In addition to the challenges of fragmented, sprawling environments in Cape Town, global issues – such as climate change, looming food insecurity, increasing fuel prices, and fossil fuel dependence – simply exacerbate the situation. When disaster strikes, the poor suffer most. We cannot continue to ignore the idea that sustainability is also about poverty reduction and working towards an equitable situation.

What does this mean for people-based urban environments and their long-term sustainability?[10] Is sustainability only about reducing ecological footprints and creating zero-carbon, zero-waste environments? Do we continue to limit our perceptions of sustainability to building technologies, energy efficiency, and the environment within a framework of economic feasibility alone? Or does it mean that we accept that we have an existing situation that, although largely dysfunctional, cannot simply be abandoned, but must be adapted over time to work towards zero-waste, zero-carbon conditions? If we only choose to prioritise zero-waste, zero-carbon built form, we create situations for wealthy opportunists to develop landscapes in the hinterland. In this way, golf estates and residential estates are built under the guise of 'sustainable development', when strictly speaking such land should be agricultural, and reserved for food security and productive purposes.

---

9   The N2 Gateway human settlement programme intends to enable approximately 20,000 to 22,000 households to be housed/re-housed, where about 18,000 people (3,600 households) of this total will be accommodated on approximately 29ha of land (to be serviced) in Joe Slovo, Langa, located approximately 14km from the Cape Town CBD. In 2003, the population size of Upper Table Valley (comprising Vredehoek, Gardens, Tamboerskloof and Oranjezicht), a wealthy area in Cape Town's inner city located within walking distance of the Cape Town Central Business District (CBD), and covering a land surface area of approximately 114ha (CCT: 'Density by Suburb' 2003 – available from the City of Cape Town (CMC Administration) Data Base), was approximately 23,857. Therefore Joe Slovo will cover 25% of the land surface area of Upper Table Valley to accommodate the equivalent of approximately 75% of the Upper Table Valley population (Mammon et al., 2008: 26).

10   The role of urban planning and design should not be underestimated in urban sustainability efforts. It is a vital component of the urban delivery process, and important as a dynamic practice of "adaptation and sustainability" (Zetter & Watson, 2006: 5), particularly in the lives of the urban poor.

Furthermore, built form is always founded in a landscape footprint that could potentially disturb the natural patterns and flows of systems (e.g. rivers) in a landscape that is by nature open and expansive; never mind the scars created from a visual impact perspective. Aside from climate change and resultant sea level rises, the same threats are posed to our ocean and marine life where claims of development rights beyond the shore lines are being made (Cape Times, Thursday 26 June 2006).[11]

Inner city Cape Town [Source: Jacqui Perrin]

The **compact city** concept is supported by what Kenworthy (2006: 68-69) terms "ten key dimensions for sustainable city development". A key element involves the return to "a compact, mixed use urban form that uses land efficiently and protects the natural environment, biodiversity and food producing areas". Public transport, walking, and cycling become key components of the city and large freeway and road infrastructure investments are de-emphasised. Burdett & Rode (2007) take this view further:

"At a metropolitan and regional scale, it is clear that more compact urban development provides the only sustainable answer to global urban growth. This is true not only because less sprawl leads to a reduction in energy use and pollution – and cities contribute 75 per cent of the world CO2 emissions – but also because dense cities require less investment in public transport, infrastructure and services to make them work."

Burdett & Rode, 2007: 22[12]

---

11 See also 'Human Shadows Leave Haunting Portrait of World's Seas' by Andrew C. Revkin, in The New York Times (The Times, Friday, March 7, 2008). The article reveals that the cumulative human impact (i.e. organic pollution from agricultural run-off and sewerage, damage from shipping, traditional fishing close to shore, construction destroying the effectiveness of continental shelves, etc.) on the ocean around South Africa is medium to high. About 40% of the global oceans are strongly affected by human impact, with only 4% pristine, but poised for change. However, the most widespread human impact is the drop of pH of surface waters as a portion of the billions of tons of CO2 added to the atmosphere from fuel and forest burning each year is absorbed in water, where it forms carbonic acid.

12 According to Burdett & Sudjic (2007), the 'Urban Age' projects by the London School of Economics and Deutsche Bank's Alfred Herrhaussen Society recognised urban connections between social, spatial, and economic sectors within six global cities: New York, Shanghai, London, Mexico City, Johannesburg and Berlin.

Burgess *et al.* also argue along similar lines to Burdett and Rode, that "sustainability can only be achieved through the concept of the compact city and a general spatial strategy of densification of the built area" (1997: 121). The ideal of a mixed-use, accessible, financially viable, and sustainable city in line with the concept of the compact city is contrary to what exists in Cape Town today.

Housing on the urban edge [Source: NM & Associates]

Some may argue that a compact city form contributes to increasing land prices (Bertaud, 2004: 8). To resolve this problem, it may be appropriate for the State to intervene in the land market by way of various densification strategies and by using the vast amounts of public land to this end. This may well assist in influencing land prices in a manner that opens up feasible development options for a range of incomes, thus integrating the diverse sectors of society. The idea of multifaceted, dense 'compact neighbourhoods' leads to the recognition of people's inherent energy, and the possibility for convergence of different cultures, generations, languages, and imagination. All these characteristics are a soft, but significant, component of sustainability.

Rural/urban edge to Cape Town [Source: NM & Associates]

The compact city moves towards achieving a positive integral and distinctive city, where multiple components result in a total living environment. These attributes of the compact city may be investigated by understanding how appropriate densification might make cities better places to live (Power, 2007); accessible places where mobility is about choice, and where there is an improved overall quality of life. The compact city presents a notion of Cape Town as a complex web of urban surfaces, interlinking its core urban nodes with a network of urban corridors and hierarchy of places and spaces as a fundamental layer reinforcing both. Towards the outer periphery, the surface reflects a zone of transition that comprises multiple green and urban layers that are edged by the rural and productive landscape. In turn, this is cradled by dominant mountain ranges on either side. These begin to define the domain of the Indian and Atlantic oceans – an endless expanse defining the south-east and north-west edges of Cape Town.

To set this landscape on a sustainable pathway, the following goals are suggested:

a. **The role of land and landscape.** A key goal in shaping urban form is to adopt a systems approach to planning and design, where land is respected as a non-renewable resource and the inherent qualities of landscape are acknowledged as an essential part of a high quality sociable living environment, rather than just spaces for sprawling, disconnected cluster developments.

b. **Mobility and land use as an inter-related potential to sustain the city over time.** The goal in this instance is to reinforce an integrated land use and public transport approach, maintaining that mobility and land use cannot be dissociated if a sustainable urban environment is to be achieved – practically, this means approving developments that are linked to public transport systems, rather than just accessible by private car.

c. **Public structure and sustainable infrastructure design.** The goal here is to create a positive urban environment comprised of multiple urban components, including a network of public transport-focused routes connecting a hierarchy of nodes and public spaces, and integrated infrastructure design between the primary requirements of the natural landscape and those of settlement.

d. **Built form.** The goal is to work towards a total living environment in neighbourhood planning and design.

Mountain to sea Cape Town [Source: Kathryn Ewing]

Table 1 concretizes these goals in describing targets that can be set to achieve them. Two points must be made clear here:

1. While targets are much better understood when expressed in quantitative terms – for example a renewable energy target of 15% by 2014, the planning and design of the compact city is as much about art and creativity as it is about science and quantum. In fact, one can argue that the latter approach has resulted in the fragmented Cape Town as we know it today.

2. The unsustainable nature of informal settlements is beyond the scope of this chapter, given their immense complexity. However, much can be learned from informal settlements about compact city elements such as space making, layout planning, resource use,[13] and so on. At the same time informal settlements should not be romanticised as they are not ideal living environments from the perspective of exposure to flooding, fire, unsanitary conditions, health risks etc. against which appropriate urban settlement making response must be investigated.

Table 1: Achievable targets and practical responses

Targets	Explanatory Notes	Diagrams/Illustrations/Examples
a. Land and landscape		
1. Set limits to urban development	Ecological and environmental constraints must inform urban development.	*Limit to urban development on edge CT* *Source: NM & Associates*
2. Conservation of land and sea as non-renewable resources	Protect natural landscape against indiscriminate development and industrial/human waste.	industrial buffer wetlands river system *Transition from industrial to natural systems*
	Fix opportunities for productive land within the urban domain e.g. 3ha urban agriculture productive land unit with basic accommodation (refer to SANRIF, 2008; also see Maas *et al.*, 1998).	*Agricultural landscape on edge of CT* *Source: NM & Associates*
	Densification and public investment in areas around nodes, corridors and high threshold points in the city (refer to CoCT, 2008 for densification strategies and tools).	*Dense settlement at Nolungile Station, CT* *Source: NM & Associates*

---

13 The poor are much more resourceful in terms of using recycling materials, albeit sometimes out of desperation (Simmonds & Mammon, 1996).

	Targets	Explanatory Notes	Diagrams/Illustrations/Examples
3.	Establish transition zones to protect valuable assets including biodiversity	Protected edges: Identification of sensitive natural areas within the development areas and along the declared biosphere reserves	*Protected nature reserve of Table Mountain* *Source: Kathryn Ewing*
		Productive edges: Positive use of transitional land which historically remained vacant	housing — veggie garden — Buffer zone – olive trees — road *Boundaries/threshold to road reserve conditions*
		Developed edges: Layer the public realm to create transitions between public and private domains.	*Layering of Belhar housing and public space, CT* *Source: NM & Associates*
4.	Make public land play a positive redistributive role	A redistributive role can be achieved through value capture which is an instrument that government can use to intervene in the land market, relying on land taxation, levies or planning instruments (see Brown-Luthango, 2006; & Gihring Thomas, 1999).	*Valuable land surrounding Khayelitsha Station* *Source: NM & Associates*
b. Transport and land use			
5.	Move to 80/20 public/private transport modal split	The greater the use of clean forms of efficient and affordable public transport (PT) the more likely the reduction in CO_2 emissions, which could be linked to carbon credits.	*Bus lanes in Pereira, Colombia* *Source: Kathryn Ewing, NM & Associates*
		The greater use of PT encourages greater movement on foot and other modes of non-motorised transport (NMT) e.g. bicycles, horse and cart.	1. 350-500m walking distance to crèche, bus station, school, play park, local store 2. 1-2km walk to PT stations, secondary schools, shops 3. 2-5km travel by NMT (cycle) to PT corridor 4. 8-10km cycle/PT (feeder service) to major PT transfer facility, institution, workplaces and travel by horse and cart — walk cycle PT *Diagram to show NMT & PT services in distances*

Targets		Explanatory Notes	Diagrams/Illustrations/Examples
5.	Move to 80/20 public/private transport modal split (continued)	Investment in PT infrastructure together with the development of land and other public infrastructure i.e. bus stops and stations, bicycle storage etc. results in greater opportunities for and improvements in economic growth and development (see Mammon *et al.*, 2008)	 *PT infrastructure in Pereira, Colombia Source: Kathryn Ewing, NM & Associates*
6.	Achieve sustainable urban mobility	Create incentives and programmes for appropriate marketing, awareness and education on the value of energy savings e.g. in Curitiba tokens for recycling enabled free PT trips for participants	 *Car free Sunday in Bogotá, Colombia Source: Kathryn Ewing, NM & Associates*
		Develop car-free weekend and public holidays and encourage safe cycling to education facilities.	
7.	Develop mixed land use opportunities along with densification to increase dwelling units/ha as defined by context and guided by the CoCT Densification Strategy (2008)	Targeting areas that are within walking distance of PT corridors and developed urban centres.	 *Mixed-use & PT Corridor in Bogotá, Colombia Source: Kathryn Ewing, NM & Associates*
		Tools for existing infill opportunities in built up areas include creating overlay zones, integrated and performance based zoning and others (see CoCT, 2008).	
		Attempt to achieve an urban gross base density of 25 du's/ha across Cape Town by 2030.	
c.	Public structure and sustainability infrastructure		
8.	Protecting public assets and places that encompass the significant roles of culture, memorialisation, heritage and celebration	Recognising the role of 'soft' aspects of sustainability through the celebration of public culture, history and memory.	 *Gugulethu Seven Memorial in Nyanga, CT Source: Paterson and Mammon, NM & Associates*
9.	Restructure the spatial environment through reclamation of public spaces for public purposes	Acknowledging the role of public elements in structuring urban space.	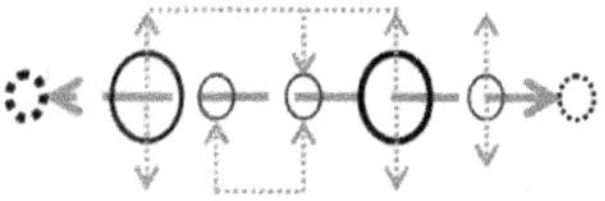 IDEA OF NODES ALONG A CORRIDOR IN RESTRUCTURING *Idea of nodes along a corridor in restructuring*
		For example, reclaiming street space for people rather than private vehicles.	

Targets	Explanatory Notes	Diagrams/Illustrations/Examples
10. All investment in public places and spaces that form a cluster of facilities that service neighbourhoods	Public expenditure should make adequate provision for investment in public structure and institutional buildings and not rely on the sale of public assets (e.g. land) to cross-subsidise such public investment.	*Philippi interchange and trading facilities, CT* *Source: Jacqui Perrin, Du Toit & Perrin*
11. Maintain intrinsic natural systems in the landscape	Hard and soft infrastructure design to respond appropriately to natural flows of systems for example, storm water management could be a lot more appropriate and less expensive when natural flows are undisturbed and carefully treated in settlement making.	housing buffer river recreation system *Transition layer from urban to river system*
12. Promote resource use conscious society	Incentives and public/professional education programmes with respect to natural and built environment relationship.	*Recycling bins in Guayaquil, Ecuador* *Source: Kathryn Ewing, NM & Associates*
13. Encourage practices that minimize the use of energy for constructing and maintaining urban settlements	Adaptation of existing public buildings and infrastructure. Using and maintaining green technology in new urban settlement with emphasis on local materials and local labour. Encourage and reinforce guidelines and practices promoted by the Green Building Council of South Africa (GBCSA)	*Bedzed 'green' urban settlement, London* *Source: Kathryn Ewing, NM & Associates*
d. Built form		
14. Promote and protect human scale urban places and spaces	Transitional spaces should always exist between private and public areas i.e. stoep, small garden or change in level/ground floor plane.	*Housing in Belhar, Cape Town* *Source: NM & Associates*
	Ensure street frontages are lively and animated by encouraging narrow housing/apartment/row housing e.g. with a traditional 6m width for a row house to achieve this condition.	

	Targets	Explanatory Notes	Diagrams/Illustrations/Examples
14.	Promote and protect human scale urban places and spaces (continued)	Ensure activated streets by perimeter block development where setbacks from the street should be no more than 2m for double storey and 3m for more than 2 storeys.	*Quinta Monroy Housing Project, Chile* *Source: Cumberlidge & Musgrave (2007: 201)*
		Orientation – wind tunnels are to be avoided (particularly in Cape Town) and all housing is to have access to natural sunlight for at least half of the day. All public spaces (streets, squares and parks) to have access to sunlight for most of the day.	*Plan to show sunlight & wind – row housing (Adapted from District 6 Architectural Guide-lines by Le Grange Architects and Urban Planners and NM & Associates, 2005)*
		All habitable rooms are to have adequate natural ventilation.	
		Vertical and horizontal circulation to be of adequate size with natural light.	
		All habitable rooms are to be of liveable sizes, where bedrooms are no less than 9m^2 and 12m^2 for a double bedroom. Comfortable habitable unit size 5 person households to be no less than 72m^2 per dwelling unit (Neufert, 1980).	
		All housing should have access to balconies and/or outdoor spaces.	
15.	Restructure the urban block to promote walkability and NMT, ease of circulation on foot, densification and access	Densify to concentrate urban settlement and minimise use of infrastructure investment and need for use of private motorised transport.	*Sketch to show perimeter block housing*
16.	Design green buildings that minimize the use of energy	Green Building Guidelines become compulsory in all new buildings and green by-laws are actively driven by local government.	*Straw Bale Hopi Nation Elder Home, USA* *Source: Architecture for Humanity (2006: 152)*

Targets	Explanatory Notes	Diagrams/Illustrations/Examples
16. Design green buildings that minimize the use of energy (continued)	Installation of water storage, solar panel heating, appropriate insulation and green building elements to existing buildings.	
	Generate involvement of communities in driving "green technologies". For example encourage community group participation in building a straw bale community centre setting precedent whilst generating active engagement and identity in neighbourhoods, such as the Hopi Nation Elder Home in Hoteevilla, Arizona, USA (see Architecture for Humanity, 2008: 150-153).	
	Encourage increased planting and maintenance of street trees and enhance public parks with tree planting programmes.	

4. Conclusion

As argued in the introductory chapter, "cities provide a unique context for realising that resource limits and degrading eco-system services are not simply constraints to development, but opportunities for redefining what development means". A consideration of urban form in Cape Town in turn gives us a unique lens through which to refine what development could be in our city.

The targets presented in Table 1 provide a practical basis for further exploration of the compact city as a key strategy for transforming Cape Town into a more sustainable city.

There is no doubt that, given the limitations of Cape Town's urban form, current attempts at densification only marginally improve sustainability. However, any debate on sustainable urban form must go beyond the traditional paradigm of environmental impact and focus instead on the relationship between densification and design. Design, while important, is not on its own an autonomous driver. Unless the many stakeholders in Cape Town recognise the crucial role that design plays in shaping workable alternatives, many of the sustainability interventions will not be realised. Design potentially provides a broad framework within which a sense of the city fuses with sustainability thinking.

References

Architecture for Humanity (Ed.). 2006. *Design Like You Give A Damn. Architectural Responses to Humanitarian Crises.* London: Thames & Hudson

Battle, G. 2007. 'Sustainable Cities', in Burdett, R. & Sudjic, D. (Eds.). 2007. The Endless City. *The Urban Age Project by the London School of Economics and Deutsche Bank's Alfred Herrhausen Society.* London: Phaidon Press.

Bertaud, A. 2004. *The spatial organization of cities: Deliberate outcome or unforeseen consequence?* Available from http://alain-bertaud.com/images/AB_The_spatial_organization_of_cities_Version_3.pdf. (Accessed on 22 April 2008)

Brown-Luthango, M. 2006. *Capturing unearned value/leakages to assist markets to work for the poor.* Development Action Group. Position Paper prepared for Urban LandMark, November 2006.

Burdett, R. & Rode, P. 2007. The Urban Age Project in Burdett R. & Sudjic, D. (Eds.). *The Endless City. The Urban Age Project by the London School of Economics and Deutsche Bank's Alfred Herrhausen Society.* London: Phaidon Press.

Burdett, R. & Sudjic, D. (Eds.). 2007. *The Endless City. The Urban Age Project by the London School of Economics and Deutsche Bank's Alfred Herrhausen Society.* London: Phaidon Press.

Burgess, R., Carmona, M. & Kolstee, T. 1997. Contemporary Spatial Strategies and Urban Policies in Developing Countries: A Critical Review. In Burgess, R., Carmona, M. & Kolstee, T. (Eds.). *The Challenge of Sustainable Cities. Neolibralism and Urban Strategies in Developing Countries.* London: Zed Books Ltd.

Cape Times (Thursday 26 June 2006). New Bill could 'be disastrous for V&A projects' by Anél Powell.

City of Cape Town. 2006. *State of Cape Town Report 2006. Development Issues in Cape Town.* Cape Town: City of Cape Town.

City of Cape Town. 2008. *Densification Strategy for Cape Town*, Draft 2. Cape Town: City of Cape Town.

Cumberlidge, C. & Musgrave, L. 2007. *Design and Landscape for People.* London: Thames & Hudson.

Department of Local Government and Housing. n.d. *The Road Map to Dignified Communities, Western Cape Sustainable Human Settlement Strategy.* Provincial Government of the Western Cape.

Dewar, D. & Todeschini, F. 2004. *Rethinking Urban Transport after Modernism – Lessons from South Africa.* Aldershot: Ashgate Publishing Ltd.

Geis, D. & Kutzmark, T. 2006. Developing Sustainable Communities: The Future is now. *Public Management magazine*, International City/County Management Association, Washington DC for Operation Fresh Start. Available from http://www.freshstart.ncat.org/articles/future.htm. (Accessed on 22 April 2008)

Gihring, T. A. 1999. Incentive Property Taxation – A Potential Tool for Urban Growth Management. *Journal of the American Planning Association*, Winter 1999, 65(1): 62-79.

Keiner, M., Koll-Schretzenmayr, M. & Schmid, W. A. 2005. *Managing Urban Futures. Sustainability and Urban Growth in Developing Countries.* Aldershot: Ashgate Publishing.

Kenworthy, J. R. 2006. The eco-city: ten key transport and planning dimensions for sustainable city development. *Environment & Urbanization*, April 2006, 18(1): 67-85.

Maas, W. & Van Rijs, J. with Richard Koek (Eds.). 1998. FARMAX – *Excursions on density.* Rotterdam: MVRDV 010 Publishers.

Mammon, N., Ewing, K. & Paterson, J. 2008. *Urban Challenges of Inclusive Cities – Towards a Spatial Realm for All.* Paper for the Development of an Urban Development Component of a Second Economy Strategy for the Office of The Presidency – Spatial Planning, Johannesburg Zoo, for Urban Landmark, 03-04 April 2008.

Mchunu, K. 2006. Planning and Sustainability in South Africa. In Zetter, R. & Watson, G. B. *Designing Sustainable Cities in the Developing World.* Aldershot: Ashgate Publishing.

Neufert, E. 1980. *Architects' Data* (2nd Edition). Oxford: Blackwell Science Ltd.

Peñalosa, E. 2007. Politics, Power, Cities. In Burdett, R. & Sudjic, D. (Eds.). *The Endless City. The Urban Age Project by the London School of Economics and Deutsche Bank's Alfred Herrhausen Society.* London: Phaidon Press.

Power, A. 2007. At Home in the City. In Burdett, R. & Sudjic, D. (Eds.). *The Endless City. The Urban Age Project by the London School of Economics and Deutsche Bank's Alfred Herrhausen Society.* London: Phaidon Press.

Robertson, L. & Dalvie, S., NM & Associates Planners and Designers. 2008. *Calculation of Cape Town's density in km² based on vacant and developed land within the CMA boundary using GIS.* June 2008.

SANRIF (South African Constitutional Property Rights Foundation). 2008. 'Owning Land can uplift the poor and clear the many ghettos', as published in *Cape Times*, Wednesday, May 28, 2008.

Sennett, R. 2007. The Open City. In Burdett, R. & Sudjic, D. (Eds.). *The Endless City. The Urban Age Project by the London School of Economics and Deutsche Bank's Alfred Herrhausen Society.* London: Phaidon Press.

Simmonds, G. & Mammon, N. 1996. *Energy services in low-income urban South Africa: a quantitative assessment.* Energy and Development Research Centre, University of Cape Town.

Republic of South Africa. 2004. *Breaking New Ground.* Pretoria: Department of Housing.

Swilling, M., De Wit, M. & Thompson-Smeddle, L. 2008. You the Planner. In Zipplies, R. (Ed.). Bending the Curve – Your guide to tackling climate change in South Africa. Africa Geographic.

Ward, S. 2008. *The New Energy Book for urban development in South Africa.* Cape Town: Sustainable Energy Africa.

Zetter, R. & Watson, G. B. 2006. Designing Sustainable Cities. In Zetter, R. & Watson, G. B. (Eds.). *Designing Sustainable Cities in the Developing World.* Aldershot: Ashgate Publishing.

Carbon Emissions:

http://www.theclimategroup.org/reducing_emissions/low_carbon_solutions/.(Accessed 8 July 2008.)

Carbon Footprint:

http://www.capetown.gov.za/en/Environmental ResourceManagement/EnergyEfficiency/Pages/CarbonFootprintCalculator.aspx. (Accessed 8 July 2008.)

Curitiba – public transport, recycling, housing, parks: Giovanni Vas Del Bello. 2006. A film entitled 'Convenient Truth – Urban solutions from Brazil'. Produced by M. Terezinha Vas. www.mariavazphoto.com/curitiba

Definitions:

http://en.wikipedia.org/wiki/Urban_design. (Accessed 3 July 2008.)

http://en.wikipedia.org/wiki/Urban_sprawl. (Accessed 2 July 2008.)

Density and socio-economic indicators:

http://www.capetown.gov.za/en/stats/Documents/2001 %20 Population%20Density%20Map.pdf. (Accessed 4 July 2008.)

http://www.capetown.gov.za/en/stats/Documents/SES_Indicators_by_2001_Suburbs_(Map)_30102006 14540_359.pdf. (Accessed 4 July 2008.)

Developing sustainable communities:

http://www.freshstart.ncat.org/articles/future.htm. (Accessed 25 June 2008.) (Refer to Geis. D. & Kutzmark, T. 2006.)

Spatial organisation of cities:

http://alain-bertaud.com/images/AB_The_spatial_organization_of_cities_Version_3.pdf . (Accessed April/May 2008.)

Sustainable Urban Resources Forum:

http://sustainableneighbourhoods.co.za/index.php?option=com_content&task=view&id=1&Itemid=2.(Accessed 1 July 2008.)

Social Justice and Sustainable Use of Natural Resources in Cape Town

Mazibuko Jara

1. Introduction

The introductory chapter notes that attention is increasingly being focused on the intersections between global warming, eco-system breakdown, resource depletion, the global economic crisis, persistent poverty, and accelerating urbanisation. As the Stern Report made clear, poorer countries will suffer first and most from the consequences of global warming even though they have contributed least to global warming (Stern, 2007).

The global economic crisis will exacerbate this suffering as the global economy shrinks, with recovery projected to take anything between three and ten years. According to the International Labour Organisation (ILO), the number of unemployed in developing countries could rise by between 18 and 51 million people by the end of 2009 over 2007 levels (cited in Barbier, 2009). When food prices rose by almost 60% during the first half of 2008, the number of people living in poverty increased by between 130 and 155 million (ibid.). The shifts in the global context mentioned in the introductory chapter provides a backdrop to the challenges that face social justice issues as they link to the matter of sustainable use of resources – as in the global context, the poor face the largest risk. This chapter highlights the many ways in which Cape Town's own legacy in the realm of social justice pose significant challenges for sustainability.

The chapter explores the concepts of sustainable livelihoods and sustainable development, and applies them to Cape Town – a city fractured by historical social injustice and inequality in a world in which social and economic inequality is worsening. Inequality is demonstrated by the vast differences in high and low levels of human development existing side by side within one city. The chapter suggests a range of integrated practical measures that Cape Town must undertake to address social injustice. This chapter explores ways in which Cape Town is characterised by social injustice, as well as the key environmental problems stemming from it, and the links between the realisation of social justice and the sustainable use of ecological resources.

The chapter is built around a sequence of questions:

- What is social justice and what brings it about? What contributes to social injustice?
- What does social justice look like in Cape Town? How is it driven and sustained?
- What is the relationship between social justice and the consumption of ecological resources?
- What are the current and historical patterns of consumption of ecological resources in Cape Town, and what drives and sustains resource consumption in Cape Town?
- How do historical and current patterns of consumption of ecological resources affect social justice, and how are these manifested?

As the chapter unfolds, it will become clear that the challenge of justice in South Africa is not merely a matter of personal guilt affecting individuals, but more a manifestation of the structural injustice that affects the entire community. In this wider systemic sense, social injustice relates to the actual and perceived unfairness or injustice of a society in the way it allocates both rewards and burdens. The idea of social justice is distinct from the legal definition of justice, which may or may not consider what is moral in practice.

It follows that social justice is the hallmark of a society in which justice is achieved in every aspect of that society, and not interpreted narrowly as the administration of law. Ideally, in a world characterised by social justice, individuals and groups receive fair treatment and an impartial approach to the division of the benefits of society. In this sense, the term describes a society where there is a greater degree of socio-economic development and equality, and less discrimination based on distinctions between class, gender, ethnicity, or culture.

The chapter then shifts to a discussion of how the concepts of sustainable livelihoods and sustainable development can be applied to the challenge of achieving social justice in Cape Town, and explores the relationship between the achievement of social justice and the sustainable use of ecological resources.

Social justice becomes a hollow catch-all phrase if it is not clearly defined and given precise meaning within a historical context. Justice refers to what is fair or reasonable relative to how individuals and groups are treated and how decisions affecting people are made. It follows that injustice is the unfair or unjust treatment of a person or group of people. When injustice operates at the broader societal level, it is normally systemic and structural – and therefore more difficult to recognise.

Another definition of social justice refers to social equality and economic justice. Social equality in a given society means that all people within that society or a particular group enjoy the same status in a certain respect. At the very least, social equality includes equal rights under the law, such as security, voting rights, freedom of speech and assembly, and the application of property rights. However, it also includes equal access to education, health care and other social securities, as well as equal opportunities and obligations. In this sense, it involves the whole society. Social equality emerges from the belief or desire that all people have both equal rights to the earth's resources and a shared responsibility for stewarding those resources. The realisation of social equality is brought to bear primarily through equal access to goods and services, and equal access to the decision-making process. Economic inequality refers to disparities in the distribution of economic assets and income, usually among individuals and groups within a society.

Indecent and intolerable living conditions, inequality, poverty, underdevelopment, lack of access to basic services, and the lack of access to assets are the very opposite of the idea of social justice. This chapter links social justice with human development. The United Nations Development Programme (UNDP) uses the concept of human development to refer to a development paradigm that is about creating an environment in which people can develop their full potential and lead productive, creative lives in accordance with their needs and interests. The notion of human development enriches a discussion of social justice by placing Amartya Sen's emphasis on the expansion of the choices people have to lead the lives that they value, at the centre of development (Sen, 1999). Building human capabilities – the range of things that people can do or be in life – is fundamental to enlarging these choices. The most basic capabilities for human development are:

- to lead long and healthy lives;
- to be knowledgeable;
- to have access to the resources needed for a decent standard of living; and
- to be able to participate in the life of the community.

Without these, many choices are simply not available and many opportunities in life remain inaccessible.

In its annual Human Development Reports (UNDP, 2008), the UNDP has consistently linked human development to the following themes:

- Social progress: meaning greater access to knowledge, better nutrition and health services;
- Economic development: highlighting the importance of economic growth as a means to reduce inequality and improve levels of human development;
- Efficiency: described as resource use and availability that directly benefits the poor, women, and other marginalised groups;
- Equity: relative to economic growth and other human development parameters;
- Participation and freedom: particularly empowerment, democratic governance, gender equality, civil and political rights, and cultural liberty, in particular for marginalised groups defined by urban/rural, gender, age, religion, ethnicity, physical/mental parameters, etc.;
- Sustainability for future generations in ecological, economic and social terms; and
- Human security in daily life against such chronic threats as hunger, as wells as abrupt disruptions, including joblessness, famine, conflict, etc.

It is not possible to make valid judgments about justice and injustice in a society without first understanding that society. In the case of Cape Town, social justice is contextualised within the histories of the country and the city. It is these histories that shape the here and now of social injustice in Cape Town, which is about material deprivation, human under-development, urban poverty, inequitable access to social infrastructure, inequitable distribution of income, inequitable access to basic services, and inequitable access to human capabilities and public goods (education, health, public transport, etc.).

3. Social injustice, inequality and human development in Cape Town

3.1 Population figures

Cape Town is the main urban centre of the Western Cape. The population of Cape Town increased by 1,6 per cent annually from 2,994 million to 3,239 million people (65,0% of the Western Cape population) in the period 2001-2006. The population is projected to grow at an average annual rate of 1,0 per cent for the period 2006-2010 to 3,368 million people by 2010. By 2014, the population is projected to grow to 3,448 million, at an average annual growth rate of 0,6 per cent. Cape Town's population as a proportion of the total Western Cape population is projected to remain stable at 65,0 per cent in 2010 and 2014 (Dorrington, 2002).

Based on the population projections for 2006, about 46,0% of Cape Town's population is classified as Coloured, 34,0% as African, 18,0% as White, and 1,0% as Asian, and the median age is 27 years. All age cohorts experienced positive growth between 2001 and 2006, except for the 10-14 and 15-19 cohorts, which declined by 0,3% and 1,5% respectively. Faster growth was registered for older age cohorts (i.e. above 40 years of age), with the highest growth in the 55-59 years age cohort, which grew by 5,01% per annum. The young age cohorts grew by less than 2% per annum on average (ibid.).

3.2 Human development indicators

The City of Cape Town (CCT) has divided the City into eight planning districts. In February 2007, the CCT's Strategic Development Information and GIS Department released 'Planning District Profiles'. These profiles compiled a set of demographic, socio-economic, housing, and crime information for each Planning District, and provide detailed statistical analysis on:

- demographic and socio-economic information (population, population projections, age profiles, work status);
- levels of living;
- a socio-economic status index;
- a service level index;
- age/gender indices;
- housing (dwelling type, household size); and
- crime statistics and patterns (for murder, rape, business crime, and drug related crime).

These indicators were developed in order to inform spatial planning with regard to future development in the City's eight Planning Districts. This information will also be used for planning purposes and to draft more comprehensive 'State of Planning District' reports. In the city's district planning profiles, socio-economic status is used to measure the quality of life of residents by focussing on income, education, and occupational status as key indicators. These indicators were combined to form an index, which represent a wider definition of socio-economic status. The index shows that Planning District F (including Khayelitsha and Mitchell's Plain) is the worst off at 54.12.

The District Planning Profiles were also used to consolidate the City Development Index (CDI) tool, which is an average of the following indices: infrastructure, health, education, and income. Overall, the City has a higher CDI of 0, 88 compared to 0, 81 for the rest of the Western Cape Province. Cape

Town out-performed the rest of the province in terms of infrastructure, income, and waste disposal. Khayelitsha, Nyanga, Langa, Gugulethu, Mitchell's Plain, and Elsies River are evidently the poorest areas, with CDI's that are below the provincial average of 0.81. These areas have the lowest levels of development in terms of infrastructure and health (which average 0.6); however, waste disposal and education indices are better (Western Cape Provincial Government, 2005).

The City also uses a Human Development Index (HDI), which is measured by averaging the following indices: health (based on life expectancy), education (based on adult literacy and gross enrolment indices), and income (based on mean household income). The City has a higher HDI of 0.82 compared with the Provincial average of 0,72, and has performed particularly well in terms of income (0.91) and education (0.88). The challenge remains the provision of health care, where both the City and the Province performed less satisfactorily, with indices below 0.7. The city's human development indicators show that the poorest areas are Khayelitsha, Nyanga, Elsies River, and Langa. These areas have the lowest health indices – averaging 0.47 – and their income indices are also below par. On the other hand, education indices are much better in these areas. Areas such as Durbanville and Melkbossstrand performed well in terms of education and income indices; however, health indices for these areas are quite low, averaging 0.69 (ibid.).

Khayelitsha, Nyanga, Langa, Gugulethu, Mitchell's Plain, and Elsies River are evidently the poorest areas, below the provincial average City Development Index (Western Cape Provincial Government, 2006). These areas have the lowest levels of development in terms of infrastructure and health. The issue of social infrastructure backlogs is critical in Cape Town, with high population density areas such as Mitchell's Plain, Khayelitsha, Gugulethu, and Langa severely affected. The backlog goes back to the enduring legacy of the segregation and apartheid periods, which skewed infrastructure development in favour of middle class white suburbs at the expense of Coloured and African dormitory townships. This ongoing legacy is compounded by the City's expanding population, which introduces secondary stress on its social infrastructure and services. In particular, education, health care, housing, and policing have been impacted.

3.3   Social infrastructure

In 2006, the Western Cape Provincial Government (WCPG) estimated the City's social infrastructure backlogs as follows:

- Schools: 156;
- Health care facilities: 100;
- Police stations: 34;
- Housing backlogs: varying estimates between 350,000 and 410,000 - 350,000 is mentioned in the City of Cape Town's 2007/8 Integrated Development Plan (IDP) (City of Cape Town, 2008), and 410,000 is mentioned in the 2007 Human Settlement Strategy of the Department of Local Government and Housing (WCPG, 2007).

The City accounts for 63 per cent of the Province's learners. However, the current number of schools only accommodates 50 per cent of learners in the City, with a high pupil-to-teacher ratio. The backlog of health care facilities constitutes the largest shortfall in the Western Cape and impacts negatively on the quality of health care services provided, given the higher population densities and concentrations of poverty and inequality in the City. In addition, the co-existence of high HIV and TB infection rates amongst the City's poorest residents have placed further strain on its limited resources (WCPG, 2006).

3.4   HIV/AIDS

The people directly affected by HIV/AIDS are some of the most vulnerable in society, and as a result the rate and depth of poverty in the city may increase. The number of working-aged adults living with HIV/Aids has implications for productivity in the work environment – and ultimately the economic growth of Cape Town. The areas most affected are Gugulethu/Nyanga and Khayelitsha, with prevalence rates of 28% and 27%, respectively. Other areas with high prevalence rates are Helderberg (19%), Oostenberg (16%), Cape Town Central (11%), and Greater Athlone (10%) (WCPG, 2006).

The number of AIDS-infected people also has large implications for health care requirements, e.g. the provision of anti-retroviral treatment. In addition, there will be increased numbers of AIDS orphans, which implies increased dependency and health care needs for both children and adults. Infant mortality per 1,000 births is very high in Khayelitsha and Klipfontein. Cape Town has a very low TB cure rate of (71%) as compared to the National target of 85%. Except for Tygerberg (82%) and the Southern region (84%), which are closer to the National target, the rest of the regions have lower TB cure rates – especially Khayelitsha, with a cure rate of 65% (WCPG, 2006).

The people already the most affected by poverty in South Africa are also those who have been the most affected by the HIV/AIDS crisis. Conversely, HIV/AIDS has also had the effect of increasing the number of households affected by poverty. The gender imbalance in HIV infections is striking, with many more women living with HIV than men.

3.5 Housing

The housing backlogs have received considerable media coverage, with contentious projects such as the N2 Gateway and Klipfontein Corridor inviting special scrutiny. While the supply of low cost housing may have increased, the economic capacity of potential first time home-owners is inadequate (WCPG, 2006). The slow delivery of housing has also not effectively overcome apartheid-era spatial divisions, which are at the foundation of social injustice in Cape Town. In this regard, post-apartheid Cape Town is really a neo-apartheid city. The peripheral location of new housing developments (Delft, Khayelitsha, Mfuleni, and other remote areas) further entrenches poverty and inequality. The combination of un-transformative housing policy and the high cost of well-located land perpetuate historical patterns of segregation. Many of the peripheral settlements also tend to have inadequate access to urban infrastructure and services, while support facilities and employment opportunities are also lacking in many cases (Landman & Ntombela, 2006). Post-apartheid Cape Town is still based on a spatial system that organises the urban population according to income groups. Hence, Cape Town's post-apartheid urban planning has not nourished urban social capital in a way that can address social injustice. Instead, the logic of one-dimensional economic growth has trumped social justice considerations.

3.6 Access to basic services and the problem of commodification

Although the City significantly improved the delivery of electricity and refuse removal services in the period 1996-2001, the number of households without refuse removal and electricity decreased by only 4,7% and 1,6%, respectively. Water supply and sanitation also remain challenges to be addressed: during this period, despite accelerated provision to the urban poor, and because the household formation rate was faster than population growth, the proportion of households without flush toilets increased by 2,1% and households without piped water on site increased by 5,4%.

Since 2001, the issues of housing backlogs, ageing infrastructure, and energy shortages have been increasingly problematic. While steps have been taken to prepare the City for the additional energy demands resulting from a growing economy, housing backlogs in particular have received considerable attention at both Provincial and National Government level. It is encouraging that both current and future fiscal budgets are addressing the shortfall.

The emerging picture of social injustice in Cape Town is also amplified by the commodification of basic services. As a result of the impact of globalisation, many publicly provided services have been commodified; much of social reproduction and other tasks the state had previously taken on have now been shunted back to the household, into the hands of women in particular. This has shifted the burden of caring for families and making up for the shortfall in both primary and secondary incomes to the household sector and the informal economy.

Such commodification does not resolve the fundamental ecological and redistribution inefficiency of the 'consumption city' model that Cape Town continues to follow. In general, there is a tendency for local government to see communities as little more than individual household 'consumers' and 'clients' of services. Not only does this bureaucratise governance, but it also fragments communities into individual households, and poverty becomes not a collective concern, but an atomised household responsibility. In her study of the livelihood strategies of the poor, Houston (2002) shows how poor households are forced not to share water with neighbours who have been cut off.

The appalling conditions that result from lack of clean water and sanitation have been well established in the literature (Hemson, 2003). Inadequate sanitation results in an environment where debilitating and life-threatening diseases can flourish. Moreover, sanitation is particularly closely related to the survival of children. The 1998 Demographic and Health Survey (Department of Health, 1998) showed that for those households that do not have piped water, the child mortality rate is twice as high, whereas it is four times as high for those households that do not have flush sanitation.

The 1994 Reconstruction and Development Programme (RDP) identified universal access to electricity as a strategy to relieve rural and urban women of the drudgery of collecting wood or using paraffin, both of which are inefficient and unhealthy fuels. Unlike the water services, there has been low consumption of electricity, typically by most of those newly connected. Electricity is used more for lighting, and less for cooking and heating. Most people find it difficult to meet normal operating costs (Hemson, 2004). By 1999, policy makers had realised that the electrification programme would not be commercially viable if it were to depend on cost-recovery. They concluded that it must be regarded rather as "fundamentally a long-term social investment programme with an indirect future return on capital" (ibid.) While the majority of municipalities in the country experience this problem, the picture is somewhat different in Cape Town, where electricity consumption is subsidised. However, despite this subsidisation, there are key concerns about the extent to which tariff levels may undermine the cross-subsidisation of the poor by the rich within the city.

3.7 Income distribution and unemployment

There is also a link between poverty and unemployment. Most people experience poverty as a result of unemployment at the household level (Bhorat *et al.*, 2001). Although economic growth has produced some income gains for poor people through job creation since 1994, the pace of job creation (especially for semi-skilled and unskilled workers) has not been sufficient to translate into the level of income creation needed. The rate of job increases is too slow relative to the growth in work seekers. The unemployment rate has risen over time; in turn, unemployment is linked to growing inequality.

Unequal distribution of income across racial groups is highly prevalent in Cape Town. In 2001, about 13% of households had no income at all, and 9% of these households were Africans, compared to Whites (1,1%) and Indian/Asian (0,1%) (WCPG, 2006). Very few households earned above R300,000 per annum (only 4,3% of the population). The unequal distribution of income is also linked to levels of employment. In 2004, Cape Town's unemployment rate was estimated at about 23%, with the number of unemployed estimated at 275,730 (Statistics South Africa, 2004). This Statistics South Africa report also shows that the proportion of households with no income in 2001 was at 13,31%, amounting to 102,062 households. The problem of unemployment in Cape Town's informal settlements also means that residents of informal settlements have the lowest share of the City's income.

A 2002 discussion paper by Houston on the livelihood strategies of the urban poor in Cape Town states:

"The Western Cape is one of the wealthiest provinces in the country; with a rate of 20% (2001) Cape Town has one of the lowest unemployment rates compared to that of other metropolitan areas in South Africa. It also has the highest proportion of Black African workers who are unskilled (31.3%), whereas the national proportion is 17.2%."

(Houston, 2002: 6)

Approximately 25% of the Cape Metropolitan Area's 3 million people live in absolute poverty (Statistics South Africa, 2001). It is estimated that 35% of households earn between R1,500 and R3,500 per month, while 45% of households earn R0-R1,500 per month. Eighteen per cent of the workforce is employed in the informal sector (Statistics South Africa, 2001).

These statistics reflect Cape Town as a place of stark contrast between rich and poor, countering misperceptions that the entire Province's poor people are found in the rural areas. Recent thinking in the field of poverty reduction acknowledges the failure of previous analyses of urban poverty to recognise the complexities that prevail in South Africa's metropolitan areas. While it is a wealthy province, with income levels that compare well with other provinces, *the Western Cape has the highest unemployment rate among black Africans (36%)*. In addition, there are also marked differences between poor communities, because, in the words of Houston (2002: 6) "... they experience poverty in different ways. There are even differences within communities where some are relatively better off than others." Despite a six-year interim, Houston's basic argument remains largely true and valid at the time of writing in 2008.

3.8   Relative deprivation

What the paper is not able to show definitively is the City's detailed profiles of inter-racial and intra-racial patterns of income distribution and consumption. Deprivation refers to unmet needs of people, whereas poverty refers to the lack of resources required to meet those needs. When reviewing deprivation numbers, the number of people experiencing a particular form of deprivation provides a clearer description of the level of deprivation within a certain ward/municipality. Available data shows that about **40 per cent of the most deprived wards in the Western Cape are within the City** (WCPG, 2006).

Cape Town's urban poor have few assets, experience social isolation and exclusion, earn low levels of income, are exposed to hazardous living conditions, have poor nutrition and high rates of HIV/Aids infection, experience social breakdown, and a general lack of infrastructure that is essential for social development. These factors undermine the poor's ability to escape from poverty through their own efforts and confine them to long-term poverty traps.

4.   The past and present causes and drivers of social injustice and inequality in Cape Town

What drives and sustains social injustice in Cape Town?

4.1   The burden of history

Critical to the urban poverty context in South Africa's metropolitan areas is the awareness of how past governments have used economic policy, spatial arrangements, and racial prejudice to contribute to social injustice for millions of people over more than a century of economic development (Houston, 2002). The way in which an economy has developed and functions can either give rise to poverty or begin to eradicate it (Heintz, 1998). South Africa's poverty is a socially constructed scarcity due to a skewed distribution of assets and incomes (Heintz, 1998). Colonial conquest, as well as gender and National oppression, shaped a particular path of economic development in South Africa (Legassick, 1977), where the economy was built on a bedrock of National and gender oppression that proved (from the perspective of White enterprise) to be an extremely profitable foundation for many decades.

Apartheid policies, in particular, were instrumental in shaping inequality in Cape Town. How did *apartheid* produce social injustice? Under *apartheid*, racially discriminatory laws and practices prevented people, other than those classified as White, from acquiring ownership and control of the land, mineral wealth, and other major means of production (Aliber, 2001). This systematically excluded Black people from wealth accumulation, and confined large numbers of them to the direst social and economic existence. Black people were also prevented from rising above subaltern positions in either State or corporate bureaucracies, condemning them to low- and semi-skilled jobs, earning low or starvation wages. At the same time, a range of racially discriminatory, oppressive laws and practices were applied to compel Black people to provide a cheap labour force to White enterprise.

As indicated earlier in this chapter, a large part of Cape Town's less affluent population live on the Cape Flats, which was relatively unpopulated until the 1960's. Since then, two waves of settlement took place:

1. The period after the 1960's saw the forceful resettlement of so-called 'Coloured' people through apartheid socio-spatial engineering.

2. In the 1980's, an illegal process of large-scale African migration from the impoverished areas of the Eastern Cape began (see Hindson, 1987; Tomlinson & Addleson, 1987).

Numerous studies (e.g. Swilling *et al.*, 1991) demonstrate how the combined effects of social engineering, spatial planning, and rural-urban migration have contributed to urban sprawl and the expansion of racialised economic geographies.

Apartheid displaced Coloureds and Africans, and maintained extreme socio-spatial marginality in Cape Town. Among the African communities, this spatial marginalisation has engendered an enduring 'outsider' complex. On the other hand, despite their own displacement in the socio-spatial marginality of the Cape Flats, there is a sense of prior entitlement and territorialism among sections of the Coloured community. In the light of the post-1994 transitional changes, this appears to be particularly prevalent among the working class, who is more vulnerable to substitution in employment and 'displacement' caused by African urbanisation. The roots of this mindset lie in the Group Areas Act in tandem with the apartheid political economy. Both were used to engineer a spatial reality that conforms to the Verwoerdian racial hierarchy. In part, the same pattern applies across the board, and especially in the African townships, where economic competition for the local market has spawned xenophobic attacks against Angolan and Somalian traders.

The uniqueness of Cape Town's urban sprawl is not confined to its recent and rapid population growth, but also resides in the fact that it reflects a nexus of extremes. Cape Town has a strong and relatively varied economy, with a mono-centric structure characteristic of South African cities. In a typical centre-periphery fashion, it represents a polarised city centre, where affluent suburbs and economic activities present a strong contrast to the overcrowded, impoverished township on the periphery. Whereas the majority of white and wealthy black people live opulent lifestyles, the majority of those on the Cape Flats live in abject poverty.

It was only towards the latter years of apartheid rule that workers' struggles compelled employers to grant basic rights, such as the recognition of trade unions. 'Cheap labour' has always meant that millions of black workers may have been employed, but never received enough income to live a decent life free of poverty and disease. Training and upgrading the human resource potential of the mass of poor and working people remain a challenge, preventing millions from overcoming poverty to this day (Van der Berg, Louw & Du Toit, 2007).

Outside employment, the urban poor have occupied well-located parcels of land since the early 1980's, thus challenging the underlying social relations and the apartheid city structure. The housing and land struggle of the people of Imizamo Yethu in Hout Bay is an interesting development, to the extent that it has the potential to really challenge the social relations that shape the structural design and composition of a neo-apartheid Cape Town. In the 1980's and early 1990's, the 'one-city-one-tax-base' demand exemplified non-racial perspectives, calling for the restructuring and transformation of the apartheid city – including specific spatial policy interventions geared at compacting and integrating development.

Paradoxically, post-apartheid Cape Town is best characterised as a 'deracialised apartheid' city – without legalised racism and discrimination, but with a continuation of working class exclusion, marginalisation, and exploitation at the hands of a deracialised elite. The major concern about the urban sprawl in Cape Town, as in other major South African cities, relates to the suburbanisation of the City's economic hubs. This is often characterised by the mushrooming of shopping centres around town houses and office blocks on the outskirts of the City. In fact, edge-city developments, such as Century City, rival the City's CBD in economic significance, despite the fact that the latter, unlike the Johannesburg CBD, has not actually decayed. Thus, morphologically, Cape Town is developing several nodes (a poly-nucleated system), where the CBD is no longer the main hub of the urban system, but is reinserted into a complex complementary and competitive hierarchy with other surrounding nodes on the edges.

Apartheid social and spatial patterns remain firmly entrenched in Cape Town, despite the unravelling class structures within black communities, supporting the official line of 'deracialising' capitalism. Thus, the exalted areas – in the shadow of Table Mountain along the western Atlantic coast, south from Green Point, east from Oranjezicht to Observatory, and south to Muizenberg – still resemble the idyllic old White suburbia. While recently there has been a proliferation of golf estates and town houses on the West Coast, the east and south-east of the city remains an expansive swathe mostly composed of slum neighbourhoods. Yet, there are signs that some integration has occurred in Cape Town (e.g. Woodstock, Plumstead), even though this is more noticeable in the former white working-class areas. In the period leading to the apartheid era that began in 1948, about a third of Cape Town lived in 'mixed' neighbourhoods.

Through the efforts of the Cape Town Partnership,[1] the CBD has been the major recipient of large-scale investments, marked by the R1 billion construction of the glamorous residential and commercial Mandela Rhodes Place complex. These developments in and around the CBD reinforce the gentrification that is already underway in the transitional zones of the city, such as the Bo-Kaap and Woodstock, where affluent locals and foreigners offer residents attractive prices for their properties. These post-apartheid processes suggest that there is no fostering of an inward and intensive development pattern based on a comprehensive and integrated spatial framework, which is crucial to restructuring Cape Town's apartheid spatiality.

We must also consider the impact of post-apartheid policies and globalisation on poverty and wider social (in)justice. Without going into a long explanation of what globalisation is, for the purposes of this paper the most important aspect of globalisation concerns the rise of production and value chains that stretch and intersect across continents and countries in search of labour, raw materials and markets. From the mid-19th century already, the South African economy had been perversely integrated into a budding global economy. The adverse integration of the South African economy into global circuits of accumulation continues to this day (Du Toit, 2005), with the inequalities of the global economic system reinforcing the underlying causes of poverty in South Africa.

It has been difficult to catalyse social and economic transformation since the 1994 democratic breakthrough that ended the apartheid political dispensation. Nevertheless, there are many areas where significant gains have been made. These include major infrastructural programmes, the provision of primary health care, a start with land reform, the transformation of the labour market, gender equality machinery, educational transformation, and the deracialisation of social grants. But the question must also be asked: have there not been strategic, subjective shortcomings on the side of post-apartheid government policies, which have continued to sustain the underlying levels of poverty? Clearly, there has been. The key policy failures can be identified in the following areas:

- The macro-economic policy framework – as formulated in the Government's 1996 Growth, Employment and Redistribution (GEAR) policy (National Treasury, 1996) – ensured macro-financial stabilisation, but did not provide a basis for sustained, job-creating growth;
- Land reform policies did not result in large-scale redistribution of land resources;

---

1   The Cape Town Partnership is a collaboration between the public and private sectors. The Partnership was formed when the City of Cape Town, the South African Property Owners Association (SAPOA), the Cape Town Regional Chamber of Commerce and Industry and other stakeholders came together to address issues of urban degeneration and disinvestment in the Central City and related social problems.

- Cost-recovery-based service delivery undermined the need for inclusive, deracialised urbanism; and
- Market-based conceptions of governance and bureaucratic incapacities resulted in a failure to build a developmental State.

How does all this manifest in Cape Town? The next section considers the particular expressions of social injustice in Cape Town.

5. Sustainable livelihoods, sustainable development and social justice in Cape Town

This section explores two separate but related concepts – sustainable livelihoods and sustainable development – and links them to a discussion of social justice and injustice in Cape Town.

Sustainable livelihoods are directly linked to vulnerability (McCaston and Rewald, 2003). All people are vulnerable to certain changes in the external environment that are beyond their control. The ability to respond to these changes, and the impact that it has on people's livelihoods differentiate rich from poor. For instance, rich people also rely on a range of livelihood strategies, but a shock to any one of these would not have as harmful an impact as it would be to a poor household. Shocks – such as death in the family, theft, conflict, gang-warfare, floods, or fire – can directly destroy the livelihood assets of poor people. Larger-scale trends influence the outcomes of livelihood strategies negatively or positively – examples include technological trends and governance trends. Seasonality is also relevant to sustainable livelihood, and refers to patterns in the social, economic, and political forces, such as price changes or employment opportunities (McCaston and Rewald, 2003).

Sustainable development was defined in the Brundtland Report of 1987 as "development that meets the needs of the present without compromising the ability of future generations to meet their own needs" (Brundtland, 1987: 54). The report made the case that "living standards that go beyond the basic minimum are sustainable only if consumption standards everywhere have regard for long-term sustainability" (Brundtland, 1987: 54).

The Institute of Development Studies defines a sustainable livelihood as follows:

"A livelihood comprises the capabilities, assets (including both material and social resources) and activities required for a means of living. A livelihood is sustainable when it can cope with and recover from stresses and shocks, maintain or enhance its capabilities and assets, while not undermining the natural resource base".

(Scoones, 2005)

According to McCaston and Rewald (2003), sustainable livelihoods are possible if people have capacities to generate and maintain their means of living, including the possibility of enhancing their well-being. Importantly, these capacities are seen by McCaston and Rewald (2003) as being crucially dependent on access to ecological, social, cultural, economic and political resources. Important here is also the extent to which there is equity, how resources are owned and the extent to which decision-making and resource allocation allow for participation by affected people (McCaston, 2004).

Therefore, sustainable livelihoods are about varied ways of living that meet individual, household, and community needs. In this context, 'needs' are understood holistically, and would include the social, economic, cultural and spiritual. A sustainable livelihood must therefore be adaptive and resilient. It should also safeguard, rather than damage, the natural environment.

Such perspectives are welcome, in the sense that they move beyond the paternalistic aid and 'poverty relief' programmes of the 1960s and 1970s. However, they generally characterise practice around 'sustainable livelihoods' as projects separate from the broader struggles to transform the accumulation regime that fosters poverty. Wealth distribution is addressed indirectly. Despite a much more progressive, inclusive, and democratic theory of development in recent years, the dominant practices associated with a 'sustainable livelihoods' approach remain on the margins of the mainstream economy. While resources are being mobilised for projects that are intended to achieve sustainable development, ironically they often tend to reproduce the marginalising features of the mainstream capitalist economy. In other words, dominant practices associated with the concept are not linked to broader development strategies, but tend to be seen as local initiatives. These typically have a more limited focus: improving living conditions for the poor without fundamentally altering the dominant economic trajectory.

The concepts and practices of 'sustainable livelihoods' are not disconnected from broader developmental perspectives. Then President Thabo Mbeki flagged the question of 'sustainable livelihoods' in his May 2004 State of Nation address (Mbeki, 2004), which led to many Government ministries developing the concept further. For example, the Department of Housing has taken up the concept of 'sustainable human settlements' as the guiding strategic concept for an important and progressive review of our approach to housing and urban development. During the first fourteen years of our democracy, Government has transferred significant social resources to the poor (both in terms of basic needs like water, sanitation, housing and electricity, and in terms of social grants). And yet, at the same time, there is a growing concern that we cannot sustain the current and burgeoning social transfers. Then Minister of Finance, Trevor Manuel, was explicit in the October 2004 Medium-Term Budget Statement debate (Manuel, 2004) about the rate of take-up on some key social grants being unsustainable.

Something 'in-between', or 'complementary to', formal sector growth and jobs on the one hand, and social transfers to the poor on the other, is clearly needed. And while both formal sector growth and jobs, and social transfers must obviously be pursued, the notion of 'sustainable livelihoods', although vague, is becoming increasingly attractive as a crucial complementary strategy. In developing a coherent conceptual and programmatic definition of the concept of 'sustainable livelihoods', there will be a need for clearer articulation about whether formal jobs should be regarded as part of building sustainable livelihoods, or whether it merely refers to livelihood strategies outside of formal jobs.

6. Cape Town: the link between social justice and the consumption of ecological resources

There is a strong link between social justice and the consumption of ecological resources. This section reviews the current and historical patterns of social injustice in Cape Town, and focuses on what drives and sustains these patterns. It also assesses the range of environmental problems caused by social injustice and highlights the relationship between poverty eradication and sustainability: If Cape Town is characterised by social injustice, what then are the key environmental problems stemming from this situation, i.e. how does social injustice affect the environment? How does it affect natural resources? And how may these negative environmental impacts best be addressed?

6.1 Trajectory of Cape Town as an apartheid 'consumption city'

Social injustice has been shaped by a particular economic development trajectory and is influenced by the fact that urban centres such as Cape Town continue to be constructed as 'consumption cities'. Essentially, a 'consumption city' is organised and structured as a site for accumulation and consumption. Globally, the 'consumption city' model was driven by the economic realities of the 20th century city, namely the need to create a mass of consumers that provide the markets for the suppliers of the basket of urban goods that are now defined as the basic elements of urban living: houses, vehicles, energy, food, leisure, household appliances, and fittings (Swilling *et al.*, 2006).

The basic building block of the 'consumption city' is the 'consuming neighbourhood', which needs to buy in the necessities for daily living – energy, water, waste removal services, building materials, food, vehicles, etc. – from the outside, and often from very distant locales (ibid.). Consumption cities' urban infrastructures had to be planned and managed to ensure that these goods and services could be supplied, transported, removed, financed, and extended. This pattern of consumption has resulted in the contemporary urban system, characterised by its spatial layout/form, function, economy, tax base, and operational requirements. As a result of the high rate of consumption, more than 80% of capital accumulation and exchange in South Africa takes place in urban centres (Kgara, 2008).

As a 'consumption city', Cape Town followed an ecologically inefficient path of development. Successive local governments failed to factor in sustainable development perspectives when it came to economic growth and development paths and strategies. The development of Cape Town as an urban system depended on vast flows of eco-system services/natural resources (water, oil, land, energy, food supplies, etc.). For a long time, these resources were cheap enough to make the development of Cape Town's urban system viable, albeit in racially skewed ways. This made it possible to subsidise the costs of middle class suburban life, at the expense of the traditional working class and rapidly expanding homeless communities (Swilling *et al.*, 2006). These natural resources were pumped through the urban system in order to feed its greedy tendency to ever-growing consumption and accumulation.

The consumption city trajectory assumed that the eco-system would continue providing natural resources forever. This assumption continues to inform the main thrust of post-apartheid Integrated Development Planning systems. They assume that the cost of water, oil, and energy will not rise and remain oblivious to the unpredictable eco-system thresholds. The net effect is to leave intact the way the urban system has been structured historically (ibid.). This also means that the historical subsidisation of extremely inefficient middle class homes by poorer households continues to be sustained. This perverse subsidisation can be seen through higher consumption rates of water and electricity, as well as larger volumes of waste streams produced in higher income homes compared to low-income homes. These are regarded as a subsidy when taking into account that water and electricity are produced from the commons (ecological resources that belong to everyone), and yet their consumption is not equitably distributed in Cape Town society. The higher volumes of waste streams from higher income homes denigrate the same commons much more than those from low income homes.

The Cape Town urban system is resource-intensive (Swilling, 2006). Every oil price rise corresponds to net increases in the amounts of cash transferred from the Cape Town economy to national and global financial circuits (Swilling, 2006). This reduces the amount of money available for circulation in the local economy. A similar situation exists for water, building materials, coal-based energy and food supplies (Swilling, 2006). When it comes to food supplies, Cape Town is a net importer of food. Food prices are directly linked to the oil price due to the chemically dependent nature of commercial agriculture in South Africa (Swilling, 2006). In analysing the Cape Town food chain, Swilling (2006) outlines some alternative links between social justice and ecological resources:

"It also needs to be noted that between 40% and 60% of the domestic waste stream is organic waste (from kitchens, garden cuttings, etc.). This is a rich source of nutrients that could be composted and ploughed back into urban agriculture. Instead, it gets combined with all other wastes and dumped into toxic landfills. In the meantime, 1,3 million tons of food are imported from a land area equivalent of 11, 000 sq. kms that stretches across the whole of South Africa, and beyond. Middle and high income households may be able to afford prices that include the costs of transporting all this food (fuel, cold storage, packaging, energy, etc.), but this is certainly not the case for poor households. Imagine the beneficial consequences for poor households if food could be made more affordable by re-using composted urban organic wastes in local urban agricultural undertakings, and then selling the product at local neighbourhood retail markets ... In one stroke, the costs of long-distance transport, packaging, cold storage, middlemen costs (wholesalers, packagers, retailers) and chemical treatments can be eliminated from the cost of each item of fresh produce. Given that some estimates put the combination of these costs at as much as 80% of the final cost before the final mark-up (which averages at 20%), the consequences for the much talked about need for 'food security', improved dietary health intake (to improve the immune system) and the need to reverse declines in expenditure on food in poor communities become obvious."

(Swilling, 2006: 33-34)

The city manages its water resources in an extremely inefficient and inequitable manner. At the time of Swilling's 2006 paper households used 37% of all water used in the City. Of this, 21,3% was used to irrigate gardens and to fill swimming pools. High income households consumed the majority of domestic water, while a significant portion of the City's poor still had no piped water supply. The ecological inefficiency of the existing water system was reflected in the fact that 61% of all water used by households in Cape Town was used to flush toilets and transport sewerage, yet 11% of the City's population had no waterborne sewerage. Only 5% of the annual 550,000 tons of sewerage produced was recycled (ibid).

Significantly, 60% of industrial waste was recycled, with only 6,5% of residential and commercial waste recycled. The bulk of the unrecycled waste went to landfills located on the Cape Flats. This effectively means that the large poorer communities on the Cape Flats hosted rubbish dumps to absorb wastes generated by a tiny minority of rich Capetonians who had one of the highest waste levels and lowest recycling rates in the world. Four years later, the situation remains largely unchanged. This is eco-inefficiency that is subsidised by nature and the poor (who often live in areas where litter is not collected often enough, because the municipality "lacks the funds") (ibid.).

An ecologically-oriented City Development Strategy for Cape Town should address the enormously costly resource inefficiencies that clearly make sustainable living choices at the household and neighbourhood level in Cape Town extremely difficult. Furthermore, poverty eradication in Cape Town can only become a realistic goal if scarce financial resources and free services from nature (water, absorption of wastes in landfills and water sinks, etc.) are not wasted on maintaining an ecologically unsustainable system that works in financial terms for the middle and high income communities, but remains too costly for those poor households that are lucky enough to be serviced (ibid.).

7. Applying the concepts of sustainable livelihoods and sustainable development to achieve social justice in Cape Town

7.1 Sustainable livelihoods and the improvement of human and social conditions

Entry points which support socially just sustainable development must exist at three levels (McCaston, 2004):

- Improving **human** conditions;
- Improving **social** conditions; and
- Creating a sound enabling **environment**.

Improving human conditions requires supportive efforts to ensure that people's basic needs are met and that they attain livelihood security to support the fulfilment of such needs. It is also about 'increasing opportunity' for current and future generations of people to meet their basic needs (ibid.). Development NGOs like CARE[2] identify productivity, livelihoods and income, accumulation of capital and assets, the development of human capabilities, the management of risk and vulnerability, and access to resources, markets, and social services as key intermediate outcomes that are necessary to lead to the improvement of human conditions (McCaston, 2004; Turner, 2007). However, none of this will be sustainable without factoring the sustainable use of natural resources into a new trajectory of urban development.

Improving social conditions is about supporting people's efforts to take control of their lives and fulfil their rights, responsibilities, and aspirations. This includes initiatives to end inequality and discrimination, and to promote mutual respect for rights and responsibilities, and equitable

---

2   CARE International ('CARE') is one of the world's largest private international relief and development organisations, and a leader in sustainable development and emergency (http://www.caresa-lesotho.org.za).

distribution of capital and assets. It also underscores the need for social inclusion and strengthening both the voice and organisational capacity of the poor (ibid.). All these imply a development approach that responds to the ecological limits to urban development.

Completing the link between social security and sustainability, Jones and Novak (2000) refer to 'popular welfare' to describe non-state and non-market types of welfare, which play an especially critical role in the lives of the poor and working classes. 'Popular welfare' then includes reliance on extended family, tribal networks, reciprocal assistance between rural migrants in the city, etc. 'Popular welfare' takes care of 'unproductive members' within the family or the community, making up for the deficit in state welfare provision. More importantly, this approach highlights the deficit in the capitalist model, which assumes that economic production is able to provide sustainable livelihoods for everyone.

In the case of South Africa, 'popular welfare' is realistically little more than a group of fragile households, providing a site to which people retreat to share the few social, economic, and natural resources they have. It is also a site for production and reproduction, attracting poorer family members in search of security. In many cases, 'popular welfare' depends on access to natural resources, and in desperate situations 'popular welfare' may well end up over-exploiting such natural resources and threatening the sustainability of this key resource. This is where the other interventions proposed above are relevant – specifically social infrastructure, better implementation of existing policies, and sustainable livelihoods.

There is a question mark around why we should continue to refer to this as 'welfare' when we could imagine a whole different kind of society – a socially just society. How could a combination of a universal social wage, sustainable growth and development, and eco-efficiency work with 'popular welfare' to create a different urban system?

Efficient and equitable provision of social infrastructure is key to laying a foundation for sustainable livelihoods for poor households. The dense and intricate layer of township and informal settlement activities – traditional activities, informal markets, spaza shops, shebeens, community projects, stokvels,[3] schools, minibus associations, church volunteer groups, neighbourhood watches, sports clubs, choirs, etc. – should not be ignored or underestimated as a potentially rich body of social capital. In this sense, ward forums/committees and local Integrated Development Plans become central to catalysing and fostering sustainable livelihoods. More dynamic synergies could be created between household subsistence programmes, for example food gardens, or baking cooperatives linked to other state interventions, such as school feeding programmes. In this way, the focus would be to buttress household sustainability with the building of locally-based economy networks, so that resources circulate actively within the community. Local economy networks can have strong multiplier effects that are different to the passive consumption of goods from outside that benefit established supermarkets and other chains. Municipal procurement policies and strategies need to be re-examined so that they support and enliven sustainable livelihoods strategies. Tendering processes tend to favour established companies, although measures have been taken to favour community-based enterprises that can increase the amount of money in local circulation. Services provided by Sector Education and Training Authorities (SETAs) could be improved and expanded so that they are attuned not just to the skills needs of the formal economy, but also to the needs of the emergent enterprises and formations in the informal economy. None of these will be possible without a range of strategic interventions by the City of Cape Town.

What could sustainable livelihoods mean for Cape Town? As we are dealing with a large urban and peri-urban reality within Cape Town's zones of social and economic exclusion, there is considerable social capital that should not be ignored. While there are positive and negative aspects, these

---

3   A stokvel is a widespread social institution in South African townships and rural villages. It is about informal collection and pooling of savings amongst a group of people that operates on a rotational basis with pooled savings normally paid out to one member at a time with all members due to benefit in a defined cycle. At times, the pooled savings are also used for bulk buying of cheap groceries which are then shared amongst members.

self-organising social groups are also a resilient product of Cape Town's human tapestry. These organisations embody a hybrid mixture of values, ranging from strong traditions of social solidarity to oppressive patriarchal customs. The exploitation of women and children in household and small enterprise activities, shack lords, gate keeping, criminal networks, and petty corruption are part of the less positive features of these kinds of networks. The challenge of building progressive sustainable households, communities, and livelihoods then includes the necessity of engaging with these contradictory elements of existing capacity.

For many of the poorest neighbourhoods, sustainability must mean investments in decent housing and the expansion and improvement in services, neighbourhood facilities, and infrastructure. However, limited household incomes mean that residents need to be protected from service systems that will be a constant drain on their finances, such as energy systems that are dependent on oil or grid electricity, transport systems that will become increasingly expensive as the oil price goes up, and sanitation options that become increasingly costly to maintain (Swilling, 2006).

The City of Cape Town could build on the initiatives that ward committees and local Integrated Development Plans are taking. Sometimes, the role played by such initiatives could be strengthened by better coordination, improved efficiency, or increased local participation. It should be possible to create the conditions in which social grant transfers are not merely consumed, or spent immediately in the formal economy, but help to trigger local-level social entrepreneurship. There is a need to ensure that social infrastructure transfers, such as housing, create viable communities and not dormitory townships, in the planning, location, character, and construction of these settlements – i.e. better location, more viable, mixed-use communities and more people-driven construction. As mentioned previously, more dynamic synergies could be catalysed to build local economy networks.

8. Social justice in Cape Town: Practical suggestions

8.1 The sustainable development option

The 'sustainable development' option is suggested by, and adapted from, Swilling (2006) as the key to social justice. This option aims to build 'sustainable neighbourhoods' as the building blocks for a sustainable urban future. Importantly for Cape Town, this option seeks to reduce the ecological footprint of the over-consumers, without fundamentally altering their lifestyles. The complementary strategy involves investing in infrastructure and housing that is designed to protect poorer houses from future ecological and economic challenges that could undermine their struggle for a better life. In this way, an urban system that reduces the cost of doing business for businesses, and the cost of living for middle and lower income households, emerges.

This approach would discourage the building of exclusive low income suburbs. By investing R75,000 to create a serviced house for a poor family in a uniformly poor neighbourhood, the value of that asset is the same – if not less – after occupation and/or sale (Swilling, 2006). The same asset constructed in a mixed neighbourhood, where there is a more active housing market, can have a market value that is three to five times the value of the initial subsidy, without any cross-subsidisation. When coupled to eco-efficiencies, this can contribute significantly to local economic growth stimulated by a virtuous cycle of access to credit, more disposable income, and higher re-investment levels back into the neighbourhood. This would retain local wealth within neighbourhoods thereby strengthening the capacity for local economic development.

The City needs to increase usage of renewable energy alternatives and promote energy efficiency. Coupled to this must be the promotion of a zero waste system, whereby waste separation at source across all households and businesses can create a new recycling industry. Sustainable transport, with a major focus on public transport, is essential. The promotion of sustainable construction materials and methods is also crucial to the success of this option.

Producing local and sustainable food is a critical component of social justice in Cape Town. The City's dependence on long-distance food supply chains from non-organic agricultural sectors results in a massive carbon footprint. This makes all Cape Town households extremely vulnerable in the medium to long term from a food security point of view. The obvious solution is the relatively low cost regulatory and investment strategy of creating neighbourhood-level spaces for food markets, where farmers and growers can sell directly to households. This stimulates the growth of local (urban) small-scale growers, who tend to be more efficient users of water and much less dependent on fuel.

8.2　Women

Women are central to the challenges of sustainable livelihoods, households, and communities. A very large number of women are engaged in economic activity in the informal or secondary economy. Stokvels and crèches are frequently run by women, and many women can be seen selling vegetables or gathering wood as a means to maintain their families. On the one hand, this shows both how vulnerable women are in the economy and yet speaks to opportunities for the development of small businesses. Often women hold the key to solving many of the developmental problems they face. The challenge is to remove the obstacles to women taking control of their own futures, and to ensure that the City of Cape Town undertakes interventions that ensure social justice for women.

8.3　Better implementation of urban development policies

The focus on traditional municipal service delivery has been partially transformed into a local government developmental approach. Although an increasing number of municipalities acknowledge their responsibility to contribute towards economic growth and job creation (Smit, 2002), this has not yet resulted in effective measures to eradicate poverty and inequality. Smit suggests the need to reshape South African urban development policies and strategies into a coherent framework that seeks to restructure apartheid spatial patterns and promote social and economic development, while responding to the problems of informal hyper growth, increased social polarisation, and the negative impacts of globalisation. In this regard, Smith suggests the following key principles:

- greater integration and coherence toward dealing with poverty and inequality;
- the promotion of participatory democracy;
- more effective coordination and policy alignment between different sectors and different spheres of government – e.g. it should be possible to create the conditions in which social grant transfers are not merely consumed, or spent immediately in the formal economy, but help to trigger local-level social entrepreneurship;
- more emphasis on partnerships between government and civil society organisations – this results in a more integrated and coherent approach to urban development, especially to the challenges of poverty and inequality;
- democratic participation by the poor and vulnerable at all levels of decision making – especially in participatory budgeting processes; and
- provision of the necessary support programmes to ensure the success of these participatory processes.

The City of Cape Town could apply the above principles in consolidating existing development policies. For example, the City could ensure that social infrastructure transfers (e.g. housing) create viable communities and not dormitory townships, both in the planning, location, character, and construction of these settlements – i.e. better location, more viable mixed-use communities, and much more people-driven construction.

8.4　A universal social wage

The structure and underlying assumptions of South Africa's social security system are still based on the fundamental premise that employment is the norm for the working-age population, and hence, social security should target the most vulnerable in other sectors of the population (the very young, the aged, etc.) (De Swardt, 2004). The current levels of poverty, inequality, and systemic unemployment all act as major restraints on any growth and development. Thus, social security and a universal social wage become critical in improving both human and social conditions. The challenge of developing a comprehensive system of social security in South Africa is then not to make nearly half the population passively dependent on welfare grants, but rather to use a comprehensive social security system as an important lever within a sustainable growth and development strategy to overcome poverty. These changes create the possibility for the majority to respond to the important challenge of respecting the ecological limits to the development which they hope will improve their lives.

The deep-seated, structural marginalisation of a large proportion of the population is a key impediment to any sustained, job-creating growth in the formal sector. Social security, and other resource transfers to the poor, could be regarded as active, transformational catalysts. They could serve to both overcome social exclusion and promote more vibrant participation in the economy (ibid.). In the second place, normal participation in the economy needs to be defined more broadly than just holding formal sector jobs. In particular, there could be reframing of both the idea and the reality of sustainable communities, and, at the household level, of sustainable livelihoods in urban, peri-urban, and rural settings. The concept of a social wage is relevant here: in developed economies, a social wage is composed of public provision of water, sanitation, electricity, transport, health, education, and other public goods. South Africa does not offer a complete and universally accessible social wage that would contribute to the goal of sustainable livelihoods and communities. However, some of the services and policies of the City of Cape Town already speak to the notion of providing a social wage, e.g. the provision of subsidies for public transport, and electricity and indigent support. The challenge here is to conceptualise these services as a social wage that can be directed at achieving social justice. If they are considered as a social wage, then the next challenge is whether such services should be commodified, or socially financed through tax rather than user-fees.

8.5 A sound enabling environment

Part of the enabling environment for a more sustainable city lies in the country's constitutional framework, which promotes the notion of socio-economic rights. Socio-economic rights are those rights that give people access to certain basic requirements (resources, opportunities, and services) necessary for human beings to lead a dignified life (Khoza, 2005). The following socio-economic rights are included in South Africa's Bill of Rights (contained in the country's Constitution, 1996) and form an enabling environment for South Africans:

- The right to a healthy environment (Section 24) – a useful foundation for the 'sustainable city' model and for addressing social injustice;
- The right of access to land, to tenure security, and to land restitution (Section 25) – in granting ownership, there is also a transfer of responsibility for sustainability;
- The right of access to adequate housing (Section 26) – may be used to ensure decent, sustainable, and environmentally-friendly housing that also undoes apartheid spatial patterns;
- The right to have access to (among other things) sufficient food and water (Section 27): sufficient food and water cannot be guaranteed in the long term if the 'consumption city' model remains intact.

Socio-economic rights are important tools for vulnerable and disadvantaged groups, who are often most affected by poverty, and who face a number of barriers blocking their access to resources, opportunities, and services in society (ibid.). The courts have confirmed that socio-economic rights are enforceable under the Constitution. In addition to these constitutional provisions, other enabling laws and policies have been put in place (local government, urban development, social development, etc.).

The concept of constitutionally entrenched socio-economic rights, accompanied by an appropriate legislative framework, transforms inequality into the social and civic obligation of addressing development as a human right. The constitutional provisions also balance socio-economic rights with due regard for ecological sustainability. In the Grootboom judgment, the Constitutional Court pointed out that the 'progressive realisation' of socio-economic rights hinges on the capacity of government to formulate and implement policy with the available resources (Constitutional Court, 2000). Unfortunately most public discourse has not included natural resources in the basket of 'available resources'.

9.   Conclusion

Natural resources, food, water, electricity, housing, transport, education, health care, and communications are all essential for a decent, tolerable, and sustainable life. As this chapter has shown, a large number of Cape Town residents do not have access to, or simply cannot afford, these essential goods and services – some of which are priced excessively high, even by international standards. The foundational constitutional values of our society – human dignity, freedom, and equality – and the socio-economic provisions of the Bill of Rights, which include the right to food, water, shelter, education, and social security, cannot be attained if people do not have access to essential goods and services. Therefore, our challenge in Cape Town is to ensure that the realisation of socio-economic rights takes place alongside ecological sustainability. This would require that the City ensures that all its policies and decisions respect, promote, and advance constitutional socio-economic rights, which in turn would require a review of a range of service delivery and urban development policies to ensure that they can withstand constitutional scrutiny. This has significant implications for the way in which a city like Cape Town balances economic growth and development with socio-economic rights, on the one hand, and sustainability on the other. The 'sustainable city' model provides a way out of this dilemma.

References

Aliber, M. 2001. *Study of the Incidence and Nature of Chronic Poverty and Development Policy in South Africa: An Overview.* Programme for Land and Agrarian Studies. Bellville: University of the Western Cape.

Barbier, E.B. 2009. *A Global Green New Deal.* Report prepared for the Economics and Trade Branch, Division of Technology, Industry and Economics, United Nations Environment Programme.

Bhorat, H., Leibbrandt, M., Maziya, M. & van der Berg, S. 2001. *Fighting Poverty: Labour Markets and Inequality in South Africa,* Cape Town: UCT Press.

Bruntland, G. (Ed.). 1987. *Our Common Future: The World Commission on Environment and Development,* Oxford: Oxford University Press.

City of Cape Town. 2008. Integrated Development Plan (IDP): 2007/8.

Constitutional Court. 2000. *Judgment: Government of the Republic of South Africa and Others v Irene Grootboom and Others, Case CCT 11/00.* Johannesburg: Constitutional Court of the Republic of South Africa.

Department of Health. 1998. South Africa Demographic and Health Survey. Pretoria: Department of Health.

Department for International Development. 1999. *Sustainable Livelihoods Guidance Sheet: Introduction.* http://www.livelihoods.org/info/info_guidancesheets.html#1. (Accessed 20 November 2008).

De Swardt, C. 2004. *Cape Town's African Poor.* Programme for Land and Agrarian Studies. Bellville: University of the Western Cape.

Dorrington R.E. 2002. Population projections for the Western Cape to 2025. Report for the Provincial Authority of the Western Cape.

Du Toit, A. & Neves, D. 2006. *Vulnerability and Social Protection at the Margins of the Formal Economy: Case Studies from Khayelitsha and the Eastern Cape.* Programme for Land and Agrarian Studies. Bellville: University of the Western Cape.

Du Toit, A. 2004. *Forgotten by the Highway: Globalisation, Adverse Incorporation and Chronic Poverty in a Commercial Farming District.* (Chronic poverty and development policy series, No. 4). Programme for Land and Agrarian Studies. Bellville: University of the Western Cape.

Du Toit, A. 2005. *Chronic and Structural Poverty in South Africa: Challenges for Action and Research.* (Chronic poverty and development policy series, No. 6). Programme for Land and Agrarian Studies, University of the Western Cape: Belville.

Heintz, J. 1998. *Poverty and Economics in South Africa.* Input paper for the 'Speak Out on Poverty' hearings organised by the South African NGO Coalition, the South African Human Rights Commission, and the Commission on Gender Equality. Available from www.naledi.org.za/pubs/1998/heintz1.html. (Accessed 24 November 2008)

Hemson, D. 2003. *The sustainability of community water projects in KwaZulu-Natal.* (Report to the Department of Water Affairs and Forestry). Pretoria: HSRC Press.

Hemson, D. 2004. *Beating the Backlog: Meeting Targets and Providing Free Basic Services.* Pretoria: Human Sciences Research Council.

Hindson, D. 1987. 'Orderly urbanisation and influx control: from territorial apartheid to regional spatial ordering in South Africa'. In R. Tomlinson and M. Addleson (Eds) *Regional Restructuring Under Apartheid: Urban and Regional Policies in Contemporary South Africa*. Johannesburg: Ravan Press.

Houston, C. 2002. *Livelihood strategies of the poor in Cape Town*. Cape Town: Development Action Group.

Jones, C. & Novak, T. 2000. *Class Struggle, Self-help and Popular Welfare*. In Lavalette, M. & Mooney, G. (Eds.). Class Struggle and Social Welfare. London: Routledge.

Kgara, S. 2008. *Neoliberal localisation: Cape Town, the Neo-Apartheid City*. (Unpublished paper).

Khoza, S. (Ed.). 2005. *Socio-Economic Rights in South Africa: Resource book* (2nd Ed.). Bellville: Community Law Centre. University of the Western Cape.

Landman, K. & Ntombela, N. 2006. 'Opening up spaces for the poor in the urban form: trends, challenges and their implications for access to urban land'. *Urban LandMark Position Paper 7*. Paper prepared for the Urban Land Seminar, November 2006, Muldersdrift.

Legassick, M. 1977. Gold, Agriculture and Secondary Industry in South Africa, 1885-1970: From Periphery to Sub-Metropole as a Forced Labour System. In Palmer, B. & Parsons, N. (Eds.). *The Roots of Rural Poverty in Central and Southern Africa*. London: Heinemann.

Manuel, T. 2004. *Address to the National Assembly on the tabling of the Medium Term Policy Statement, the Adjustments Appropriation Bill and the Revenue Laws Amendment Bill*. 26 October 2004. Pretoria: National Treasury.

Mbeki, T. 2004. *State of Nation Address to the National Assembly: Address to the First Joint Sitting of the Third Democratic Parliament*. 21 May 2004. Pretoria: The Office of the President.

McCaston, M.K. 2004. *Moving CARE's programming forward. Summary paper: unifying framework for poverty eradication and social justice and underlying causes of poverty*. Atlanta, US: CARE.

McCaston, M. and Rewald, M. 2003. *A Conceptual Overview of the Underlying Causes of Poverty*. Johannesburg: CARE.

National Treasury. 1996. *Growth, Employment and Redistribution: A Macro-economic Strategy*. Pretoria: Republic of South Africa.

Nattras, N. & Seekings, J. 2006. *Class, Race and Inequality in South Africa*. Pietermaritzburg: UKZN Press.

NEDLAC Community Constituency. 2006. *Position Paper on the Informal Sector*. National Economic Development and Labour Council (NEDLAC): Johannesburg.

NEDLAC Labour Constituency. 2003. *Draft Position Paper to the Growth and Development Summit*. National Economic and Development Council (NEDLAC): Johannesburg.

PCAS. 2006. *A Nation in the Making: Macro-Social Trends in South Africa*. Policy Coordination and Advisory Services (PCAS), Pretoria: The Presidency.

Republic of South Africa. 1996. *Constitution of the Republic of South Africa*. Pretoria: Government Printers.

Scoones, I. 2005. *Sustainable rural livelihoods: a framework for analysis*. IDS Working Paper 72. London: Institute of Development Studies (IDS).

Sen, A. 1999. *Development as Freedom*. Oxford: Oxford University Press.

Smit, W. 2002. 'The urban development imagination and realpolitik'. *Development Update*, 5(1): 53-80.

Statistics South Africa. 2001. *Census 2001 Report*.

Statistics South Africa. 2004. *Quarterly Labour Force Survey*, September 2004.

Stern, N. (2007) *The economics of climate change: the Stern review*. Cambridge, Cambridge University Press.

Swilling, M., Humphries, R. & Shubane, K. (Eds.). 1991. *Apartheid City in Transition*. Cape Town: Oxford University Press.

Swilling, M. with Davids, R., Ward, W., Wetmore, S., Jackson, N., Paschke, S., Moosa, S. & Khan, F. 2006. *Cape Town 2025: A City of Sustainable Neighbourhoods*. Paper commissioned by the Isandla Institute. Stellenbosch: Sustainability Institute.

Tomlinson, R. and M. Addleson (Eds.). 1987. *Regional Restructuring Under Apartheid: Urban and Regional Policies in Contemporary South Africa*. Johannesburg: Ravan Press.

Triegaardt, J. 2006. *Reflections on Poverty and Inequality in South Africa: Policy Considerations in an Emerging Democracy*. Paper presented at the 2006 Annual Conference of the Association of South African Social Work Education Institutions (ASASWEI), 18-20 September 2006, University of Venda: Thohoyandou.

Turner, S. 2007. *The Underlying Causes of Poverty in Lesotho*. Johannesburg: CARE.

UNDP. 2008. Human Development Report 2007/2008: Fighting climate change: Human solidarity in a divided world. New York: United Nations Development Programme (UNDP)

Van der Berg, S. Burger, R., Louw, Y. & Yu, D. 2005. *Trends in poverty and inequality since the political transition*. Stellenbosch: Department of Economics, University of Stellenbosch.

Van der Berg, S., Louw, M. & Du Toit, L. 2007. *Poverty Trends Since the Transition: What We Know*. Stellenbosch: Department of Economics, University of Stellenbosch.

Western Cape Provincial Government (WCPG). 2005. *Measuring the State of Development in the Western Cape – May 2005*. Cape Town: Western Cape Provincial Government.

Western Cape Provincial Government (WCPG). 2006. *Socio-economic profile: City of Cape Town*. Cape Town: Western Cape Provincial Government.

Western Cape Provincial Government (WCPG). 2007. *Isidima, the Western Cape Sustainable Human Settlement Strategy*. Cape Town: Department of Local Government and Housing.

Economic and Industrial Development

Options facing Cape Town from a 'Sustainable City' Perspective

Mazibuko Jara

1. Introduction

This chapter opens by acknowledging the impact of rising oil prices on the global economy and the affect it has on economic and industrial development in Cape Town. Cape Town's economy is in trouble. For the city to grow in such a way that it can sustain a tax base that can finance a more sustainable urban infrastructure, it is clear that Cape Town's economy can neither afford the continued decline of manufacturing, nor remain reliant on ongoing dependence on debt-financed consumerism.

The challenges of economic development become clear in this chapter, as we sketch out the ecological limits to growth in Cape Town. There are positive responses that are emerging in addressing regional economic development. The 'sustainable city' model is suggested as a viable alternative, when compared and contrasted with other possible future scenarios: 'business as usual', the developmental state option, and the 'deep green' option. This is followed by an analysis of the structural elements of Cape Town's economy, and trends in unemployment. Using a 'sustainable city' model, this chapter explores possibilities for sustainable economic and industrial development for the City of Cape Town, in a global context characterised by continuing economic challenges such as peak oil, energy constraints, rising food prices, environmental pollution, and permanent structural unemployment. The chapter builds the case for an alternative economic development path for Cape Town that respects and integrates the ecological limits to economic and industrial development. Practical proposals are suggested. A central challenge is how to redirect investments into the kinds of activities that will enhance the emergence of a more sustainable city. The chapter concludes by offering a wide range of actions and policies designed to enable Cape Town to become sustainable.

2. Implications of the Global Economy for Cape Town

It is no longer possible to think about the economic growth of developing economies within a globalised economy without taking into account the effects of oil peak on oil prices, and the way in which rising oil prices will blow holes in oil dependent economies (Swilling *et al.*, 2006). The linkage between climate change and human activity is now an uncontested fact. There are similar concerns about the impact on food supplies, and the relationship between population growth and the transition to a majority urban world. How will all this be affected by oil peak, climate change, and eco-system degradation? What are the implications for municipal policies on local economic development, service delivery, revenue generation, and the provision of infrastructure?

What does all this mean for the economic development path that a city such as Cape Town must follow? Should it aspire to be part of global competition and find a place for itself in the globalised economy? The dominant answer to be found in city development strategies is 'yes' to the global integration of the City – draw foreign investors to the City; let there be a shopping mall for local townships; let the property market work to promote investment, etc. (City of Cape Town, 2006a). The implications of such an approach are reduced capacity for local production because of the flooding of local markets by cheaper commodities from other countries displacing local goods. But is this desirable and sustainable, or is there an opportunity to imagine a local economy that meets basic needs through local resources?

3. Cape Town recognises the challenges

The concepts and solutions proposed in this chapter are already embedded in policy debates within the Cape Town municipality. For example, the City of Cape Town (CCT) *Updated Growth Scenarios* (2007d), the *Economic and Human Development Strategy* (EHDS) (2006a), the 2007-2012 Integrated Development Plan (IDP) (2007c), and other key strategy documents underline the importance of economic growth as "the primary vehicle through which to address the development goals of reducing poverty and unemployment" (CCT, 2007a: 6). In all these key strategy documents, the emphasis on economic growth has also sought to integrate environmental sustainability, which is identified as one of the "5 stars" of the EHDS strategy. Impending climate change and the imperative for Cape Town to continue to work on positioning itself as a 'green' destination to attract investment have driven the city's strategic thinking on economic growth scenarios (CCT, 2007b). Background research used by the CCT strongly states that "global evidence is increasing in its consensus that future economic growth will indeed depend on economically positive ecological preservation" (CCT, 2007b: 2).

By December 2007, the CCT recognised that "future economic growth in Cape Town is dependent on environmental sustainability" (CCT, 2007a: 4). This perspective is contained in a discussion document released by the CCT's Department of Economic and Human Development entitled *The Economic Imperatives of Environmental Sustainability*. This document underlines the following policy perspectives:

- The CCT is increasingly identifying environmental sustainability as a strategic priority: "the economic opportunities inherent in the threats of climate change should be immediately exploited, benefiting both the economy of Cape Town and the local community, and the environment" (*ibid.: 6*);
- The CCT recognises the economic benefits of environmental sustainability for firms and individuals alike, and that these would require changing current consumption and production patterns;
- The CCT recognises that current models and measurement tools do not take into account the very real negative environmental, opportunity and blackout costs for economic activity – in acknowledging this, the City is considering using policy tools and planning models that address this gap;
- A major determinant of current economic activity in the City and South Africa as a whole is the fluctuating prices of exported oil – recognising this, the CCT now believes that it is essential that all future policy decisions are mindful of the need to reduce the economy's dependence on oil;
- The CCT recognises the prudence of investing in renewable energy technologies and related benefits, such as the creation of a private market in renewable technology;
- The 'greening' of Cape Town will reduce pressure on basic service infrastructure capacity and attract investment, which will grow the economy and contribute to the environmental preservation of the city at the same time;

Several key documents have been developed, e.g. the *Draft Energy and Climate Change Strategy* (CCT, 2006b) and a draft report entitled *Renewable Energy and the City of Cape Town: an Economic Model of Environmental Costs* (CCT, 2007b).

The Sustainable Development Summit, which was organised by the multi-stakeholder Provincial Development Council (PDC) on 21 and 22 June 2005, adopted a 'Declaration of Intent' that has far-reaching implications for Cape Town. It sets a framework for positioning the Western Cape as the Province that will lead the way both nationally and internationally with respect to regionally-based sustainable development strategies, focusing on:

- Promoting greater equity in wealth and land distribution;
- Satisfaction of fundamental human needs;
- Job creation through economic development and investment strategies;
- Incorporation of biophysical and ecological limits in planning and decision-making;
- Conservation and sustainable use of biodiversity and other natural resources;
- Sustainable living within peaceful, economically viable communities;
- Internationalisation of environmental and social costs in the design and operation of production and consumption systems; and
- Promotion of environmental, social, and economic justice.

From these documents and processes, the CCT has drawn the conclusion that the current rate of natural resource exploitation will retard economic growth when it reaches a supply ceiling. Resource use occurs at a highly correlated level to economic growth (CCT, 2007a). This recognition is an important platform from which to take forward action on ecologically-informed economic growth and development. However, there is no information available regarding the extent to which this recognition is translated into practice, and whether it informs the investment and production decisions of industry and other economic players. Another challenge has been how to ensure that paradigms that inform economic growth and development plans integrate the provisions of the above documents. Not much success has been achieved in this regard. Therefore, this chapter seeks to consolidate these processes by putting forward concrete proposals for a 'sustainable city' model that require action by the City.

4.   The 'sustainable city' model

The solutions suggested in this chapter rely on the very useful framework provided by Swilling *et al.* (2006). This framework is implied, but not explicitly stated, in current internal CCT policy debates and recommendations. According to Swilling and his colleagues, the 'sustainable city' model requires the Provincial, City and National governments to realise early on that *sustainability provides the means to achieve equity and integration*. Massive savings are generated from investments in smart infrastructures, designed by engineers and financial experts who are briefed to learn from the thousands of working models – found locally and elsewhere in the world where poverty eradication has been achieved via sustainable resource use strategies. More equity is achieved by removing the subsidies and reducing the costs of the middle class lifestyle via restructuring the techno-infrastructure that supports it.

Swilling *et al.* (*ibid.*) contrasts this 'sustainable city model with three other development scenarios:

- *Business as usual*: This scenario is basically a continuation of the status quo into the future. The consequences are increasing inequalities and deepening poverty; increasing race and class divisions in a context of state-subsidised urban sprawl; and persisting highly inefficient urban systems, which will cause eco-system services to further degrade, and rapidly rising costs of energy (in particular oil), water, refuse removal, transport, food, and shelter, which, in turn, will exacerbate inequalities reinforced by spatial disintegration. Ultimately this will lead to a general state of decline, represented most graphically by the continued expansion of gated communities on the one hand, and sprawling informal settlements on the other, with corresponding, totally disconnected lifestyles.
- *Developmental state option:* In this scenario, Provincial government and the City manage to agree on a package of developmental interventions in the land market, in particular in order to achieve greater integration and equity. At best, they would give a rhetorical commitment to sustainability. Publicly-owned inner city land is secured for socially mixed integrated developments that result in densification, new local economies, and the promotion of a kind of 'melting pot' lifestyle. Infrastructure funds would be spent on expensive sewers, rather than much cheaper biogas digesters; public transport is expanded, but remains dependent on fossil fuels. The middle class is allowed to build energy-inefficient houses that are not forced to separate wastes, and nothing is done to promote local food markets. Unfortunately, the choice to ignore eco-system thresholds comes home to roost. This includes the wrong assumptions about cheap water, oil, energy, waste management, sewerage treatment, food, and eco-tourism dollars. The net effect would be that there would be fewer resources for poverty eradication, and that ultimately integration would mean that the rich buy their way out of unsustainability by actually reversing real integration. Islands of eco-efficiency do emerge, but the continuation of public subsidies of eco-inefficiencies (that the emerging black middle class require to make it into the middle class) undercut the resources required to fully realise the poverty eradication aims of the developmental state option.
- *'Deep green' option:* In this scenario, sustainability becomes an end in itself. *Instead of using sustainability as a means to achieve sufficiency and integration via the notion of sustainable resource use, 'greening' becomes the overriding focus.* Such focus may result in integration via densification, but not necessarily across racial boundaries, because race will not be seen as such a problem. This scenario is most likely to happen later, after a 'business as usual' scenario has been allowed to run its course, ending in a severe crisis in ten or fifteen years' time.

5. Cape Town economy: a continuing 'consumption city'

This section surveys the major structural features of the economy of Cape Town in relation to the broader Cape Town Functional Region and the rest of the Western Cape economy. The city's insertion into the global economy and future growth and employment prospects are also highlighted. Applying Swilling's development scenarios (Swilling *et al.*, 2006), the analysis indicates that Cape Town's current and future economic growth and development trends display a dominant 'business as usual' approach, although there are nascent elements of the 'developmental state' option. The dominant 'business as usual' approach perpetuates established production and consumption patterns.

6. Size, sectors and growth

The economy of Cape Town is hardly small and insignificant in relation to the South African economy as a whole. In the year 2000, Cape Town produced goods and services worth more than R85.9 billion (PGWC, 2005). Cape Town's 2006 Gross Geographic Product (GGP) is estimated at R123 billion. The City constitutes 78% of the Western Cape economy at large, and 11.58% of South Africa's Gross Domestic Product (GDP) (Ashley, 2008). This is many times larger than Lesotho or Swaziland, and is essentially the same size as the Tanzanian economy, a country of 35 million people.

Table 1: Operating and capital budgets for the City of Cape Town, 2004-2009

Financial year	Operating budget	Capital budget
2004-05	R11.3 bn	R1.53 bn
2005-06	R13.97 bn	R3.8 bn
2006-07	R11,3bn	R1,94bn
2007-08	R14,15bn	R4bn
2008-09	R15,75bn	R3,91bn

[Source: CCT Budget: 2008/2009-2010/2011]

The most important characteristic of Cape Town's economy and its economic development is its broad, diversified base. The key sectors of the economy (manufacturing, trade, and financial and business services) are each well diversified, thereby reducing the risk of over-dependence on any single industry. In addition, despite the fact that it is not primarily agricultural, the Cape Town economy also includes a significant agro-processing component. The City's economy is also strengthened by a well-developed and diversified tourism sector, and relatively strong and dynamic construction, fishing, professional services, higher education, and transport sectors. Within manufacturing, the major activities are agro-processing, metals and engineering, oil and petro-chemicals, and clothing and textiles. The absence of large mines and capital-intensive mineral processing, as well as heavy industry complexes, accentuates the dominance of small and medium-sized enterprises (SMEs) in Cape Town's economy (PGWC, 2005).

The infrastructure and economy of Cape Town is far more established than any other settlement in the Western Cape. Since 1996, the Cape Town economy has grown by 51% (CCT, 2007a). Between 1995 and 2006, the fastest growing service sub-sector per annum was communications (16.39%) (Ashley, 2008). Second was finance and insurance (10.82%), followed by construction (8.02%), then wholesale, and lastly the retail trade (7.03%). The 1995-2006 growth period of the service sector was built on an earlier period of expansion that goes back to the early 1980's (CCT, 2001), and by 2001, the fast growing service sub-sectors had increased their contribution to Cape Town's economic output from 35% to 42%. Other contributors to the 1990's period of growth were the transport, communication, and construction sectors, which are expected to continue to be significant contributors to local economic growth. An important pillar in Cape Town's growth pattern was the positioning of export products as the central driver of local economic growth, increasing from R6.6 billion in 1996 to R9.7 billion in 1999. Between 2001 and 2005, the average annual growth in exports from the Western Cape as a whole was 5% (Wesgro, 2006). Cape Town contributes a significant part of this growth.

Textiles have declined to 3.5% of exports in 2005 compared with 6.1% in 2001 (*ibid.*). A number of other manufacturing sectors saw overall declines: food and beverages, non-metallic minerals, and wood, paper, and publishing. Some of the decline is due to world trade commitments to lower trade tariffs against imports, which affected the clothing and textile sector in particular. The general decline in manufacturing was paralleled by growth in the informal sector, which by 2001 employed more than 22% of the labour force, and produced about 12% of local economic output. The increasing export orientation of the local economy was actively encouraged by the National, Provincial and local spheres of government (see discussion below on globalisation and its impact on the City).

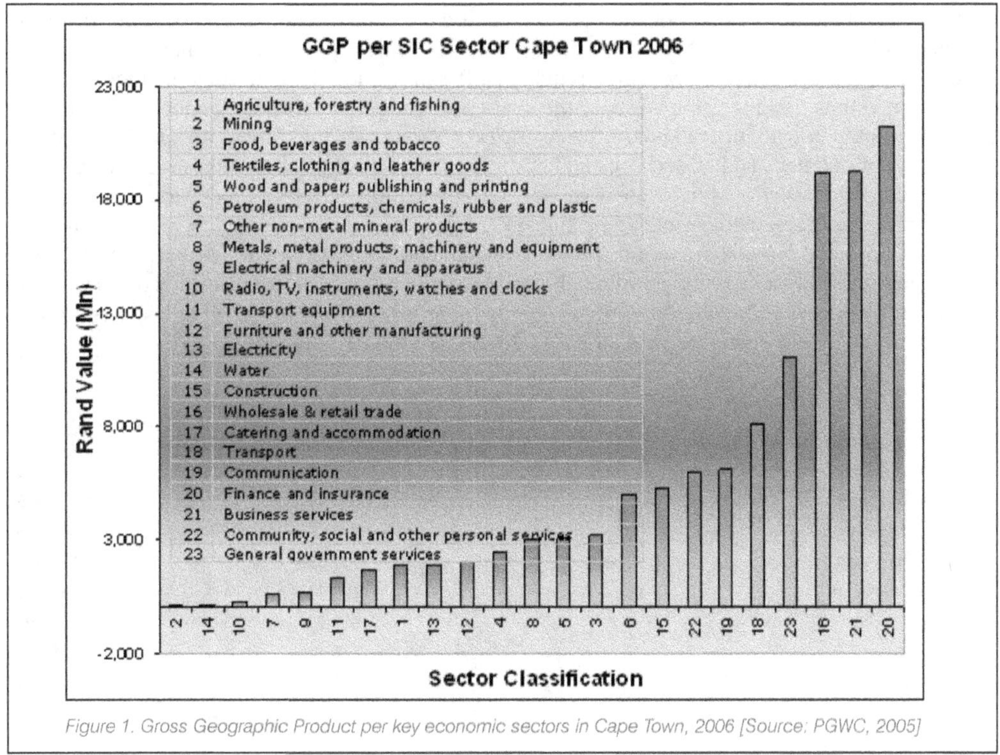

Figure 1. Gross Geographic Product per key economic sectors in Cape Town, 2006 [Source: PGWC, 2005]

In line with higher rates of economic growth, fixed investment spending in the Western Cape also grew more rapidly than the National average between 1995 and 2005 – 4.2%, as compared to 2.2%. Most of this investment has gone into economic sectors located in Cape Town. Financial and business services played a major role in this increased investment; investment in the service industries is less cyclical. Transport and communication, government and internal trade, and catering also contributed to the growth in fixed investment spending. The only manufacturing sectors that made any significant positive contribution were petroleum products, chemicals, and transport equipment (PGWC, 2005).

But GGP is in fact only a partial measure of the actual City economy, as it excludes the 'second economy' and does not measure the ecological impact of growth – i.e the running down of natural capital that is not factored in when progress is measured in terms of GDP growth. The informal economy has become an important source of livelihood for many, employing an estimated 22% of the labour force, and contributing 12% to economic output (CCT, 2007a). 93% of Cape Town's formal businesses are small, contributing 50% of total output and 40% of total employment (PGWC, 2005). These figures would be even higher if informal business were included.

Despite the rates of growth reported above, formal unemployment still grew from 16% in 1996, to 18% in 2000 (see Table 2). In terms of the Labour Force Survey (Statistics South Africa, 2005), unemployment in the city had grown from 13% in 1997, to almost 23% in 2004, although a drop to 20.7% was recorded in 2005. In line with national and international trends in economic growth, the City experienced both 'jobless' and 'job-loss' growth in the 1990's (CCT, 2001). The decline in manufacturing has largely been responsible for the rise in unemployment, driven largely be the decline of blue collar jobs in the textile sector as Government removed the protective tariff barriers.

What are the underlying causes of the economic and employment trends? The legacy of *apartheid* has meant that a large portion of the labour force has limited skills that are not suitable to an increasingly globalising economy, which requires soft skills. There is a shortage of these skills in the economy, thus limiting the growth of the domestic market. Other factors which explain these economic and employment trends are the historical structure of the South African economy, the impact of globalisation, and deepening capital intensity. Historically, the South African economy as a whole has been based on a dependent capitalist accumulation path that relies heavily on primary product exports – particularly from mining and agriculture – and is excessively dependent on imports of capital goods and many commodities required in manufacturing. Therefore, economic growth and development have been exceedingly vulnerable to global fluctuations, and particularly to movements in commodity prices. The structure of the economy retains the basic features of an export-oriented and import-dependent economy, with an under-developed domestic market.

Table 2: Employment and unemployment between 1995 and 2004

	1995	2004
Population aged between 15-64	1,644864	1,983,916
Economically active	1,014,102	1,178,436
Not economically active	630,762	805,480
Employment: formal	727,538	715,505
Informal employment	111,412	187,201
Total employment	838,951	902,706
Unemployed	175,152	275,730
Unemployment %	17,27	23.4

[Source: Quantec Research, using data from Statistics SA's 2004 Labour Force Survey, cited in PGWC, 2006]

Table 3 shows that, for all the growth sectors mentioned above (petro-chemicals, transport, communications, and financial services), there is a disjuncture between growth in fixed capital formation and employment creation: significant increases in capital investment were not accompanied by any net employment growth. Employment declined in almost all sectors – the major exceptions being automotive, retail and wholesale trade, and business and community services. Employment decline was particularly sharp in manufacturing, at 3.2% per annum, compared to a National average decline of 0.8% per annum. Between 1995 and 2005, employment in financial services grew by 1.1% per annum, while output growth was 6.1% and investment growth 5%. Rising fixed investment, accompanied by falling employment, is a clear indication of capital deepening and labour displacement. In line with this strong tendency to capital deepening, employment creation in the Western Cape (in particular in Cape Town) is significantly skills-biased. In all sectors, even in the unskilled areas, the share of unskilled workers has tended to decline. Between 2000 and 2003, two-thirds of jobs created accrued to those with a Matric certificate or higher. Another factor in growing unemployment has been the net in-migration to the city from the poor and rural parts of the Eastern Cape, which has seen a massive growth of unskilled labour in the city (PGWC, 2005).

Table 3: GGP Growth and Associated Employment Growth by Skill Category, 1995-2006

Rank	Sector	Average GGP growth p/a 1995-2006	Highly skilled	Skilled	Semi & unskilled	Total	Highly skilled % growth	Skilled % growth	Semi & unskilled % growth
	1995-2006		**Skills Demand Average**				**Skills Demand % Growth p/a**		
1	Communication	16.39%	1195	6629	2821	**10646**	0.68%	-3.62%	-3.45%
2	Finance & insurance	10.82%	15787	35874	1414	**53075**	1.79%	-0.35%	3.74%
3	Construction	8.02%	2882	6923	47824	**59628**	-2.89%	-2.86%	-3.94%
4	Wholesale & retail trade	7.03%	18080	74627	20190	**112895**	3.30%	3.81%	1.48%
5	Transport	5.69%	1779	13753	4165	**19698**	-2.75%	-3.56%	-3.83%
23	Mining	-5.80%	400	379	2019	**2797**	0.21%	0.23%	-2.32%
	Total average	4.79%							

[Source: CCT, 2007a]

The above trends also raise serious questions about the efficacy of economic growth as a vehicle for job creation and poverty alleviation. The challenge, therefore, is not solely to raise the rate of growth, but also to foster growth that takes a more labour-absorbing path, and more especially to increase the demand for unskilled and semi-skilled labour. This underlines the importance of SMME development, cooperatives, entrepreneurial development, skills development, and public works programmes.

6.2 Cape Town in the global economy

One of the most significant spatial dimensions of global growth is that coastal economies have tended to grow significantly faster than hinterland economies (PGWC, 2005). The principal factor leading to the enhanced economic performance of coastal regions is that their location allows for easier engagement in the global economy. This is particularly true in the case of the Western Cape, as it sits astride established trading routes. For the Western Cape, the rise in exports has exceeded the growth rate of the regional economy by a significant margin. The growth in Western Cape exports in the period 1996-2003 has been estimated as being between 8% and 11.4% per annum, compared to a provincial growth rate of 3.9% over the same period. Much of this export growth comes from the Western Cape's top three export categories: fruit, alcoholic beverages (wine), and iron and steel. These three categories account for a little under half of all Western Cape exports. Cape Town benefits specifically from the growth in exports in the following sub-sectors: boats and ships, furniture, chemicals, wood products, automotive components, tobacco, and paper (PGWC, 2005).

The global economy is seen by both the CCT and the Provincial Government as offering several advantages for Cape Town:

- The global dispersion of production offers new export opportunities to manufacturers – but only if they can meet the exact technical requirements, operational parameters, and pricing structures specified by their global value chains. However, the other side of export opportunities is vulnerability to global economic fluctuations, and the threat of cheaper imports wiping out domestic market share.
- The global importance of knowledge-intensive processes (branding, marketing, design, research and product development, logistics, organising sourcing, managerial skills, financial services, etc.) as high rent yielding activities, which has implications for fostering specific knowledge activities/ sectors, and attracting skilled high income earning and spending personnel.
- The segmented nature of global tourism and the advantages Cape Town has in facilitating the synergies between natural and cultural resources.

What implications do these globalising tendencies have for sustainable and all-round growth and development in Cape Town? In the current efforts to build the Cape Town Functional Region as a globally competitive city region there is inadequate consideration of ecologically sustainable resource use, equity, and ways of overcoming under-development. There is also limited consideration of the negative impacts of an over-reliance on exports in an unstable global economic environment.

Furthermore, globalisation has had negative consequences for the Cape Town economy already. Global economic developments are significantly determining policy options at the local scale, to which most of the economic activity is increasingly confined (Kgara, 2008). Yet, in the South African context, the macro parameters set in the governance, spatial, and housing policies have ensured that the National scale significantly frames local development trajectories, even though there may be some room to pursue alternative paths. This can be seen in how interest rates, fiscal austerity, and trade liberalisation have affected the local economy.

According to Kgara (*ibid.*), the impact of globalisation on urban spaces and city economies has resulted in –

– a polarised and fragmented urban morphology, defined by de-industrialisation and urban sprawl;
– competition for prestige and investment by increasingly global cities and 'city regions'; and
– profound changes in the management of municipalities, provision of infrastructure, and service delivery.

"Polarised and fragmented urban morphology" refers to the simultaneous flourishing of elite enclaves at the urban cores (CBD, some inner-city zones and edge-cities)[1] and the increasing expanses of decaying working class neighbourhoods. De-industrialisation is one of the underlying factors in the development of this deepening socio-spatial polarisation and fragmentation. This has caused decays, especially in cities that were anchored by the old 'smoke-stack' industries. The private sector-led urban sprawl further contributes to fragmentation on the fringes of the city, through the expansion of town house estates, shopping centres, and industrial and techno parks. The Century City, Grand West Casino, and Cape Town International Convention Centre (CTICC) developments are examples of this global trend. In essence they are emerging as key sites where prime multi-use property is allocated alongside complementary developments. These do not contribute to spatial integration and equity. The combination of de-industrialisation and urban sprawl reproduce zones and circuits of inclusion and exclusion. This reinforces unequal access to infrastructure and economic activities. The net effect is that Cape Town remains one of the most unequal cities in the world, with 36% of households (1,2 million people) living in poverty (Ashley, 2008); it maintains a Gini coefficient of 0.67.

6.3 Future Growth and Employment Prospects

Wesgro has identified several growth sectors within the province, one of which is the film industry, estimated to be worth R2,65 billion per year. Investment in this sector is perceived to generate positive spin-offs to the hospitality, clothing, and carpentry industries. According to the Provincial government and the CCT, the outlook for a number of rapidly growing sectors is seen as very favourable (PGWC, 2005). These include:

• Financial and business services;
• Transport and communications;
• Property and retail; and
• Wholesale trade and catering.

A key question is whether these sectors will grow the economy while creating jobs. Will they overcome under-development? Will they integrate ecological aspects into economic growth and development? Financial and business services, as well as communications, require a certain level of skills – a factor that does not take into account the low skills base of the majority of the unemployed labour force in the City. As long as transport relies heavily on imported oil, its contribution to future growth is

---

1   Inner city refers to places just outside the CBD, such as Sea Point or Hillbrow, whereas the edge-city refers to the new nodal developments on the outskirts of cities partly caused by capital flight from the CBD. Edge-cities tend to emerge around the elite residential landscapes, e.g. Century City, Sandton, etc.

likely to be limited. Property and retail may contribute to growth, but their capacity to contribute to job creation is limited. This can be seen in the already existing property and retail mega-investments that are mainly focused on the development and expansion of commercial and residential property, e.g., Century City, Grand West Casino, the CTICC, the V&A Waterfront, expansions (particularly in the Cape Town CBD), and other new developments located in other nodes. These create construction jobs, but the number of permanent jobs created in the sectors these investments support grow at rates that are lower than the economic growth rates that they stimulate. One needs to ask whether investments of a similar scale in manufacturing jobs would not generate greater output and more jobs per R1 invested. This requires further research.

Even though manufacturing does not feature strongly in these projections, there is a latent assumption in CCT and Provincial government discussions that the manufacturing sector is likely to be favourably affected by strong domestic demand. Again, this requires further research to substantiate.

The proposed 'sustainable city' model could boost manufacturing to meet local demand in a way that mainstreams sustainable resource use in economic planning. However, there are some limiting factors and potential threats:

- The strong Rand will have short-to-medium-term implications that directly impact on a wide range of Cape Town exports, and dampen the opening-up of new export niches (e.g. yacht building and film production);
- Stark and increasing income and wealth inequality in the region will limit local markets, and potentially threaten social cohesion;
- Sizeable net in-migration, with about two-thirds of the 50,000 net immigrants per year coming from poor, rural communities in the Eastern Cape;
- Steady annual out-migration of relatively well-skilled (young) people to the northern hub and other centres of the country and abroad;
- The challenge of quota removals and other global trade liberalisation measures on hitherto relatively protected local industries (such as clothing).

Further, the uncertainty over guaranteed electricity supply and the massive rise in electricity tariffs may dampen growth prospects – for the 2008/9 financial year, the CCT raised electricity tariffs by 38%, and future increases will be informed by the extent to which ESKOM raises tariffs to meet its huge capital expenditure programme. Given that 5% of businesses employ approximately 90% of the local labour force (CCT, 2007a), the Cape Town economy is distorted and potentially vulnerable to external shocks, which, even if they only affect a few firms, may affect a large part of the labour force. A collapse of key sectors could therefore be disastrous for the local economy.

6.4 Critique of the ongoing 'consumption city' model

Globally, the 'consumption city' model was driven by the economic reality of the 20th century city, namely, the need to create a mass of consumers that provide markets for the suppliers of urban goods that are now defined as the basic elements of urban living: houses, vehicles, energy, food, leisure, household appliances, and fittings (Swilling *et al.*, 2006).

The CCT has recognised the importance of ecologically sustainable economic growth and development (as discussed above). However, investment patterns confirm that the main economic development trajectory is for Cape Town to remain a 'consumption city'. Cape Town is essentially organised and structured as a site for accumulation and consumption.

According to Swilling and his colleagues, the basic building block of the 'consumption city' is that the 'consuming neighbourhood' needs to buy in the necessities for daily living from the outside (often from very distant locales) – energy, water, waste removal services, building materials, food, vehicles, etc., which were all provided cheaply and abundantly to make it all possible. This pattern of consumption, overlaid with the distorted logic of separation from the *apartheid* and colonial periods, has resulted in a particular urban system with its own spatial layout/form, function, economy, tax base, and operational requirements. Historically, the development of the apartheid city, in particular its white core, was at the expense of its black and working-class appendages. Post-apartheid social and spatial relations in many South African cities have often reproduced the key features of an *apartheid* city structure, where largely black poor and working people are perversely included (through simultaneous integration and marginalisation) into the city's economy and spaces. The CCT has also acknowledged this tendency in Cape Town (CCT, 2007a).

How do *apartheid* spatial relations combine with this consumption and accumulation model? They ensure that contemporary urban policies link urban land use to economic growth in a way that reduces the potential for sustainable resource use, equity, and integration. Contemporary urban policies perpetuate the view that urban areas are the primary engines of growth. In the South African context, local governments have thus been encouraged to plan and promote economic growth through developing efficient cities (Pieterse, 2003). The assumption that economic growth must be the outcome of competitive urban land markets (Marx, 2006) clearly underlies the approach of the CCT. The result is that many local, Provincial and National government policies are oriented to ensure that everything is done to make land markets contribute to economic growth. No critical questions are asked. There is no inquiry into the potential for sustainable land and natural resource use; there are no questions about who benefits most from economic growth.

The geographical location of economic growth is a key outcome of this framework: economic growth takes place in 'consuming' enclaves. Within these 'enclaves', economic growth is considered to possess an inherent, organic logic that will inevitably increase the standard of living and welfare of the general population. In the search for economic growth, this dominant approach is blind to its own ecological inefficiency. Another aspect of this logic is how it situates poor people's economic activities in the household, rather than the economy at large. This limits the understanding of the possibilities for outputs from poor people's economic activities. This approach inherently denies the possibility of seeing the location of poor people's economic activities as productive and contributing to urban economic growth. In doing so, it simply freezes poor people in their current locations. A sustainable resource use approach to economic growth and development would reintroduce ecological considerations and also integrate poor people into economic growth in a way that ensures growth, human development, integration, equity, and sustainability.

To conclude this critique, despite the recognition by the CCT of the need for a sustainable resource use approach, the pervasiveness of the 'consumption city' model has meant that there is yet insufficient room for alternative economic policy and investment decisions. For example, the provincial Micro-Economic Development Strategy ignores the economic advantage and potential of a Cape Town city region that specialises in knowledge-based systems, technologies, products, and value chains, and investments that seek to promote alternative/sustainability based industries (Crane & Swilling, 2008). This is the case with most other investment and economic policy decisions. The absence of mechanisms to measure progress made in areas where the City has taken action in this direction is a further problem.

What are the options for sustainable economic and industrial development for the City of Cape Town in a world characterised by multi-fold economic crises? Can there be another option for local economic development that delinks from a crisis-ridden path of development? The next section considers proposals for sustainable models of economic growth.

7. Towards a 'sustainable city' model for economic growth and development

7.1 Equity, integration and sustainability

This chapter builds the case for an alternative economic development path for Cape Town that would respect and integrate the ecological limits to economic and industrial development. In this way, the 'consumption city' model would be gradually replaced by a 'sustainable city' model that envisages the decoupling of improving living standards from rising (and increasingly unsustainable) resource consumption (Swilling *et al.*, 2006). Equity, integration and sustainability lie at the core of the 'sustainable city' model.

Equity involves significantly reducing the inequalities that exist between classes. Such inequalities are typically expressed in the City in spatial terms. A precondition for addressing such inequality is the provision of urban infrastructure, including housing, which many poor communities lack. Integration inevitably means both densification and a social mix across race and class boundaries (*ibid.*). According to several authors (Smajgl & Larson, 2007; Crane & Swilling, 2008; Prato, 1998; and Lakhani, n.d.), the concept of sustainable resource use is pivotal and integrates the following elements:

- Ensuring that our rate of consumption of natural resources does not damage the environment. This involves rethinking which natural resources we use and how we use them – by reducing, reusing and recycling;
- Reducing the quantity of natural resources used as material inputs into production and consumption processes (inputs here would include fossil fuels, wood, water, metals, clay, and food supplies produced from the exploitation of soils, sea, and water resources);
- Improving the efficiency of through-puts as they go through production and consumption cycles over a given period; and
- Eliminating entirely unproductive wastes by turning all outputs into inputs in a complete cycle in which waste is thought of as a 'residual product or a 'potential resource'.

These elements combine to define sustainability as an approach to resource use that ensures that every citizen has sufficient resources to have a decent quality of life, without consuming more than his/her fair share of the eco-system services that are required by this and future generations to live a similar or better quality of life (Swilling *et al.*, 2006; UNEP, 2002; Brundtland, 1987; and Burgess & Carmona, 1997).

7.2 Elements of the 'sustainable city' model

The basic building block of the sustainable city is the 'sustainable neighbourhood', namely the neighbourhood that generates more energy than it consumes, generates zero waste (both liquid and solid), meets most of its basic food requirements from local sources, requires little or no fossil fuels to transport people, and releases minimal amounts of CO_2 into the atmosphere (Swilling *et al.*, 2006). The 'sustainable neighbourhood' helps to rebuild eco-systems and mitigates the risks associated with the rising costs of fossil fuels as these non-renewable resources run out. The 'sustainable neighbourhood' is economically more efficient, because, in theory, it should be able to pay lower rates and taxes. Because it makes fewer demands on externally provided services, it becomes a more attractive operational space for households and businesses (Crane & Swilling, 2007, and Swilling *et al.*, 2006). The biggest transformational challenge for cities is to put in place frameworks and tools by building up 'sustainable neighbourhoods' from below.

In the Cape Town context, key elements of the 'sustainable city' model would include:

- Overcoming *apartheid* spatial relations through densification and multi-income settlements – with positive spin-offs for public transport routes and location close to work opportunities;
- An approach to economic growth and development in which process, means, and benefits are shared, democratised, and equitably redistributed among all sections of society. There is an explicit bias towards those who are marginalised and excluded from current patterns of growth and development;
- Addressing energy security and natural resource constraints through ensuring that recycling and increased usage of renewable energy lie at the centre of economic growth and development. This would include shortening logistics lines and value chains (not least through local, National and regional economic development) whereby opportunities for local value-adding are optimised, while those for long-distance hauls of supplies are minimised as far as possible;

- The re-conceptualisation of the entire economic sphere in such a way that poor people are perceived as inherently having a role to play in economically productive activities;
- Universal access to free basic services (discussed in Chapter 5); and
- Consolidating food sovereignty – this would require a combination of sustainable urban agriculture, land reform, small-scale production for household consumption, and local markets.

The model focuses on building self-sustaining communities and stimulating local economies. This would include stimulating local consumption and trade, the building of cooperatives, fostering training and skills development, and public works programmes. Important by-products of this model would create the conditions conducive to valuing authentic cultural diversity and a sense of community. A participatory culture celebrating diversity could enliven approaches to health, well-being and soulfulness (UNEP, 2002). In this 'sustainable city' model envisaged for Cape Town, the economic development path would follow policy measures, infrastructure investment, and sector specific programmes. Ideally, the intention would not be to seal off the local economy, but to ensure as best as possible a sustainable and balanced economic growth path that responds appropriately to both the challenges of global competitiveness and the often contradictory requirements of local, National, and regional development.

The 'sustainable city' option recognises that local government has the capacity, via land-use, investment, and service delivery to influence the way urban infrastructures evolve. Such strategies could result in different outcomes for households, either producing more or less equity, more or less integration, and more or less sustainability.

The logic of what is suggested in this section is absent from current 'consumption city' paradigms that inform Cape Town's economic growth and development. This explains to a great extent why the commitments made by the City to sustainable ecological resource use have not been realised: 'sustainable city' opportunities cannot be fully recognised in an economy oriented mainly towards consumption.

7.3 Overcoming apartheid spatial relations

The logic of 'one-city-one-tax-base' implies the spatial restructuring of the *apartheid* city linked to sustainable urban development. Key principles here include:

- Compacting development of the City in order to stem urban sprawl and foster intensive and inward-oriented spatial development;
- Mixed activity corridors along public transport routes (rail and bus), which break with the inherited mono-functional design of land uses with the view of enhancing sustainable socio-economic development, characterised by high-density, mixed land uses, and activity corridors, such as in Salt River and Observatory; and
- Densification, especially in the development of low-income dwellings – not only to maximise the use of space, but also to build the necessary economic threshholds to support local economic development, and public amenities in fostering sustainable urban development.

The above principles begin to make 'sustainability' accessible to the poor. In this sense, sustainability must mean investments in decent housing, more and better de-commodified services, and neighbourhood facilities and infrastructure (both built and natural). A 'sustainable development' approach has to discourage the building of exclusively low income suburbs. There is also an economic logic to sustainability: by investing R60,000 to create a serviced house for a poor family in a uniformly poor neighbourhood, the value of that asset is the same – if not less – after occupation and/or sale (Swilling *et al.*, 2006). The same asset constructed in a mixed neighbourhood, where there is a more active housing market, can have a market value that is three to five times the value of the initial subsidy, without any cross-subsidisation. Simply by being creative with space, zoning, by-laws and planning instruments, the same investment can generate much greater returns for poor households. This, when coupled to eco-efficiencies, can contribute significantly to local economic growth, stimulated by a virtuous cycle of access to credit, more disposable income, higher re-investment levels back into the neighbourhood, and reduced leakage as the benefits of eco-efficiencies kick in and local food markets reduce the costs of healthy eating.

The legacy of spatial segmentation along racial lines undermines alternative and integrated approaches to urban design that would focus on the public realm through sufficient public and private investment. According to Landman & Ntombela (2006), the characteristics of an integrated urban design approach would include mixed land use and the externalisation of public facilities and amenities along accessible roads. There would be activity corridors and mixed use nodes, as opposed to centralisation inside residential neighbourhoods. The integration of different urban areas and smaller neighbourhoods would be fostered through integrated routes, a well-functioning public transport system, and a continuous 'open space' system. Such an integrated urban design approach would be a better model to ensure wider access to multiple social ad economic benefits, markets, and services, including land, housing, jobs, safety, mobility, and opportunities for individual development.

When it comes to the reform of tenure systems in ways that can benefit the City's poor, consideration has to be given to using housing policy as an intervention in land tenure systems and housing markets. Key here is the practical creation of human settlements that can be regarded as assets by the beneficiaries, contributing to their quality of life. Among the contributing factors to the asset value of human settlements are the location of the land and the level of infrastructure and services available. The City will have to work with the Provincial government in this regard.

There must be a policy commitment to increasing the supply of subsidised formal public and private housing. This includes the intention to both improve the quality and speed up the rate at which delivery is made. Viable implementation must ensure that it is located in thriving economic zones. Another necessary intervention would be to introduce regulations on inclusionary housing, where developers building more than a certain number of houses at a time would be obliged to build a certain proportion (around 20-30%) in the affordable housing band (Napier, 2007). Such socially mixed settlements would provide a platform for the poor to enter profitable economic activities.

These combined recommendations would go a long way towards ensuring security of tenure and economic revival. The critical element in housing delivery is whether land and housing (with secure tenure) can be used as a platform for accumulation. Ideally, this would entail linkages to a socially mixed income milieu as the basis for generating household savings, income generation, and gradual capital formation. The challenge would be to align these activities with sustainable resource use. In this scheme, integration would be achieved via densification and the use of public land to build numerous socially mixed and mixed-use areas that are designed to reduce the distances between home and work and increase use of public transport by the middle class. In these ways, a sustainability approach can release large amounts of funds that would otherwise be spent on the subsidisation of the current ecologically inefficient urban systems.

7.4 Shared, democratic, equitable and redistributive economic growth and development

A strategic approach to the City's economic growth and development is called for. This must rest on local job creation as a major goal of economic growth and development. This can be done through policies that promote businesses that either operate at high levels of labour intensity or can generate substantial employment multipliers. This must go together with policies that address the tendency to 'informalise' employment and would require a more balanced growth and development strategy, going beyond the inherited economic structure with all its distortions and blind spots. Elements of this balanced strategy must include:

- Developing an effective industrial policy that focuses, in particular, on ensuring that the labour-intensive manufacturing sector is developed into a much more vibrant and dynamic sector of the economy.
- Actively linking industrial policy with support for major infrastructure development;

- Well-resourced and strategically directed education and training would be necessary to overcome the massive skills distortions in the local economy – a sustainable resource use approach would be instructive here, thus ensuring that large numbers of unemployed unskilled labour can be skilled for the purposes of a changed economy;
- Reconceptualising the economic sphere of the poor by integrating them into economic growth – this would need to involve innovative thinking about how urban land markets can be related to supporting networks of innovation and productivity, valuing land in terms of its ability to create new economic opportunities for the poor, and other measures to ensure that a sustainable resource use approach integrates poor people in economic growth; and
- Balanced development and effective industrial policy integration with other parts of the province and country.

The above elements must be integrated into the conceptualisation of economic growth and development. This would mean a review of the existing policy frameworks and plans. The CCT's challenge is to work out local industrial development strategies in collaboration with a range of stakeholders. This must start with ensuring that the ***economics of sustainable resource use*** are factored into all planning, in particular the costs of known constraints such as finite water supplies and rising oil prices; the economic job-creating potential of eco-efficient technologies such as waste recycling; and the renewable energy economy and urban agriculture (including food markets). It would require the creation of an entirely new 'green economy', driven by a new generation of 'GEEs' – 'Green Economic Empowerment' entrepreneurs, with massive investments in renewable energy, wind farms, the solar power sector, carbon trading, new inner city eco-designed office blocks, sustainable mariculture businesses, organic farming, and the fastest growing sectors of all, namely sustainable construction, recycling, public transportation, and authentic eco-tourism (Swilling *et al.*, 2006). A local industrial strategy must speak to all these potential opportunities.

7.5 Cooperatives

Cooperatives have a central role to play in an economic growth and development path that is informed by a commitment to democracy, equity, and redistribution. Although they are not necessarily going to solve all the ills of society, they show simply that there is an alternative way of doing business and really making a difference. Cooperatives are even more relevant to the present challenges facing society – employment creation, wider distribution of surplus, fighting financial exclusion, local economic development, building community buying power, and enhancement of broad-based economic empowerment. When actively promoted, cooperatives can achieve sufficient concentration in a particular geographic area, and perhaps more than any form of enterprise, can contribute significantly to local economic development. They can also contribute to mobilising community buying power and broad-based economic empowerment.

Because surplus profit remains with the members of the cooperative, money potentially stays and grows in the local community. In times of economic crisis, cooperatives do not readily close down or relocate, because they are controlled by members residing in the affected communities, hence they are then well placed to protect the sustainable development and livelihoods of many communities, especially in rural and urban areas.

Currently, cooperatives form a very small of part of the local economy in Cape Town. Systematic steps will have to be taken to ensure that cooperatives can grow and develop in ways that can change the accumulation logic of the local economy and meet the goals of equity, sustainability, and integration. The following practical measures are proposed:

- A process to identify, mobilise, and engage with existing cooperatives about their needs, challenges, and interests, and proposals on their role in the local economy;
- Municipal policy and strategy on the promotion of cooperatives;
- Procurement spend on cooperatives by the municipality;

- Municipally supported programmes for the business promotion, support, and incubation of cooperatives;
- The integration of cooperatives in the IDP, the Economic and Human Development Strategy, and other key strategies of the City;
- Providing incentives for cooperatives; and
- Building institutions for the support and promotion of cooperatives – an interesting example here is the Nelson Mandela Metropolitan Municipality's piloting of the establishment of a municipality-linked Cooperative Development Centre. The CCT should consider this option.

These proposed measures on cooperatives would build on what the City is doing already. Cooperatives would consolidate these City initiatives into demonstrable interventions, particularly if they are also linked to supporting economic activity in the townships.

There is ample international evidence to show that cooperatives have a long and successful tradition around the world, and have proved to be flexible and resilient in meeting a wide variety of social and economic human needs. International experience shows that countries that have achieved economic development have a vibrant and a dynamic cooperative sector, contributing substantially to the growth of those economies. For example, in Kenya, co-operatives contribute 45% of the Gross Domestic Product (GDP), and 31% of the total national savings and deposits. Cooperatives control 70% of the coffee market, 76% of the dairy market, and 95% of the cotton market (ICA Report, 2007). With a supportive and enabling legislative environment and supportive economic context, cooperatives in South Africa have the potential to contribute to economic growth, development, job creation, and a fairer distribution of surplus. The challenge for the CCT lies in demonstrating an explicit and shared commitment to growing cooperatives into a significant player in the economy of Cape Town.

The CCT can explore successful practices within the Ekurhuleni, Nelson Mandela, and eThekwini Municipalities. The Western Cape Provincial Government has also completed drafting its strategy and policy on cooperatives, which could provide a foundation for what the City undertakes in this regard. The challenge would be to ensure that the promotion of cooperatives is linked to sustainable resource use.

7.6 Trade

Equity and fair trade at all levels (global, regional and local) could change the way Cape Town is inserted into the global economy. Although increasingly common across many value chains at the local and global levels in many other parts of the world, there is little evidence that the local Cape Town discussion about ways of growing the economy have factored in a value chain approach that takes into account equity and fair trade issues (Swilling *et al.*, 2006). The 'Proudly South African' campaign is clearly a key reference point when it comes to branding, but this way of thinking does not seem to have penetrated into the discussion of micro-enterprise development, which remains fixated on the traditional concerns of this sector – namely micro-entrepreneurs and how to secure finance and opportunity for them. An equity and fair trade perspective looks at value chains creatively and explores how these can be restructured to advantage smaller players and build local economies. There are inspiring examples of employment creation that is independent of new investment. Other parts of the world have successfully introduced local credit systems. Or they have developed sophisticated ways of linking waste streams from certain industries to the input streams of others. Local food markets could well become the key driver for an equity and fair trade focus. This approach does not see globalisation in a unilinear way, which emphasises producing for, or importing from, global markets at the expense of local markets. The CCT already recognises that it can play an enabling and catalytic role to encourage equity and fair trade practices amongst Cape Town businesses. However, even existing initiatives undertaken by the City are not done to sufficient scale, and are still subordinate to the dominant economic growth and development paradigm. The recommendations made below seek to break out of this logjam.

Crane and Swilling (2008) have called on the CCT to instigate the formation of an 'energy consortium' to mobilise research and investment to pilot massive renewable energy and energy efficiency schemes, including wind farms on the West Coast, massive solar/geo-thermal power stations, the installation of solar roof tiles in all new houses together with incentives to install them on existing structures, and ensuring that the Cape Town CBD has energy efficient buildings.

On solar water heaters, the City must mobilise investors, local business, National government and other players to put resources on the table in order to inject capital into an industry that is vital to environmental sustainability. Current demand for solar water geysers is low, because of high start-up costs. A massive subsidy or capital investment would induce a shift in the demand curve, thus preceding future economic growth. Furthermore, and most importantly, it would provide the kind of output volume opportunities on which firms depend to achieve increasing returns to scale.

Competition and scale in the market of solar geyser production and installation will also bring the initial start-up costs down, thus incentivising middle and upper income households to invest in the sustainable, medium-term cost saving option. All this is contained in several recommendations already before the City. However, the bias is to seek leadership from, and reliance on, the private sector. Energy is a strategic resource for City- and Nation-wide social and economic well-being and development. In this sense, energy is a public good. Viewing energy in this way points to a necessary discussion about the role of the local state vis-à-vis the role of the private sector in generating and transmitting renewable energy and the provision of renewable technology into other loops of the renewable energy value chain. Also relevant is the important international experience of energy production by cooperatives which are not the same as the private sector.[2] It is therefore proposed that the CCT itself must take the lead in ensuring sustainable energy security. This does not mean that the private sector should be excluded. Instead, a multi-layered approach to the creation and development of markets in renewable energy and related technologies that could include the local State and other public agencies, as well as the private sector and new cooperatives, which will need to be promoted, supported, and developed.

7.8   Biogas promotion

In February 2008, the *Engineering News* magazine reported on a National biogas feasibility study, which was completed in November 2007 for the Department of Minerals and Energy (Tzoneva, 2008). Biogas is a cheap form of energy that can be used for cooking, lighting, and heating. The feasibility study showed that a rural biogas pilot project covering 360,000 households would have 15% and 67% financial and economic internal rates of return, respectively. The study recommended the roll-out of the first phase of a National biogas programme, which would involve the construction of 20,000 biogas digesters at a cost of R238 million over five years. For a densely populated metropolitan area like Cape Town, biogas production can be more cost-effectively produced.

7.9   Consolidating food sovereignty

Cape Town depends on long-distance, energy-intensive food supply chains to import 1,3 million tons of food per year. This makes all Cape Town households extremely vulnerable from a food security point of view in the medium to long term (Swilling *et al.*, 2006). Consideration has to be given to the relatively low cost regulatory and investment strategy of creating neighbourhood-level spaces for food markets, where farmers and growers can sell directly to households (*ibid.*). This could reduce prices for the consumer and increase the returns for farmers. It could also stimulate the growth of local small-scale growers, who tend be much less dependent on oil and more efficient users of water.

---

2   The US and Germany has a few case studies of renewable energy cooperatives, which, unlike private producers, are locally embedded and can be locked into long-term socio-economic development planning.

The Abalimi Bezekhaya story is relevant here. The group is comprised of more than 200 local farmers, who use small pieces of land in Khayelitsha, Langa, and Phillipi. Over time, they have built production systems and relational markets that not only provide them with their own food, but also with surplus production for sale throughout the City of Cape Town. This initiative can be supported and adapted to several other settings in the City. The urban agriculture policy of the City does not go far enough in this regard. Absent in the policy are the following elements:

- Up-scaling of urban agriculture to develop a critical mass of producers providing a significant portion of food for local markets;
- Provision of infrastructure and the extension of financial and market support for urban agricultural producers at the required level, scale, and regularity; and
- Adapting other policies and by-laws to ensure the success of urban agriculture – e.g., the City ban on mobile marketing of agricultural produce may hinder the success of urban agriculture.

There are many examples around the world of localised and decentralised food systems. For example, in the densely populated cosmopolitan city of Shanghai, 80% of the land is reserved for crop growing. The city is self sufficient in vegetables, rice, pork, chicken, duck, and carp. Urban agriculture is part of city planning, and local production is integrated into city plans. Another example is the city-state of Singapore, which is self-sufficient in meat production and produces 25% of its vegetable consumption locally. An example closer to home is Bamako, Mali, which is self-sufficient in vegetable production and meets 50% of its poultry needs through local production. There are also many examples where urban families have food gardens and meet local needs through their own production. In Dar es Salaam, 67% of families are involved in farming, while in Moscow, 65% of families are engaged in farming.

8. Conclusion

In critically analysing the features and prospects of the Cape Town economy, this chapter has made a case for the introduction of a new paradigm for economic growth and development. This new paradigm has been elaborated as the 'sustainable city' model. The CCT has already recognised the need for this paradigm shift, but it has not been fully internalised, nor optimally implemented in practice. By providing several concrete proposals, this chapter seeks to enable the City to take the next step. Many of these concrete proposals arise from already existing recommendations that have been published in documents either directly commissioned by the CCT or related stakeholders. Others come from initiatives taken by citizens in the City and elsewhere. Can the City of Cape Town make the necessary choices to become a 'sustainable city'? Whether it will take the steps required to shift away from the current unsustainable models of economic growth and development remains to be seen. Cape Town now has an opportunity to choose a new path. The City clearly has taken some initial steps to recognise the direction it needs to go in order to bring about sustainable economic development.

References

AGAMA Energy. 2007. *Integrated Biogas Solutions*. Available from www.agama.co.za. (Accessed 29 September 2008).

Ashley, B. 2008. *An economic overview of Cape Town*. Presentation to a seminar of the Don Davis People's Centre, Cape Town, 21 June 2008.

Burgess, R. & Carmona, M. 1997. *Challenge of Sustainable Cities: Neo-liberalism and Urban Strategies in Developing Countries*. London: Zed Books.

Jenks, M. & Burgess, R. 2000. *Compact Cities: Sustainable Urban Forms for Developing Countries*. New York: Taylor & Francis Ltd.

Bruntland, G (Ed.). 1987. *Our Common Future: The World Commission on Environment and Development*, Oxford: Oxford University Press.

City of Cape Town. 2001. *Towards an Economic Development Strategy for the City of Cape Town: a discussion paper*. Cape Town: City of Cape Town.

City of Cape Town. 2006a. *City Economic and Human Development Strategy*. Cape Town: City of Cape Town.

City of Cape Town. 2006b. *Draft Energy and Climate Change Strategy*. Cape Town: City of Cape Town.

City of Cape Town. 2007a. *The Economic Imperatives of Environmental Sustainability*. Cape Town: City of Cape Town.

City of Cape Town. 2007b. *Renewable Energy and the City of Cape Town: An Economic Model of Environmental Costs*. Cape Town: City of Cape Town.

City of Cape Town. 2007c. *Five Year Plan: Cape Town 2007-2012*. Cape Town: City of Cape Town.

City of Cape Town. 2007d. *Updated Growth Scenarios*. Cape Town: City of Cape Town.

City of Cape Town. 2008. *City of Cape Town Budget: 2008/2009-2010/2011*. Available from www.cityofcapetown.gov.za. (Accessed May 2008.)

Crane, W. and Swilling, M. 2008. 'Environment, Sustainable Resource Use and the Cape Town Functional Region – An Overview'. *Urban Forum*, 19: 263-287.

Deininger, K. 2003. *Land policies for growth and poverty reduction: A World Bank Policy Research Report*. Washington DC: The World Bank.

Department for International Development (DFID). 2001. *Meeting the challenge of poverty in urban areas*. Department for International Development UK, UK.

Department for International Development (DFID). 2005. *Making market systems work better for the poor: An introduction to the concept*. Discussion paper prepared for the ABD-DFID 'learning event', ABD Headquarters, Manila. Department for International Development UK, UK.

Gasson, B. 2007. *The Ecological Footprint of Cape Town: Unsustainable Resource Use and Planning Implications*. Available from www.saplanners.org.za/SAPC/papers/Gassonpaper.pdf. (Accessed 7 September 2008.)

International Cooperative Alliance (ICA). 2007. *General Assembly Edition*. Geneva: Review of International Cooperation 100(1). International Cooperative Alliance. n.d. *Global 300*. Available from *http://www.ica.coop*. (Accessed 29 September 2008)

Kgara, S. 2008. *The Neoliberal localisation: Cape Town, the Neo-Apartheid City*. (Forthcoming.)

Lakhani, M. n.d. Zero Waste: An Introduction. *Biophile, Issue 9*. Available from www.biophile.co.za. (Accessed 29 September 2008)

Landman, K. & Ntombela, N. 2006. Opening up spaces for the poor in the urban form: trends, challenges and their implications for access to urban land. *Urban LandMark Position Paper 7*. Paper prepared for the Urban Land Seminar, November 2006, Muldersdrift. Available from www.urbanlandmark.org.za. (Accessed 29 September 2008)

Marx, C. 2006. Conceptualising the 'economy' to make urban land markets work for the poor. *Urban LandMark Position Paper 3*. Paper prepared for the Urban Land Seminar, Muldersdrift, November 2006. Available from www.urbanlandmark.org.za. (Accessed 30 September 2008)

Monbiot, G. 2006. *Heat, How to stop the planet burning*. London: Zed Books.

National Assembly. 2003. *Municipal Finance Management Act*. Act 56 of 2003. Parliament: Cape Town.

Napier, M. & Ntombela, N. 2006. Towards effective state interventions to improve access by the poor to urban land markets. www.urbanlandmark.org.za (Accessed 30 September 2008).

Napier, M. 2007. Making Urban Land Markets Work Better in South African Cities and Towns: Arguing the Basis for Access by the Poor. Johannesburg and Washington: Urban Landmark and World Bank.

Pieterse, E. 2003. Unravelling the different meanings of integration: The Urban Development Framework of the South African government. In Harrison, P., Huchzermeyer, M. & Mayekiso, M. (Eds.). *Confronting fragmentation: Housing and urban development in a democratising society*. University of Cape Town Press: Cape Town.

Prato, T. 1998. *Natural Resource and Environmental Economics*. Ames, IA: Blackwell Publishing Professional.

Provincial Government of the Western Cape (PGWC). 2005. *The Micro-economic Development Strategy Synthesis Report: Version 1: July 2005*. Cape Town: PGWC.

Provincial Government of the Western Cape (PGWC). 2006. *Socio-economic profile of Cape Town*. Cape Town: PGWC.

Provincial Government of the Western Cape (PGWC). 2007. *Cape Shared Growth Initiative*. Cape Town: PGWC.

Smajgl, A. & Larson, S. 2007. *Sustainable Resource Use: Institutional Dynamics and Economics*. London: Earthscan.

Statistics South Africa. 2005. *Labour Force Survey*. Pretoria: Statistics South Africa.

Swilling, M., Davids, R., Ward, W., Wetmore, S., Jackson, N., Paschke, S., Moosa, S. and Khan, F. 2006. *Cape Town 2025: A City of Sustainable Neighbourhoods*. Paper commissioned by the Isandla Institute. Stellenbosch: Sustainability Institute.

Tzoneva, D. 2008. Rural biogas programme could benefit 20 000 households. *Engineering News*. Available from www.engineeringnews.co.za. (Accessed 29 September 2008).

United Nations Environment Programme (UNEP). 2002. *Melbourne Principles for Sustainable Cities*. Nairobi: UNEP Division of Technology, Industry and Economics.

Urban Sector Network and Development Works. 2004. *Scoping study: urban land issues – final report*. Prepared for DFID-SA (South African office the British government's Department for International Development). Braamfontein: USN.

Wesgro. 2006. *Western Cape Trade and Investment Trends 2006*. Available from www.wesgro.co.za. (Accessed 30 September 2008).

Wesgro. 2007. *Survey of Constraints to Investment Constraints in Cape Town*. Available from www.wesgro.co.za. (Accessed 30 September 2008).

Water and Sanitation in the City of Cape Town

Sustainability Uncertain

Kevin Winter

This chapter focuses on the challenges and opportunities facing the water and sanitation sector in developing a more sustainable Cape Town. Major concerns are raised about the lack of attention by the City to policy imperatives which specifically focus on sustainability issues. Failure to see and respond adequately and in a timely way to factors such as increased population, migration to the city, climate change and the uncertainty of meeting the ecological requirements, place residents of Cape Town, and particularly the poor, at risk. The central questions are:

- *How can the City best provide water and sanitation services, while conserving the resource?*
- *How can we ensure a sustainable supply of water, while minimising pollution of the natural environment?*
- *Are there are other ways of managing the flows of sewage, water and storm water in our city that are cheaper, more innovative, less resource-intensive, more environmentally friendly, and easier to operate?*

The discussion begins with a broad overview of water and sanitation in the City, outlining limits to the supply of water and the elements shaping the increasing demand for water. This is followed by a discussion on what sustainability means in relation to water services. To some extent, key barriers to sustainability are social and institutional in nature, rather than technical. A key theme in the chapter is the idea that the integration of environmental, social, economic, and institutional imperatives remains one of the main challenges in the management of resources such as water. Although practical suggestions are offered for water and sanitation, the conclusion is that however uncertain the transition towards sustainability for the City of Cape Town may be, sustainability is no longer a matter of choice. It is an imperative that calls for commitment and dedication from all stakeholders in the water and sanitation sector.

1.   Introduction

Reports commissioned by the City of Cape Town (CCT) in the past few years on water and sanitation largely provide a technical analysis of water and sanitation services, but fail to offer a clear statement as to whether the City is managing these resources sustainably or not. An unequivocal statement on sustainability is required not only to inform consumers, but also to inspire public confidence and trust in those responsible for the management of water resources and water related services. It is therefore sobering that measures for determining sustainability are not discussed in two of the most comprehensive water reports of recent years: the Water Services Development Plan for the City of Cape Town (WSDP) (CCT, 2007) and the Western Cape Water Supply System Report (DWAF, 2007).

Without a doubt, the City's Water and Sanitation Services Department faces a host of critical challenges. These include:

- Meeting the increasing water demand;
- Eradicating the backlog of basic sanitation services;
- Complying with waste water effluent standards;
- Managing ageing infrastructure and deteriorating assets;
- Extending infrastructure to meet the developmental needs;
- Ensuring financial sustainability of services; and
- Building the human capacity capable of addressing these pressing challenges (CCT, 2007).

A failure to make substantial progress in these issues will seriously impact on the water security and health of the city. The consequences of the failure to make sufficient progress could easily result in widespread conflict and socio-economic instability. These challenges are exacerbated by the fact that the City has limited control over the influx of an estimated 7,700 households per annum; an average population growth rate of 2.5% (2001-2006) that is likely to continue at a rate of 1.3% to 2016 and beyond; and an expected population of 4 255 847 by 2031 (*ibid.*). These factors alone compromise the City's efforts to realise its self-proclaimed vision of becoming a leader in the provision of "equitable, sustainable, people-centred, affordable, and credible water services" (*ibid.*:17). In essence the City is involved in a struggle to control the consumption of a finite resource amidst unreasonable demand, where the biophysical environment is no longer able to absorb pollution or to dilute waste water adequately. These factors are even more alarming when one considers that the investment in fixed assets and human resources is severely constrained by a limited budget and human capacity.

2. Troubled waters: an overview of the City's water supply and demand

Surface run-off in South Africa comprises approximately 9% of the total precipitation, while the remaining 91% either evaporates or infiltrates. The conversion ratio of rainfall to run-off for the country as a whole is among the lowest in the world compared to the global average of 35% (Shultz, 1997). Currently, surface water resources for the City represent 440.5 Mm³/year, or a 97% total yield. Between 70% and 75% of the City's raw water requirements are obtained from dams owned by the South African Department of Water Affairs and Forestry (DWAF)[1] all of which fall outside the boundaries of the City's catchment area. Only 15% of the raw water requirements are met from sources within the Cape Metropolitan Area (CMA). It is projected that existing water resources, including the new Berg River Water Scheme, will meet Cape Town's water demand until approximately 2013, provided a low water demand projection is maintained (see Figure 3 and the discussion later). While groundwater might be a further option to consider, it is estimated that the total yield is little more than 6.64 Mm³/year, representing only 1.5% of the total yield. Of critical importance is the relationship between groundwater and surface water. Groundwater can only be abstracted on a sustainable basis at a rate less than, or equal to, its long-term average recharge of rainwater. Over-extraction will result in the drying of rivers and wetlands, a reduction in subsurface flow, and groundwater failure.

In April 2007, the input into the reticulation system was estimated to be 268 Mm³/year; the non-revenue demand being 62 Mm³/year, and real losses of 47 Mm³/year. The distribution of this water demand for 2006/7 is represented by different categories of consumers (Figure 1). Of note is the relatively large proportion of consumption in the domestic clusters (combined total of 52%), and also a relatively high proportion of non-revenue water demand. It is also worth noting that water use in sectors such as urban agriculture and tourism are not disaggregated. This highlights the need to improve understanding and regulation of water-use in these and other strategic sectors of the urban economy.

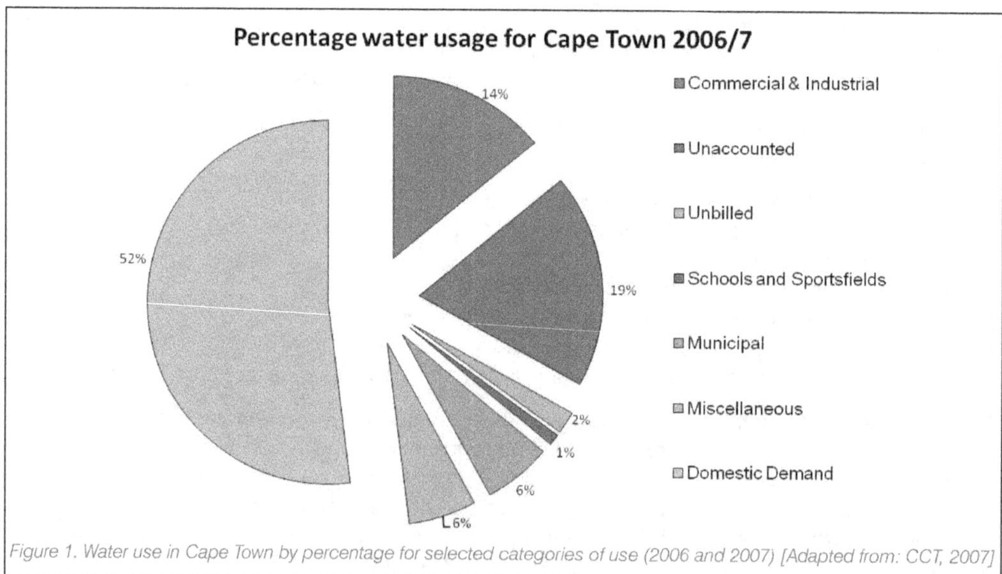

Figure 1. Water use in Cape Town by percentage for selected categories of use (2006 and 2007) [Adapted from: CCT, 2007]

The average water use per capita in the City is approximately 230 litres per person daily, which indicates a slight decrease from 270 litres per person per day in 2000 (CCT, 2008). This decrease is attributed to the introduction of water restrictions in 2001, and serves to illustrate the response to a

---

1 Renamed Department of Water Affairs in 2009.

water demand management strategy – although the demand increased soon after these restrictions were lifted (Figure 2). In 2003/04, the daily average water demand was 850 Ml/day. Again, after severe Level 2 restrictions were imposed in 2004/05, the demand dropped to 737Ml/day, but only to rise again in 2005/06 to 796Ml per day, and 845Ml per day in 2006/07. Clearly, consumers respond well to periods of imposed water restrictions, but consumption rates return to previous highs soon after restrictions are lifted. It can be concluded that water demand management is perceived as an administrative exercise that is inconvenient for consumers for a short period, but has limited impact in influencing long-term adaptation and behavioural change.

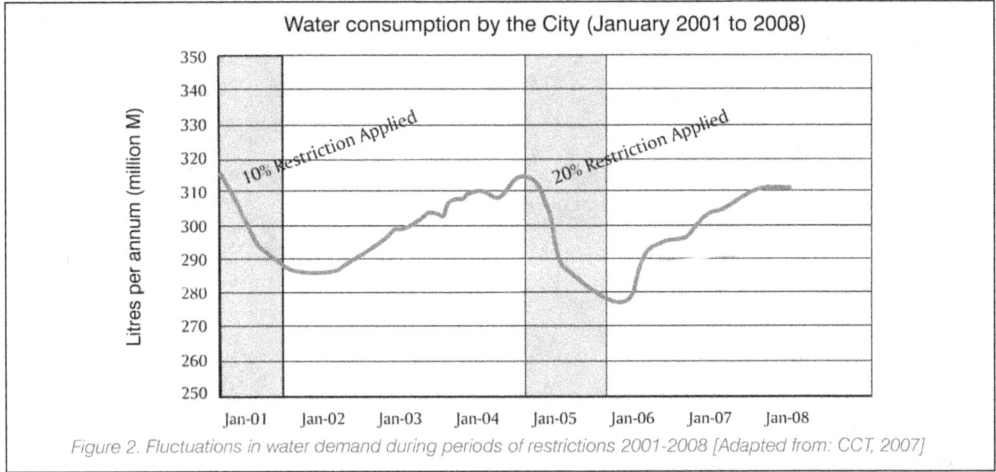

Figure 2. Fluctuations in water demand during periods of restrictions 2001-2008 [Adapted from: CCT, 2007]

The targeted savings for 2006/7 was 18.7 Ml/day, or 6.84 million m³ per annum, representing approximately 2% of the total demand. Demand is currently 25.5% below the unconstrained water demand (Figure 3). This is better than the DWAF requirement of 20%. Despite this reduction, the reality is that population growth remained at a high level of 2.9% as at December 2007(DWAF, 2007). According to the Integrated Development Plan for the City (IDP), the City is committed to reducing the actual demand for water by 20% from the projected unconstrained demand scenario by the year 2010 (Figure 3). The extent to which water management strategies incorporate exposure to climate change and the release of an estimated 10% of the total surface yield to the ecological reserve remains uncertain.

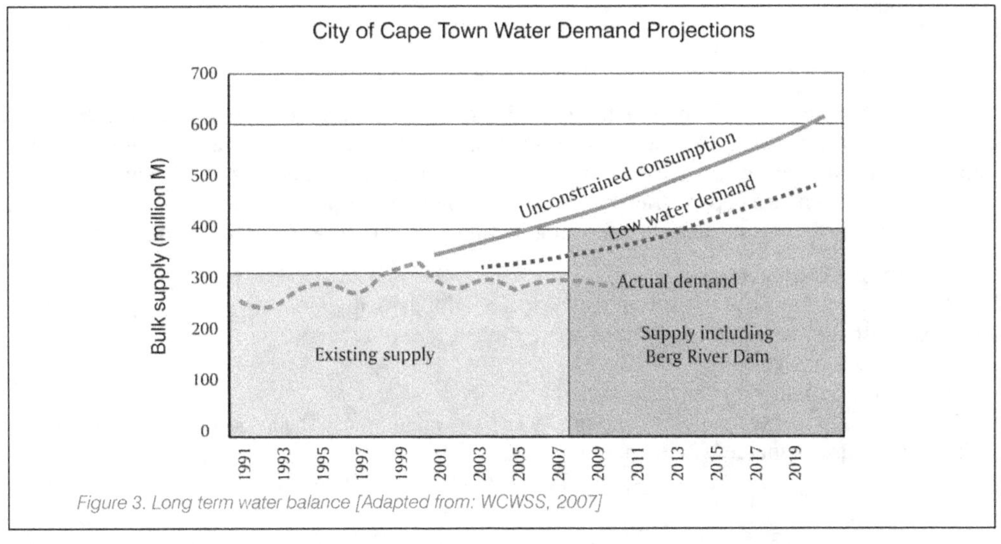

Figure 3. Long term water balance [Adapted from: WCWSS, 2007]

After 2013, a number of scenarios have been suggested to meet the water demand up to about 2030. These scenarios include comprehensive water demand management and water conservation (WDM/WC) initiatives, extraction from aquifers, and desalination. Augmentation of water in the future will need to focus on reducing the dependence on surface water. Table 1 outlines the most comprehensive scenario (Scenario 10) – as described in the WCWSS (DWAF, 2007) – in which a number of schemes and strategies are outlined, including that of meeting the ecological reserve and the accommodation of climate change. The real cost of Scenario 10 interventions has a net present value of R2,7 billion. By 2023, these interventions will require an annual cash flow of R450 million, which is ten times more than the current cash flow. The most cost effective intervention for the largest proportion of water saving remains that of WDM/WC (combined total of 106 Mm^3/year for implementation during the next 8 years).

Table 1: Proposals for the Western Cape Water Services Scheme. Water Demand Management is likely to be effective only until 2016, thereafter new augmentation schemes will be required

Intervention	Year of 1ˢᵗ Water Saving	Yield million m³/year
Water Conservation and Water Demand Management strategies including tariffs, private boreholes and wells, detection and leaks, and education	2007-2016	126.1
TMG Aquifer Scheme 1	2017	19.0
Voëlvlei Phase 1	2018	29.8
Raise Lower Steenbras dam	2019	21.3
Lourens River diversion	2020	16.2
TMG Aquifer Scheme 2	2020	47.5
Upper Molenaars diversion dam	2022	23
Re-use potable	2023	180
Desalination	2028	66

[Source: WCWSS, 2007]

The financial sustainability of water services will need to ensure full cost recovery and debt management at a fair and equitable tariff in order to finance the required capital investments. Currently the loss of income of uncollected revenue amounts to R205 million per annum (approximately 20% of total demand) (WSDP, 2007). Given the deterioration of assets and shortages of human resources in the water sector, it seems obvious that the real cost of water services is not being achieved either in the collection of revenue or the current unrealistic tariff structure in which the cost of water represents a substantial under recovery.

3.   Imperatives of sustainability

The pursuit of sustainability is not a matter of choice; it is an imperative that requires commitment and dedication. This imperative includes a high level of commitment to a range of socio-economic, administrative, and resource management processes. Valentin & Spangenberg (2000) illustrate these obligations in a useful conceptual framework in which they identify processes that should be pursued. The suggested processes include the following mutually reinforcing and interdependent goals:

- To establish fair access to resources;
- To create conditions and opportunities for meaningful participation;
- To establish a society that is willing to share the burdens of others;
- To ensure social justice;
- To take responsibility for the environment; and
- To use eco-efficient technologies and practices that will minimise the impact on the environment and the natural resource base (Figure 4).

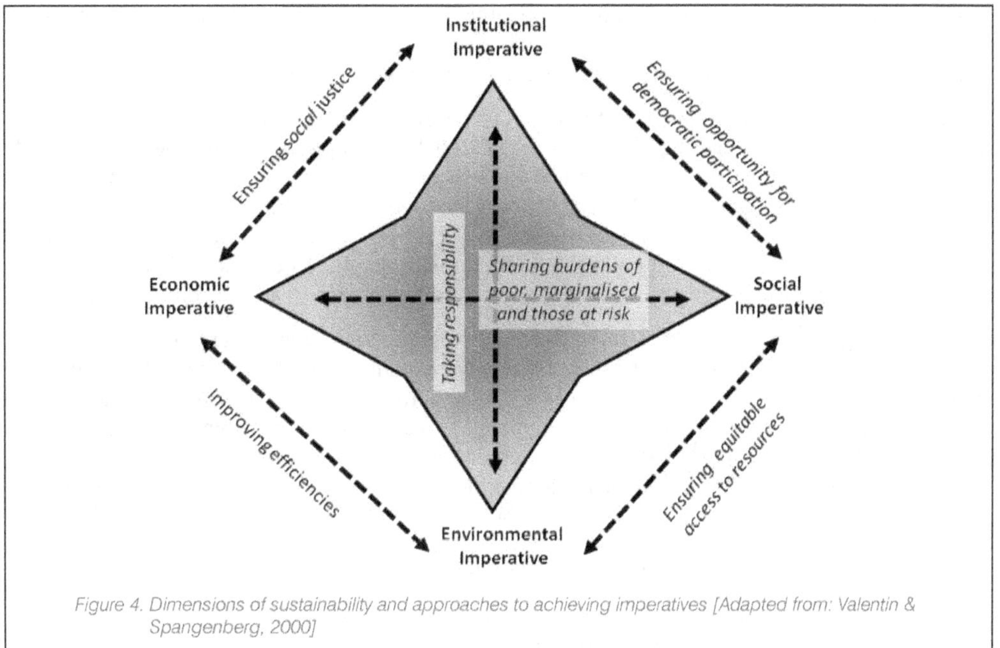

Figure 4. Dimensions of sustainability and approaches to achieving imperatives [Adapted from: Valentin & Spangenberg, 2000]

The intention of this framework is to provide a conceptual understanding of sustainability. It also offers the potential to develop holistic indicators of sustainability for measuring the performance of the City's water and sanitation services.

Integration remains one of the key challenges in the management of environmental resources. Key reasons underlying the lack of integration include the following factors:

- Efforts to integrate multiple dimensions of complexity often fail;
- Resource managers are poorly trained in the theory and practice of integration (and may also be unconvinced of the value of such an approach); and,
- Ideas and processes that inform integration are limited by financial budgets and time constraints.

Nevertheless, integration remains fundamental to addressing the imperatives of sustainability. One of the critical lessons learnt by the United Kingdom's Department for International Development (DFID) projects, for example, is that it is no longer possible to solve water supply and sanitation problems with concrete pipes alone. What must be acknowledged is that solutions depend on the integration of political, social, technical, and institutional approaches, and changing established practices (Department for International Development, 1998). Integrated urban water management (IUWM) is described simply by the United Nations Environment Programme (UNEP) (2003) as the management of water in the urban environment. Such integrated management specifically minimises the impact on rural water resources (quality and quantity), and maximises its utility within the town or city. In broad outline, the operational management of water and sanitation services will need to include the following:

- Water demand management (appropriate levels of service, improved technology, pricing, leak detection, etc.)
- Waste management (appropriate levels of service for different types of residential, industrial and commercial users, improved technology, pricing, immediate re-use of grey water, minimisation of waste, etc.)
- Urban drainage (minimise the risk of flooding, especially by reducing run-off rates through local storage and infiltration; maximise the use of storm water as a local source of water, improve the quality of storm water run-off, protect and/or remediate urban rivers, etc.)

- Bulk water and waste water treatment (minimise costs, maximise effluent quality, promote re-use, etc.)
- Local water supplies (groundwater, rainwater harvesting, evaporation control, etc.)

Furthermore, effective management of the urban water cycle must recognise the interrelationships between sanitation, water, drainage, and hygiene (Table 2). For example, an improvement in the supply of water may fail to improve public health if there is inadequate provision for the removal of waste water. This failure is widespread in the City's informal settlements, where improved access to water results in increased volumes of grey water, which, when mixed with black water, forms a hazardous transmission route for disease and pollution. In part, this example represents a failure to integrate services that might otherwise have minimized the risk to human health and the environment.

Finally, an integrated approach is characterised by the involvement of end-users in decision-making and management. All too frequently, projects are implemented largely on the felt or perceived needs of users, rather than on wide consultation and the use of participatory methods. As a consequence, services often fail to reflect user preferences and result in poor maintenance of service technologies linked to inappropriate interventions. Now, more than ever, interventions should be socially acceptable and efforts should be made to achieve a 'genuine commitment to partnership and empowerment' (Department for International Development, 1998: 41).

Table 2: Broad elements of sanitation, hygiene and water management

Sanitation	Safe collection, storage, treatment and disposal/re-use/recycling of human excreta (faeces and urine)
	Management/re-use/recycling of solid waste (rubbish)
	Collection and management of industrial waste products
	Management of hazardous wastes (including hospital wastes, chemical/radio-active and other dangerous substances)
Hygiene	Safe water storage
	Safe hand-washing practices
	Safe treatment of foodstuffs
Water management	Drainage and disposal/re-use/recycling of household waste water (also referred to as 'grey water')
	Drainage of storm water
	Treatment and disposal/re-use/recycling of sewage effluents

[Source: Evans, 2005]

4. Cape Town: the unsustainable city

It is often easier to identify unsustainable indicators than to find empirical evidence and observations to inform these judgments. This section identifies a list of water and sanitation service indicators for the City that could be considered as 'unsustainable' and these have been linked to broad categories shown through the prism of sustainability (Figure 4). The problems highlighted in Table 3 are not usually the result of intention, but much rather *inattention* – i.e. the failure to see and respond adequately and in a timely way to factors such as increased population, migration to the city, and climate change, and the uncertainty of meeting the ecological requirements. The list highlights the urgent need to change current thinking and practices so as to prevent the continued risk to human and environmental health, and to avoid a looming water crisis.

Table 3: Links between indicators and evidence of 'unsustainable' urban water services[2]

Component	Indicator	Evidence of 'unsustainable' water and sanitation services
Social	Access to water supply	
	Access/use of sanitation facilities	Basic sanitation services backlog.

---

2   Note: Empty cells indicate that there is no evidence or that the evidence is insufficient.

Component	Indicator	Evidence of 'unsustainable' water and sanitation services
Social (continued)	Levels of service	Limited access to adequate water and sanitation services – 300 000 to 400 000 households are limited to basic and/or emergency access to water and sanitation.
	Vulnerability to disasters	
	Health (morbidity and mortality)	
	Education/awareness	Residents in formal areas are 'over' satisfied with water services, while those in informal settlements are increasingly dissatisfied.
Economic	Capacity to pay/access services	
	Cost recovery	Excessive water loss as 'unaccounted-for' water (UFW) – average of 19.3% (June 2007) represents a loss of 47 million m^3 pa. Full cost recovery is not being achieved – non-revenue demand is 62 million m^3 pa or 23%. Outstanding debt of R1.77b places a constraint on infrastructure and development.
	Investment levels	Ageing reticulation systems – number of bursts per 100km has deteriorated from 6 to 14 in the last 20 years, due to a reduction in the number of kilometres of pipeline replaced per annum. Further investment in bulk water supply water schemes is inevitable – new large bulk water supply schemes will need to be implemented anywhere between 2011 (no WC/WDM) and 2019 (comprehensive WC/WDM). Unaffordable rising costs caused by diffusion of the city – urban sprawl increases the unit cost of piped water, sewers and drains.
Environmental	Fresh water resources	Water demand exceeds surface yields – surface water represents 97% of the total yield. Demand is likely to be exceeded by 2019, even at the low water requirement trajectory. Daily per capita usage is unsustainably high.
	Sustainability of water source	Major uncertainties in the future – available supply depends on the effects of climate change and the implementation of ecological reserves for existing schemes.
	Use (resource distribution per sector)	
	Waste water management	Effluent from waste water treatment works (WWTW) fails to comply to DWAF water quality standards – mean compliance for discharge into freshwater bodies was 81.3% (Jun 2007), with 83% being the target for 2008/09. Increase in volume of waste water sent to WWTWs – from 470ml/day in 2000 to 565Ml/day in 2007.
	Stormwater management	Blue Flag beach status is threatened along with the coastal marine ecology – 24% of sample points exceeding 80th percentile water quality guideline in 2004/2005.
	Compatibility of water system with surrounding environment	
	Compatibility of sanitation system with surrounding environment	
	Environmental stresses	DWAF water quality standards too lax for water quality being discharged into freshwater resources – given the deterioration of freshwater resources within the city, the current 1984 Water Quality standards/targets should be re-evaluated
Political	Governance	
	Compliance with policy	
Institutional	Institutional and technical capacity	Limitations of current WDM strategy on conservation of water resources. Targeted savings for 2008 are 18.7 Ml/day or 6.84 million m^3 per annum, approximately 2% of demand. Limited critical personnel to manage key technical services.

[Source: Carden, Winter & Armitage, 2009]

Sustainability offers an approach to designing policies and strategies that integrate social and biophysical resources. If taken seriously, a sustainable approach could significantly influence policy, plans, and practices in the long term. *Sustainability principles recognise that all elements of water provision, water use and services are interrelated, and this includes supply and demand, sanitation, sewage (treatment), sewerage (reticulation), grey water (excluding input from toilets), and storm water drainage.* There is increasing evidence that this is being integrated into water and sanitation policies by the City. However, the existing institutional structure is problematic. For example, Water and Sanitation Services are separate from storm water drainage and road services. This already places a severe strain on efforts to achieve sustainable integrated urban water management (IUWM), as suggested in the first point by Brown *et al.* (2007) listed below.

The bulleted list outlines some of the foundational principles of the philosophy of sustainable IUWM:

- All parts of the water cycle should be considered as an integrated, inter-connected system, which includes protecting and restoring the health of water sources.
- Multiple purposes for water use (human and environmental) need to be accepted and flexible, and multiple solutions need to co-exist.
- Context matters: all perspectives need to be considered (environmental, social, cultural, and institutional).
- Public participation in planning and decision-making is vital.
- Programs, projects and policies need to be considered over long-term timeframes guided by a common vision.
- Inter-disciplinary approaches are required (e.g. engineers, environmental scientists, social researchers, economists, planners, etc. all working together).

In general, IUWM thinking is applied more readily to the management of water in formal settlements, but it should be equally replicable in informal settlements. To this end, it is imperative to measure performance elements within the urban water cycle for the city as a whole, rather than adopt a separate approach for the formal and informal settlements of the city. *It is also worth noting that barriers inhibiting the adoption of sustainable practices in the city are social and institutional, rather than technical.* While a range of technologies are well known worldwide, human behaviour and governance remain major challenges in the City as a result of a complex socio-political history and inertia to change. Attention should therefore be given to exploring social and institutional indicators of sustainability, rather than focussing exclusively on technology.

There is also a strong argument for creative and innovative management of water resources that is dependent on strengthening neighbourhood and 'community' social structures. This assumes that the greatest resource for change lies in the human and social capital of the City. This concept is not new. The UN Global Report on Human Settlements recommends that informal settlement upgrading comprises "improvements which can be undertaken cooperatively and locally among citizens, community groups, businesses and local authorities", and suggests that such actions could include the installation or improvement of basic infrastructure (2003:165). However, participation remains largely dependent on the existence of some form of social structure, such as an interest group, a street committee, or some representative council. In addition, social structures can become the catalyst for establishing a 'community'. In this instance, the concept of 'community' refers to a social context in which individuals relate to each other for the purpose of achieving shared goals (Warburton, 1998), but for a variety of reasons the idea of community is largely absent in informal settlements. Community structures in informal settlements in and around the city are frequently found to be weak and fragmented. There is an argument therefore that building and supporting robust social structures is integral to effective water and sanitation services. It is based on the premise that citizens are more likely to take greater responsibility for managing scarce water resources and services if they are able to participate meaningfully. The task of setting up representative social structures and providing support for communities should not be performed by the City alone. Considerable lead support must be acquired from NGOs that are better placed to provide this support and to facilitate partnerships between the City and residents.

Finally, human capacity to govern water resources is fundamental in order to achieve a sustainable approach. The Water Services Development Plan raises numerous concerns about the capacity and training of personnel, and concludes that it is grossly under-resourced in the water and sanitation sector (City of Cape Town, 2007). The report alludes to the constraints that exist due to long overdue vacancies, and asserts that low staff morale is counter-productive to the high expectations of service delivery. There is clearly an urgent need to give attention to human resources within the City. This argument is reiterated by Brown (2008) who states that a commitment to local leadership and organisational learning by way of capacity building is important for enabling a sustainable urban water management approach.

6. Water and Sanitation options

6.1 A Pro-poor priority

The flow and exchange of water in informal settlements is a daily struggle involving fetching and carrying water and selecting disposal sites for waste water. Poor hygiene, exposure to water pollution and the uncomfortable fact of living in damp, wet conditions are commonplace. Poor drainage, rising levels of groundwater, and flood risks affect the lives of many. It is not surprising that those living in these under-serviced settlements want nothing less than household access to potable water and on-site toilets so that they may have greater control and management of their water and waste.

The Water Services Development Plan (WSDP) claims that non-access to basic water provision was effectively eradicated in the period 2005/06, when 35 000 informal households were granted access water. The report continues by claiming that 17 050 informal households were provided with basic sanitation during this period, but approximately 30 000 households still have no sanitation. However, all too often these facilities become inoperable soon after installation due to vandalism, the general failure of the technology, or both (Carden *et al.*, 2008). The cost and capacity for fixing these broken facilities is taking its toll on the City budget and its officials. Broken systems often remain in this state, because residents have an expectation that the City is duty-bound to repair this equipment. Steps to address the issue may lie in an integrated strategy in which the available human and financial capital is maximised. Consultation and local level user involvement remains crucial.

A range of options are available to address the critical concerns identified above, including:

(a) Communal bathrooms and toilets

The City recognises the general failure of existing communal facilities. In part, failure can be attributed to the limited services offered by these facilities, the poor design and location of these facilities, and the fact that the responsibility and arrangements for maintaining these facilities falls largely on the City. Studies elsewhere (Maili Saba Research Report, 2005) report on the popularity and success of communal facilities, but only because these facilities offer more than just a toilet. Appropriate sanitation includes toilets, but also extends to washing (having a safe, private place and sufficient clean water); cleaning of clothes and keeping homes, latrines and bathrooms clean; and better drainage to avoid dirty water from pooling in the streets. Warm showers, basins and toilets must be located optimally and managed seven days a week. NGOs and SMMEs should become essential partners in this service provision and in turn offer employment for local residents. Most importantly, the monitoring of water and energy resources, the cost recovery from consumers, and the management of communal facilities are suggested as crucial to success.

(b) Dry Sanitation Units

eThekwini Municipality is a leader in the field of dry sanitation in the peri-urban areas of South Africa. They have installed over 30 000 urine diversion toilets. Urine is absorbed into the ground while the faecal matter collects in a chamber with a storage capacity of one year. The household needs to take responsibility for disposing the contents of the chamber. The success of this intervention and household practice is still uncertain. While dry sanitation might be an option, the density of Cape Town's informal settlements and the high water table in many parts of the city is likely to make this an inappropriate option for most locations. Further research is required.

(c) Condominial Sewer Systems

The use of a shallow sewer network in informal settlements can provide an incentive to encourage residents to connect their waste water and sewage to a pipeline that will discharge into a treatment works, rather than the storm water drainage. There are obvious risks involved, but the potential exists to support such arrangements through education of users and community involvement, and in the partnership of NGOs and the City. The concept of shallow sewerage condominial systems involves laying pipes from one house or dwelling to another, rather than using a conventional approach of laying pipes at some depth alongside roads or similar linear structures. Most importantly, a condominial system involves a system of user agreements for managing blockages and leaks so as to avoid disorder on the land and to reduce the potential risk to human health and the environment.

(d) A household water and sanitation unit: a pipe dream?

This option runs counter to the current national housing policy and approach to building RDP houses. Three assumptions are forwarded:

- Effective sustainable water and sanitation services strategies are dependent on some form of land security or tenure;
- All reticulated services have the potential to create planned order on the land and organisation of space; and
- Water and sanitation is fundamental to social upliftment, and is an enabling factor necessary to break the cycle of poverty.

This option recommends that greater attention should be given to services, rather than attempting to provide costly, inappropriate RDP houses for all. Furthermore, it suggests that providing water and sanitation is not enough. Water and sanitation services must form an integrated package that includes energy and communication services. The inclusion of electricity not only offers the potential to monitor resource usage at a household or neighbourhood level, but also offers a range of management possibilities that could be extended to include cost recovery mechanisms. In addition, it would offer a communication link (either by wireless or cable) allowing householders an entry into the information age and the possibility of accessing education and public services using the internet. A prefabricated unit is proposed of approximately 6m^2, which could provide a secure weather-proof environment for incorporating the services mentioned above and form the basis of the housing structure. This suggestion has not been found elsewhere in the world, but may well merit further discussion, simply because it seeks to offer an integrated approach to address the social, economic, and environmental imperatives of sustainability as discussed earlier, and may also offer a means to break the cycle of poverty through structured learning programmes being made available directly to the dwelling.

6.2 Limits to growth in demand: managing the middle-upper income consumers

The Western Cape Water Supply Reconciliation System (DWAF, 2007), the Water Services Development Plan for City of Cape Town (City of Cape Town, 2007), and the Integrated Analysis: Water and Sanitation Baseline Report (Sustainability Institute, 2007) all emphasise that water conservation/water demand management (WC/WDM) is the single most important means of ensuring that water demand is kept well below the low water demand projections as shown in Figure 3. Consumption patterns currently suggest that further augmentation of new water sources could be delayed until at least 2029.

While it is salutary to note a reduction in actual water demand, the continued increase in economic growth of the City will require additional water and sanitation infrastructure, and additional water resources. Given that current measures are inadequate, a water conservation and demand management strategy will become increasingly necessary. To this end, water demand management strategies should be strengthened and expanded to include:

- Incentives – Informative billing to provide knowledge and information on consumption patterns over time; interactive flow limiter devices that provide a digital display of household water use (currently being tested by CCT); rebates for households and institutions able to use grey water technologies in controlled, safe environments; rebates for properties equipped with rain tanks and rainwater harvesting facilities; subsidies for the purchase of approved water saving and retrofitting devices; reduction in sewage disposal charges for off-grid sewage treatment (e.g. dry sanitation of private households, small bore bioreactor services); financial incentives to encourage NGOs and community-based groups to support water resource management;
- Disincentives – A tax on swimming pools; increasing the cost of water significantly for households consuming over 10 000 litres per month; establishment of a water 'Scorpions' (replaced by the 'Hawks' in July 2009) division to ensure compliance and cost recovery of water services;
- Annual water audit report on consumption and effluent releases of large institutions and industry;
- Reduction of water losses and Unaccounted-for Water (UFW) to meet targets progressively and extending the Water Leaks Project and efficiency throughout the City;
- Encouraging research initiatives in water resource management at a City scale. It should be noted that a fundamental prerequisite for IUWM is the availability of appropriate data in order to be able to examine individual components and understand the interactions between them (Fletcher & Deletic, 2008).

A reduction in water demand must continue through a combination of a variety of incentives and disincentives, and an approach to education that is likely to motivate users to consider behavioural change. There is no reason to suggest that the trend of reduced consumption should not continue. Ongoing education and information dissemination is required (City of Cape Town, 2007).

While informal settlements and low income households are frequently chosen as sites of experimentation and research, greater attention should be given to regulating the demands of large consumers. Moreover, if technologies become socially acceptable, then these technologies are far more likely to be accepted widely, especially by consumers who perceive that there are no alternatives to conventional technologies. A case can thus be made for research that tests the viability of biogas digesters in new 'greenfields' development. Biogas reactors, working together with secondary treatment systems, might be one of the most important technologies to reduce the volume of solid waste entering waste water treatment works (WWTWs), and moreover has the potential to supply households with gas for cooking either in the form of compressed biogas (tanks) or via a gas reticulation system. There is potential to use biogas production as part of integrated treatment systems to close the loop on the generation of sewage, while at the same time providing SMME opportunities. It is recommended that the City needs to pilot such projects in formal residential areas first, rather than frustrate the poor with failure.

7. Conclusion

Water and sanitation services in Cape Town face numerous critical challenges highlighted in this chapter. Arguably, integration is fundamental to managing the sustainability of water resources and water-related services. This approach will give more attention to human health, well-being and productivity, and a healthy environment. Most of all, integration should improve the provision of sanitation and water services through better co-ordination. In this way, more attention might be given to sanitation and water services in relation to livelihood, cultural and gender issues, and to an acknowledgement that appropriate sanitation is more than simply providing toilets and tap stands.

Ongoing research and experimentation must be directed towards understanding the impacts of future activities, such as the implementation of biogas digesters, abstraction of water from aquifers, and development of desalination plants. While technological advances worldwide suggest that cost-effective sanitation systems, treatment works, and desalination processes are increasingly being considered, scientific research must inform future choices and options in the context of the City and its needs. Brown (in Fletcher & Deletic, 2008) notes that community and government should jointly develop visions of a future sustainable water environment, and use science and other relevant inputs to determine targets to achieve this vision.

Sustainability, and a clear understanding of how to achieve the principles and vision of sustainability, is fundamental to the development of Cape Town, but it is unclear how this is being achieved in practice. As a result, there is a widespread perception among academics and researchers that the failure to commit to sustainability is placing the future of the City in jeopardy. To overcome such uncertainty, we will need to mobilise courage and invite commitment to alternative visions of a sustainable future for water and sanitation.

Acknowledgements

The constructive comments of Kirsty Carden (UCT) and an anonymous reviewer are gratefully acknowledged in developing this chapter.

References

Brown, R., Farrelly, M. & Keath, N. 2007. *Summary report: perceptions of institutional drivers and barriers to sustainable urban water management in Australia.* Report no. 07/06, National Water Governance Program, Monash University, Australia. ISBN 978-0-9804298-2-4.

Brown, R. 2008. Local institutional development and organisational change for advancing sustainable urban futures. *Environmental Management*, 41: 221-233.

Carden, K., Armitage, N., Winter, K., Sichone, O. & Rivett, U. 2008. The management of greywater in the non-sewered areas of South Africa. *Urban Water Journal*, 5 (4): 329-343.

Carden, K., Winter, K. & Armitage, N. 2009 *Sustainable water management in Cape Town, South Africa – is it a pipe dream?* Paper presented at 34th WEDC International Conference, Addis Ababa, Ethiopia.

City of Cape Town, 2006. *State of Cape Town Report 2006: Development issues in Cape Town.* Cape Town: City of Cape Town Municipality.

City of Cape Town, 2007. *Water Services Development Plan for City of Cape Town 2007/08-2011/12.* Cape Town: City of Cape Town Municipality.

City of Cape Town, 2008. *Water Services Development Plan for City of Cape Town 2008/09-2012/13.* Cape Town: City of Cape Town Municipality.

Department of Water Affairs and Forestry. 2007. *Western Cape Water Supply Reconciliation System, Report No. 3.* DWAF: Directorate National Water Resources Planning, Pretoria.

Department for International Development (DfID), 1998. *Guidance manual on water supply and sanitation programmes,* Water and Environmental Health at London and Loughborough (WELL), Water Engineering and Development Centre (WEDC), Loughborough University, UK.

Evans, B., 2005. Securing Sanitation: the compelling case to address the crisis. Report produced by the Stockholm International Water Institute (SIWI) in collaboration with WHO and commissioned by the Government of Norway as input to the Commission on Sustainable Development. http://www.who.int/water_sanitation_health/hygiene/securingsanitationpref.pdf

Fletcher, T. & Deletic, A. (Eds.). 2008. *Data requirements for integrated urban water management,* Volume 1 in Urban Water series. – UNESCO-IHP Publishing: Paris, France. ISBN 978-0-415-45345-5.

The Intermediate Technology Development Group. 2005. *Livelihoods and Gender in Sanitation Hygiene Water Services among Urban Poor.* Maili Saba Research Report. http://www.odi.org.uk/resources/download/2959.pdf.

Morris, J. 1996. Water Policy: Economic Theory and Political Reality, in Howsam, P. & Carter, R.C. (Eds.). *Water Policy: Allocation and Management in Practice.* London: E & FN Spon.

Schultz, R. 1997. South African Atlas of Agrohydrology and Climatology. Pretoria: Water Research Commission. Report TT82/96.

Sustainability Institute. 2007. *Integrated Analysis Water and Sanitation Baseline Report.* SI UNDP Cape Town Project. Stellenbosch: Sustainability Institute.

United Nations. 2003. *Global Report on Human Settlements.* London: Earthscan.

Valentin, A. & Spangenberg, J. 2000. A Guide to Community Sustainable Indicators, *Environmental Impact Assessment Review, 20(3):381-392.*

Warburton, D. 1998. A Passionate Dialogue: Community and Sustainable Development. In Warburton, D. (Ed.). *Community and Sustainable Development: Participation in the Future.* London: Earthscan.

In This Changing City

Exploring Urbanisation in Cape Town

Luke Metelerkamp

Baden Powell Drive, September 2009 [Image supplied by CDNGI ©]

I picked up a book once about the transformation of agrarian landscapes in Europe which left a lasting impression on me. The author's work engrained itself in my brain due to the skilful way in which he condensed the long term, incremental process of urbanisation into seventy pages. It was a book of photographs, with no words, which focussed solely on the green field construction of a single mall in a small German town. Each week from start to finish the photographer returned to his vantage point over the site of the new mall and recorded what had taken place, even if nothing appeared to have changed. The seventy weeks which he recorded and condensed served not only as a reminder of what had been before, but also of the speed with which a total and irreparable transformation of space can take place. The collection of images in this chapter takes inspiration from this book and the transformation of a pasture into a strip mall which it so eloquently captured.

However, this collection also tries to peel back what could become an inanimate reflection on the change in urban form or a dualistic battle between cul-de-sacs and green fields, by bringing the eyes and voices out from within the smarty box satellite images. The photographs draw people out from within the maps. What I suppose emerges is obvious but perhaps easy to forget; chiefly that the changes captured by the satellite represent more than a transformation of place, there was also a transformation of lives. And, conversely, that changes in people's lives and aspirations are changing the face of Cape Town. These changes are happening more rapidly and at a grander scale than I think most people realise, and that it is us, the residents of Cape Town both old and new, who are driving this change.

Largely missing from this collection are the voices of that which has been replaced to make way for the new form the city is taking. The voices of these people, animals, plants, soils and insects are for you, the viewer, to hear.

Delft, August 2008 [Image supplied by CDNGI ©]

List of Images

Page 116 Baden Powell Drive:
September 2009 [CDNGI]

Page 118 Delft:
August 2008 [CDNGI]

Page 120 The corner of Sheffield and New Eislenben Road:
August 2002 – February 2009 [CDNGI]

Page 121 The corner of Sheffield and New Eislenben Road:
August 2002 – February 2009 [Metelerkamp]

Pages 122 & 123 Meeting of Milnerton Lagoon and Atlantic Ocean:
July 2009 [CDNGI]

Page 124 The corner of Waterville and Sunningdale Road:
January 2001 – February 2009 [CDNGI]

Page 125 The corner of Waterville and Sunningdale Road:
January 2001 – February 2009 [Metelerkamp]

Pages 126 & 127 Bush between Symphony Way and Cape Town International Airport:
August 2008 [CDNGI]

Page 128 The corner of Landsdown and Amsterdam Road:
March 2001 – February 2009 [CDNGI]

Page 129 The corner of Landsdown and Amsterdam Road:
March 2001 – February 2009 [Metelerkamp]

Pages 130 & 131 Border between Durbanville and Durbanville Winelands:
July 2009 [CDNGI]

Page 132 The corner of Xorana and Faure-Klipfontein Road:
July 2005 – February 2009 [CDNGI]

Page 133 The corner of Xorana and Faure-Klipfontein Road:
July 2005 – February 2009 [Metelerkamp]

[Source and copyright: Images used with kind permission from Chief Directorate: National Geo-spatial Information (CDNGI)]

The corner of Sheffield and New Eislenben Road: August 2002 – February 2009 [Images supplied by CDNGI ©]

The corner of Sheffield and New Eislenben Road: September 2010

"I'm just renting this house, I came here from Joburg,
but I have been here a long time now. I didn't have
family down here, I just came for work."

— Petrie Mayaphi

Meeting of Milnerton Lagoon and Atlantic Ocean [Image supplied by CDNGI ©]

The corner of Waterville and Sunningdale Road: January 2001 – February 2009 [Images supplied by CDNGI ©]

The corner of Waterville and Sunningdale Road: July 2010

"We moved here after we got married, y'know. It's nice and quiet here, the kids can play outside. There's also nice schools and things around here."

– GW Pilkington

Bush between Symphony Way and Cape Town International Airport, August 2008 (Image supplied by CDNGI ©)

The corner of Landsdown and Amsterdam Road: March 2001 – February 2009 [Images supplied by CDNGI ©]

The corner of Landsdown and Amsterdam Road: July 2010

"I live near by, next to the Phillipi train station. I moved there from Landsdown Road after they cleared that area and built government houses."

– Thamiso George

Border between Durbanville and Durbanville Winelands, July 2009 [Image supplied by CDNGI ©]

The corner of Xorana and Faure-Klipfontein Road: July 2005 – February 2009 [Images supplied by CDNGI ©]

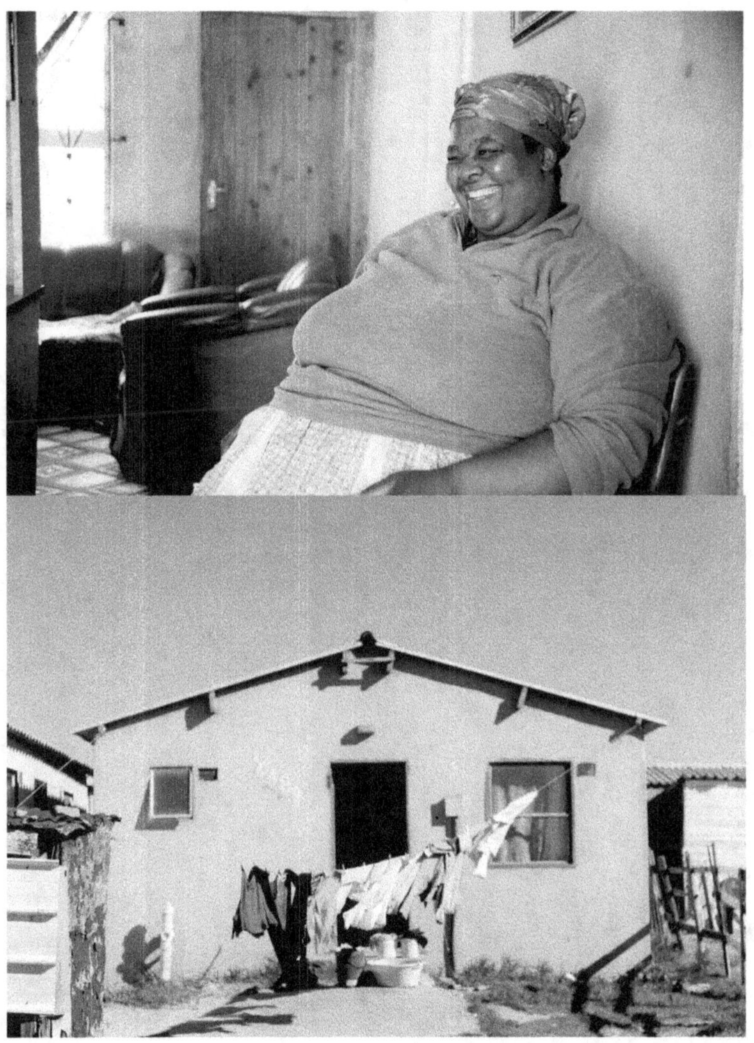

The corner of Xorana and Faure-Klipfontein Road: August 2010

"My sister's daughter used to be at school in the
Eastern Cape, but the the schools are better here so we
came to Mfuleni for her."

– Fundiswa Nqaiqa

Sustainable Energy

Frank Spencer

"We have an opportunity over the decade ahead to shift the structure of our economy towards greater energy efficiency, and more responsible use of our natural resources and relevant resource-based knowledge and expertise.

Minister of Finance, Trevor Manuel, 20 February 2008

1. Introduction

The City of Cape Town faces vast and compelling challenges when it comes to energy. Practically all the energy consumed in the city comes from outside the city's boundaries or control. The main sources supplied to Cape Town are high in carbon emissions. In the light of climate change, this is not sustainable in the long term. As we imagine a Sustainable Cape Town, we need to consider what might be the ideal mix of energy types. Can we move to low-carbon fuels and reduce our energy intensity without compromising growth? What new principles might underlie the kind of energy policies which would support a sustainable Cape Town? These arc the items we will explore in this chapter.

The chapter discusses the current context of energy in Cape Town, starting with the national, regional and municipal polices that shape the use of energy.[1] It demonstrates the limitations of current resource use approach towards energy, highlighting the ways in which this makes neither financial nor ecological sense over the long term. Cape Town electricity is supplied by the City of Cape Town and ESKOM, with little cooperation between the two. This makes city-wide energy planning difficult. At the same time, Cape Town's economic growth is constrained by supply limits. This explains why many key actors are seriously considering investments in renewable energy and energy efficiency. The chapter clarifies the necessity for the city to shift policy and practice to a sustainable resource use approach. We explore and illustrate some of the challenges and the opportunities facing the city, and conclude with a scenario outlining goals and targets to underpin best sustainable practices. The chapter will argue that, with a little bit of determination, much can be accomplished to bring more energy security to Cape Town, while also protecting our environment. The chapter begins with an outline of the policies which shape the energy sector.

2. Energy Policy and Cape Town

The impact of government policy on the direction of energy can never be underestimated, since policy is what gives investors and project developers direction and surety. The table below summarises national, provincial, and City of Cape Town policies without going in to them in detail.

During the apartheid years, energy policy was driven by a need for the economy to be self-sufficient. As a consequence, South Africa is now dependent on dirty, inefficient fuels, and the energy sector is dominated by a few large players. Furthermore, poor communities have inadequate access to affordable and safe fuels. While current policy represents a shift in direction towards improving access to energy, particularly for previously disadvantaged communities, sustainability in the energy sector has yet to emerge as a key national or local priority.

From a policy perspective, national government is still the major player, and still has strong links with the two biggest energy players – ESKOM (which it owns) and SASOL (our main provider of liquid fuels). It is also responsible for developing integrated resource plans that determine the strategic future of South Africa. Although there have been some attempts to promote Renewable Energy

---

1   This chapter draws heavily on the report by Sustainable Energy Africa commissioned by the Sustainability Institute (see Sustainable Energy Africa, 2007).

(through a target of 10000 GWh/annum of Renewable Energy by 2013), and a proposed Renewable Energy Feed-in Tariff (REFIT), to date these have largely been a failure.

There has also been the ESKOM Demand Side Management Solar Water Heater programme, which is an incentive scheme providing a subsidy of R2000-R4000 (approximately 10-20%) per SWH (solar water heater/heating) system installed in a residential house. However, this is probably not sufficient to drive large-scale adoption of SWH. Moreover, it is a paradoxical initiative, coming from a company whose mandate it is to *sell* electricity as opposed to reducing its use.

Provincial governments have little mandate in the energy space, except to coordinate between national and local governments. Nevertheless, the Western Cape government has started a process to look at sustainable energy supply, and has proposed a 15% Renewable Energy target for the Western Cape by 2015.

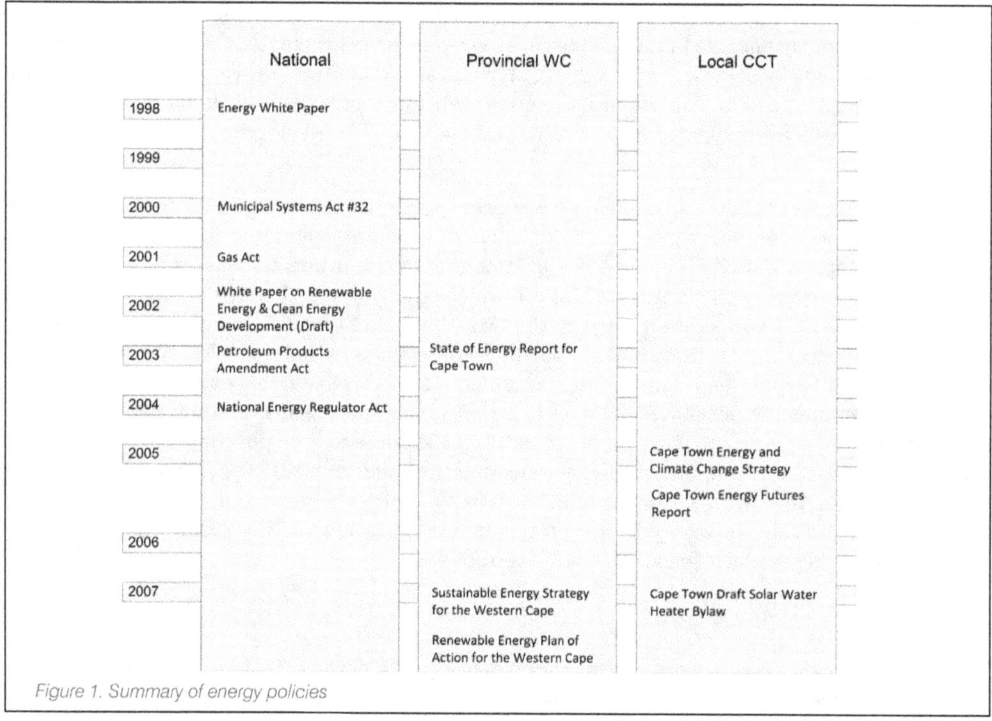

Figure 1. Summary of energy policies

Local governments are responsible for the distribution of electricity. In addition, municipalities have a significant responsibility in the shift towards the provision of initiatives to support sustainable energy. Cape Town currently purchases 'green electricity' for resale from the small Darling Wind Farm, and has investigated legislation around compulsory installation of SWH, but this has not yet emerged as a bylaw.

Thus, by 2007/8, very little has been practically implemented across the National, Western Cape and Cape Town government spectrum.

3.   Current Cape Town Energy Scenario Baseline

Most households are currently serviced with electricity, although about 29,700 households (approximately 145,000 people) are not (City of Cape Town, 2007).

Since energy is a major creator of wealth and employment in Cape Town,[2,3] historically there has been a direct relationship between sources of energy and economic development. In order to move forward with a low-carbon green economy, ways of reducing electrical consumption – while still delivering the same economic services – will need to be found. Whatever energy is consumed will also need to move towards low carbon emissions.

Cape Town's major sources of energy – and the energy sectors that consume them – are shown below. The tables and figures show that the major energy sources are electricity and liquid fuels, while transportation, commerce and industry consume most of this energy.

Table 1: Summary of Cape Town's Energy Consumption (2006)[4,5,6,7]

User Group		Households	Industry & Commerce	Local Authority	Transport	Total	Total %
Electricity*	PJ	17.97	24.76	1.75		44.47	29%
Petrol	PJ			0.12	42.29	42.41	28%
Diesel	PJ		13.16	0.23	14.34	27.73	18%
Heavy Furnace Oil	PJ		4.70			4.70	3%
Paraffin	PJ	2.59	0.44			3.03	2%
Jet Fuel	PJ				13.62	13.62	9%
LPG	PJ	0.55	2.72			3.27	2%
Coal	PJ	0.04	10.79			10.83	7%
Wood	PJ	0.36	0.56			0.92	1%
Total	PJ	21.51	57.12	2.10	70.25	150.98	100%
Total %		14%	38%	1%	47%	100%	

*   Note 1 PJ = 278 GWh.

2   Sustainable Energy Africa, *Energy Baseline Analysis*, iii.

3   Ibid.

4   *Cape Town State of Energy Report.*

5   *Household Numbers in Cape Town – Discussion Document.*

6   SEA, *SA State of Cities Report.*

7   *Cape Town Energy Futures Report.*

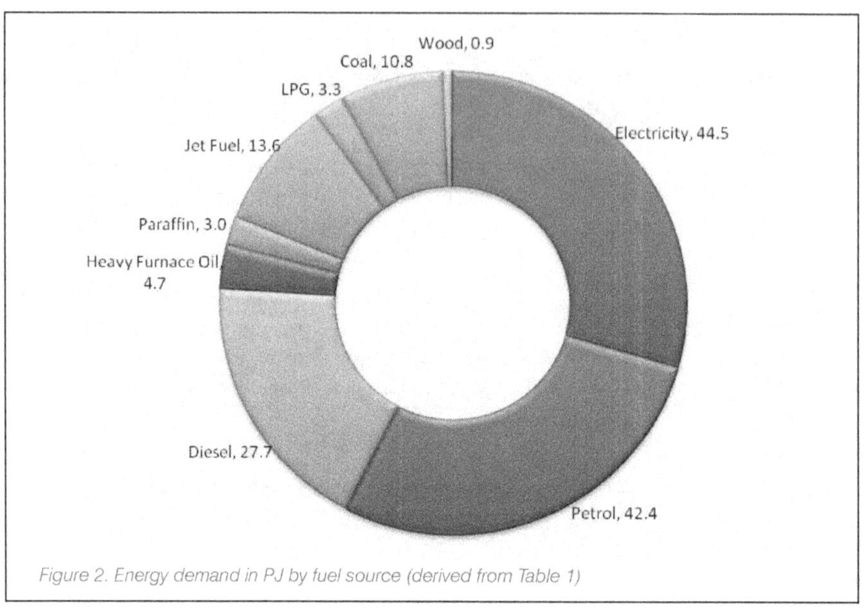

Figure 2. Energy demand in PJ by fuel source (derived from Table 1)

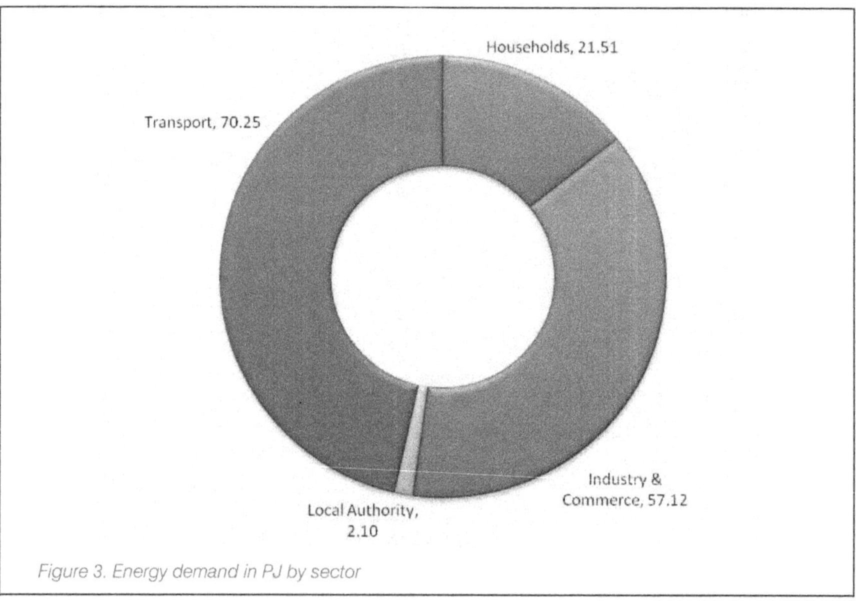

Figure 3. Energy demand in PJ by sector

a) Electricity

Cape Town consumed approximately 12,000 GWh (44PJ, or 3,090 kWh, per person per year in 2006, which contributed 50-59% of the City's CO2 emissions. The average retail price was 27c/kWh. Some power is supplied directly by ESKOM, and some through the City's Electrical Department.

The graph below shows the revenue split between ESKOM and Cape Town for July 2006 to June 2007.

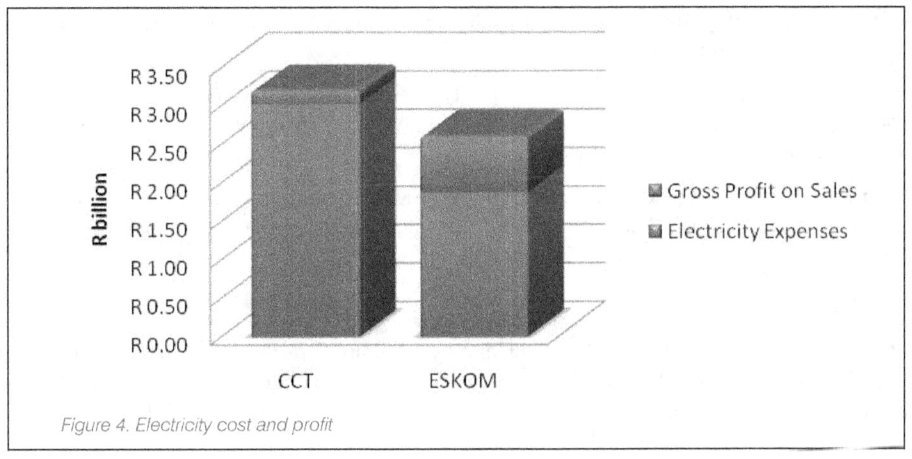

Figure 4. Electricity cost and profit

The installed electricity supply capacity is summarised as follows:

Table 2: Summary of installed electricity supply capacity[8]

Producer	Type	MW	%
ESKOM	Transmission lines (Coal)	2,600	50%
ESKOM	Nuclear	1,800	34%
ESKOM	Palmiet Pumped Storage	400	8%
CCT	Steenbras Pumped Storage	168	3%
ESKOM	Acacia Gas Turbine	171	3%
CCT	Roggebaai & Athlone Gas Turbine	80	2%
TOTAL:		5,219	

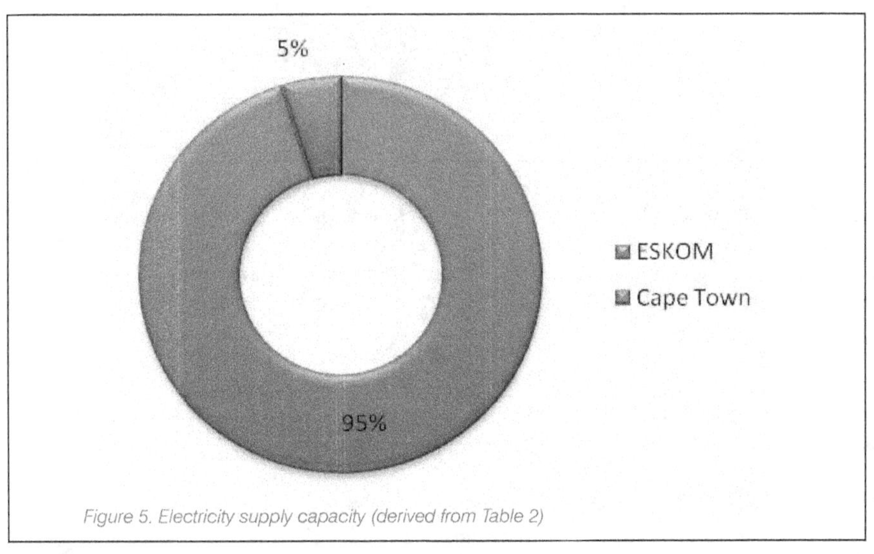

Figure 5. Electricity supply capacity (derived from Table 2)

---

8 Kenny, "ESKOM CCT Electricity Dept Presentation."

All of Cape Town's electrical power is brought in over the ESKOM grid. The Steenbras Pumped Storage facility does not produce any power; it only stores power, and helps level out the difference between low and high demand. The Open Cycle Gas Turbines are extremely expensive to operate compared to the ESKOM supply price.

South Africa is currently experiencing an electricity crisis, primarily due to capacity issues (demand outstripping supply) and the quality of electricity supply from ESKOM, which directly impact on the service that Cape Town is able to provide.

Although the bulk of electricity is purchased from ESKOM, a small portion of 'green' electricity is purchased from the Darling Wind Farm (5.2MW installed capacity) – at a premium price – which is also wheeled over the ESKOM grid to Cape Town.

b) Liquid Fuels

Liquid fuels currently constitute 60% of all energy used in Cape Town, primarily for transport, industrial heating, and household cooking and heating. The dominant liquid fuels are petrol, diesel, and jet fuel. Cape Town's average demand for refined liquid fuels is approximately 46,200 barrels of oil per day (SAPIA cited in Sustainable Energy Africa, 2007).

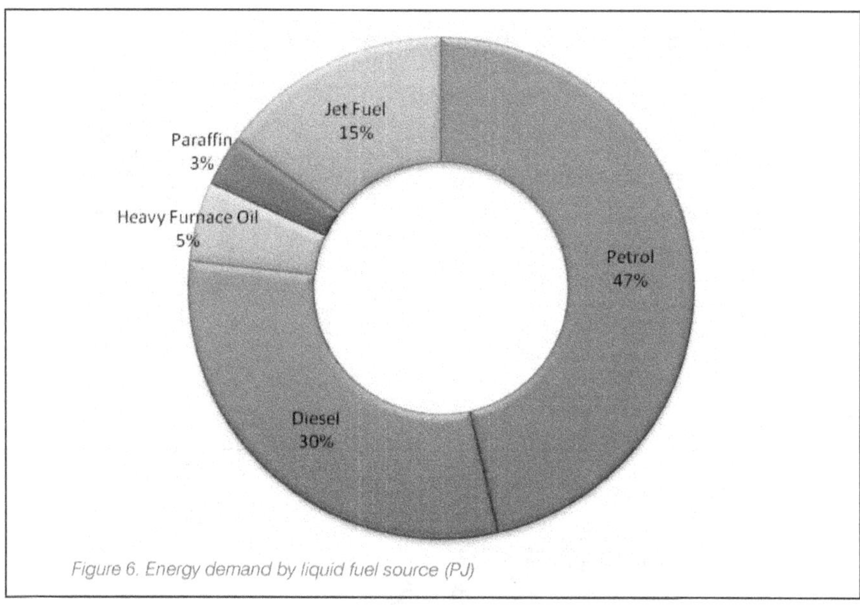

Figure 6. Energy demand by liquid fuel source (PJ)

Calref, the Cape Town-based, Chevron-owned oil refinery, provides Cape Town with most of her liquid fuels. The refinery processes over 100,000 barrels per day (SAPIA cited in Sustainable Energy Africa, 2007), and would cost about USD1 billion to replace. The quantity of imported refined fuel is expected to increase in the future, as South Africa's demand for fuel exceeds its ability to refine.

3.1 Sustainability issues with current scenario

There are a number of fundamental flaws in the current energy scenario, which are discussed below. We begin with electricity.

a) Electricity

The following concerns exist for the ongoing supply of electricity:

- An increasing demand for services with a supply that is currently constrained due to –
  - the actual transmission capacity into Cape Town;
  - the ability of ESKOM to meet national peak demands; and
  - the ability of ESKOM to build new power stations;
- An increasing energy demand for water, due to increasing water scarcity;
- An increasing demand for HVAC (heating, ventilation and air-conditioning), refrigeration, and water heating, often without using the most efficient technologies;[9]
- The dramatically increasing price of electricity supplied by ESKOM (more than 30% pa);
- The setting of targets by ESKOM for reduction in demand, with steep penalties for non-compliance (the Power Conservation Programme, or PCP);
- Tariff increases that historically have been below inflation;
- High carbon and other emissions from coal-based electricity, especially when considering the losses from long transmission lines;
- Dwindling coal reserves internationally, which will place pricing pressure on local supply to ESKOM; and
- Due to massively rising costs,[10] nuclear energy (including PBMR [pebble bed modular reactor] power) is considered an unsustainable future supply source.

b) Liquid Fuels

The following concerns exist for the ongoing supply of liquid fuels:

- The ongoing international pressures on the supply of oil;
- The probable shortfall by 2012 in the ability to produce enough liquid fuel (SAPIA cited in Sustainable Energy Africa, 2007)
- Tourism may be negatively impacted, because of the likelihood of a reduction in air travel due to carbon taxes; and
- Paraffin for cooking and heating is a dangerous and inefficient fuel choice.

c) Transport

In terms of sustainability, transport has the following implications for the energy sector:

- Energy consumption per km per capita is high, because of the high usage of cars with low occupancy;
- Busses (Golden Arrow) are (aerodynamically speaking) poorly designed and have inefficient engines;
- Rail – the most sustainable form of transport, due to its low energy consumption per km per capita – is not used as extensively as might be possible.

d) Carbon Emissions

Due to the use of fossil fuels, high energy usage normally translates directly into high CO_2 emissions. CO_2 (and equivalent emissions) are considered to be the primary cause of accelerating climate change. Both electricity and liquid fuels produce a high amount of CO_2 emissions, as shown in the table below:

---

9   In commenting on the growth in energy demand in China, Lovins & Sheikh observes that "... half the demand growth is due to largely wasted air-conditioning and refrigeration" Lovins and Sheikh, "The Nuclear Illusion," 39.. Similar wastage occurs in South Africa.

10   "... Nuclear is so costly and slow relative to its winning competitors that it will retard the provision of energy services" Ibid., 3.

Table 3: CO$_2$ emissions in Cape Town (2006/7)[11, 12, 13, 14]

User Group		Households	Industry & Commerce	Local Authority	Transport	Total	Total %
Electricty	CO$_2$ (t)	4 362 575	6 648 614	469 273		11 480 462	59%
Petrol	CO$_2$ (t)			8 768	3 123 417	3 132 185	16%
Diesel	CO$_2$ (t)		968 720	17 254	1 055 332	2 041 306	11%
Heavy Furnace Oil	CO$_2$ (t)		346 309			346 309	2%
Paraffin	CO$_2$ (t)	189 346	32 475			221 821	1%
Jet Fuel	CO$_2$ (t)				996 658	996 658	5%
LPG	CO$_2$ (t)	30 085	149 498			179 582	1%
Coal	CO$_2$ (t)	4 080	1 014 072			1 018 152	5%
Wood	CO$_2$ (t)						0%
Total	CO$_2$ (t)	4 586 085	9 159 688	495 295	5 175 406	19 416 474	100%
	%	24%	47%	3%	27%	100%	

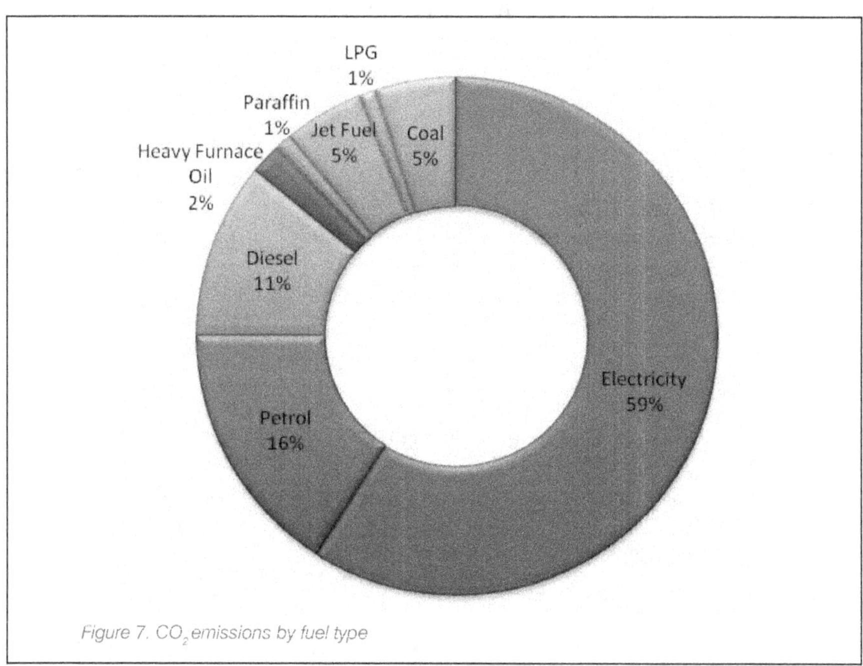

Figure 7. CO$_2$ emissions by fuel type

Cape Town has a per capita CO$_2$ emission toll of 6.4 tons per year.[15, 16]

Cement and steel – some of the primary building resources – consume a large amount of energy in their production, and thus the embodied CO$_2$ emissions for these materials are extremely high.

---

11 *Cape Town State of Energy Report.*

12 *Household Numbers in Cape Town – Discussion Document.*

13 SEA, *SA State of Cities Report.*

14 *Cape Town Energy Futures Report.*

15 PDG, *State of Energy Report for the City of Cape Town 2007*, 25.

16 Ibid.

Moreover, as cement and steel are used extensively in the city, these CO_2 emissions are imported into the city.

A significant number of goods, including food, are imported from far away from the city, increasing the imported CO_2 emissions, whereas some of these goods could be grown more locally.

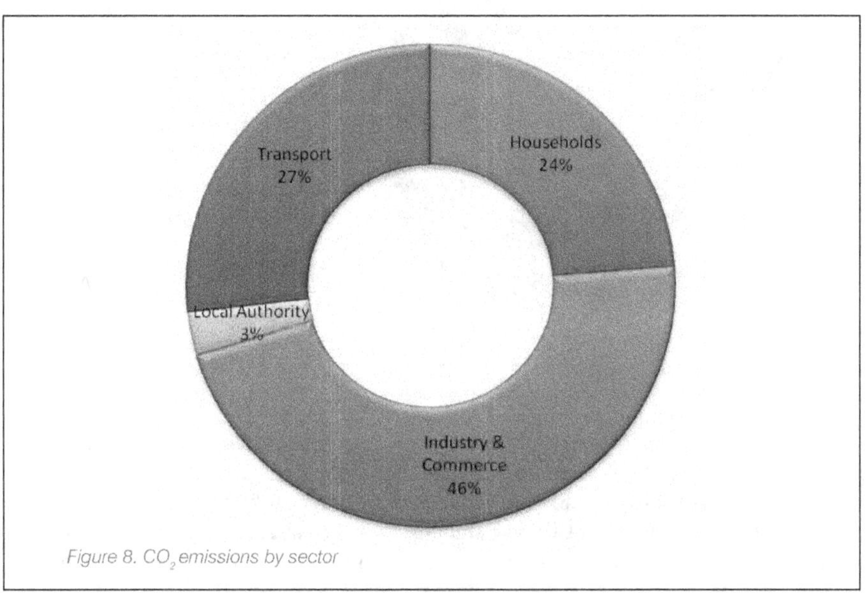

Figure 8. CO_2 emissions by sector

4.   Current Projections for a Future Scenario

4.1 Future total energy requirements

Assuming a 4% growth in all sectors in a "Business as usual' scenario, the 2050 requirements for energy will be around 529 PJ, broken down as follows:

Table 4: Energy requirements by sector by 2050 ('Business as usual' scenario)

User Group		Households	Industry & Commerce	Local Authority	Transport	Total	Total %
Electricty	PJ	93.31	128.55	9.07	0.00	230.93	30%
Petrol	PJ			0.62	219.62	220.24	28%
Diesel	PJ		68.34	1.22	74.45	144.00	18%
Heavy Furnace Oil	PJ		24.39			24.39	3%
Paraffin	PJ	13.43	2.30			15.74	2%
Jet Fuel	PJ	0.00	0.00		70.70	70.70	9%
LPG	PJ	2.84	14.11			16.96	2%
Coal	PJ	0.23	56.02			56.25	7%
Wood	PJ	1.86	2.91			4.78	1%
Total	PJ	111.67	296.63	10.91	364.77	783.98	100%
Total %	%	14%	38%	1%	47%	100%	

The total electricity demand is equivalent to 64,000 GWh.

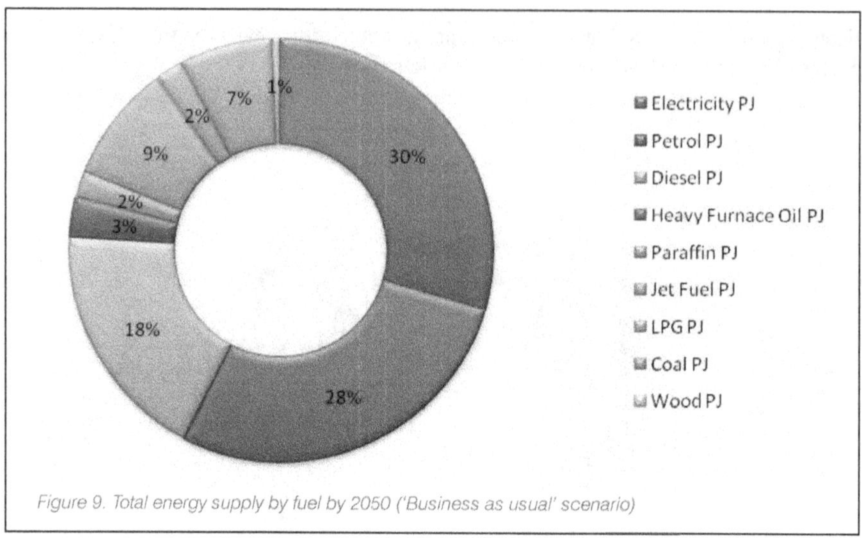

Figure 9. Total energy supply by fuel by 2050 ('Business as usual' scenario)

It is highly unlikely that Cape Town will be part of a world where green house gasses – and CO_2 in particular – will be emitted into the atmosphere without being taxed in order to finance mitigation and adaptation to the resulting change in climatic conditions. South Africa introduced carbon taxes in 2008, and these are most likely to grow steadily into the future. A 'Business as usual' scenario should, therefore, be seen in economic terms as the equivalent of a commitment to systematically bleed Cape Town's economy of desperately needed investment capital. In the end, CO_2 emissions mean more outward flows of capital, and therefore, fewer jobs. This triggers a negative feedback loop that will become increasingly difficult to break as time goes by.

4.2 Future renewables projection

South Africa has large potential for Renewable Energy and job creation from Renewable Energy, as illustrated by the following graphs taken from the *White Paper on Renewable Energy*:[17, 18]

---

17 DME, "White Paper on Renewable Energy."

18 Ibid.

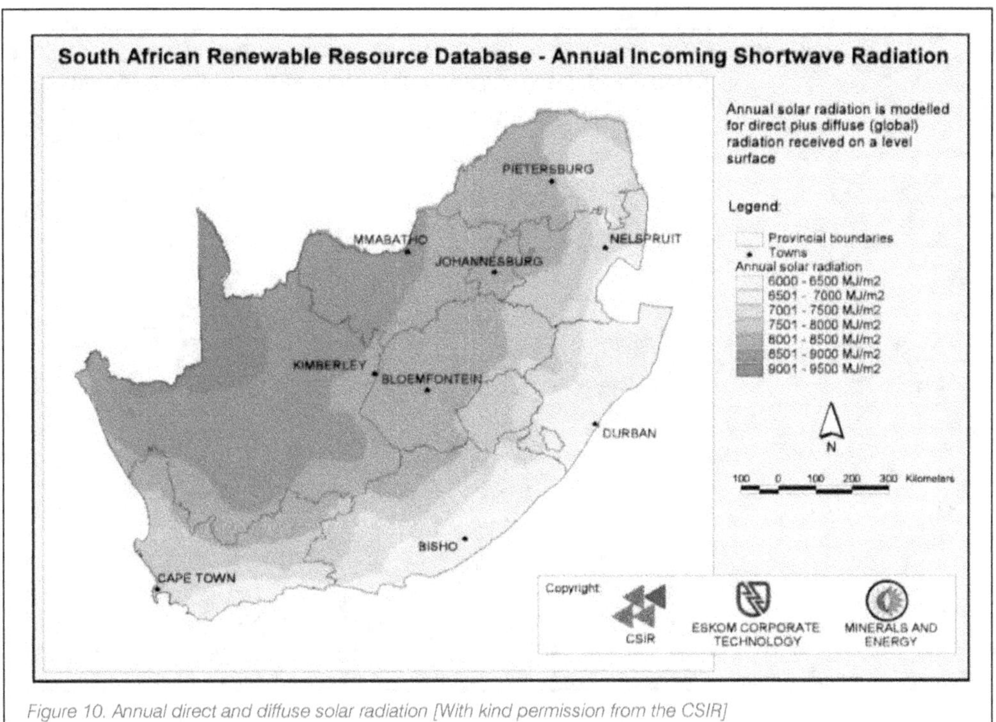

Figure 10. Annual direct and diffuse solar radiation [With kind permission from the CSIR]

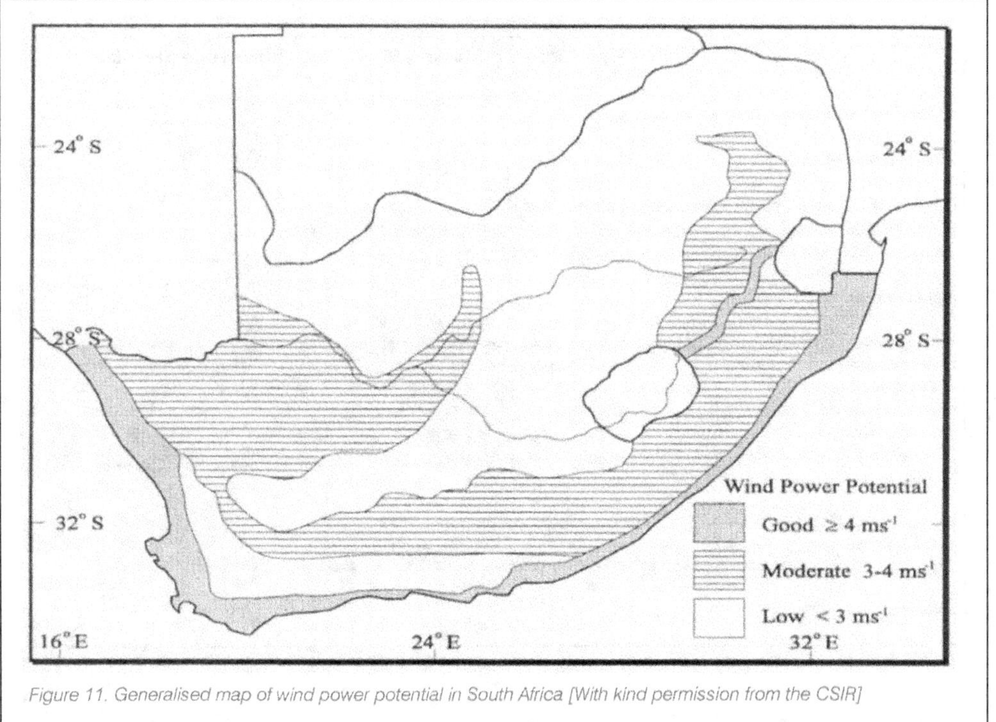

Figure 11. Generalised map of wind power potential in South Africa [With kind permission from the CSIR]

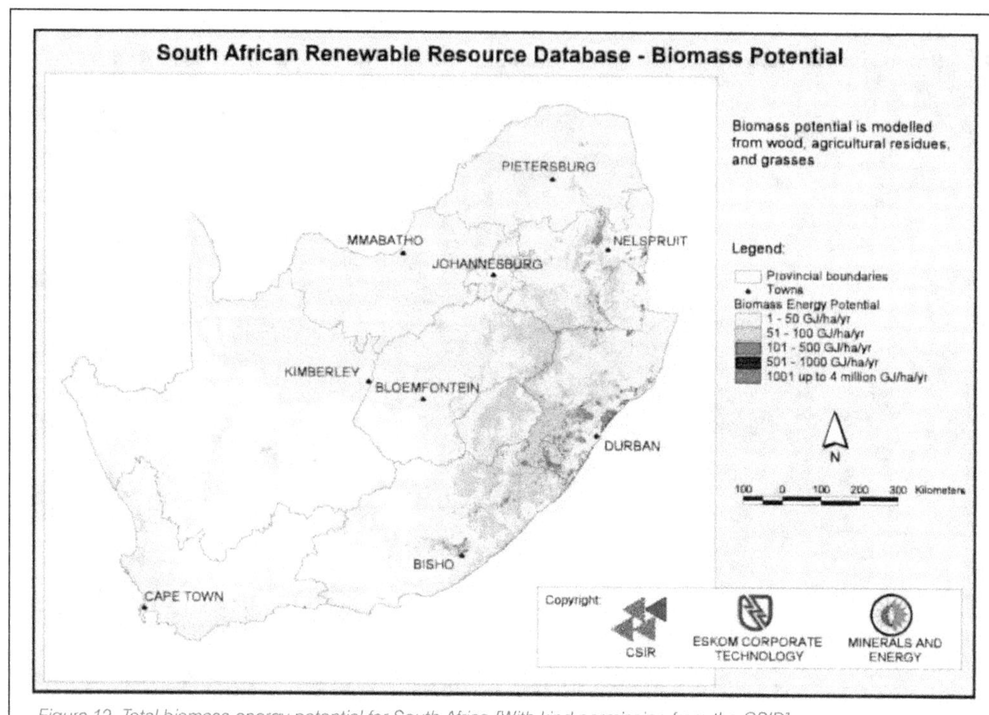

Figure 12. Total biomass energy potential for South Africa [With kind permission from the CSIR]

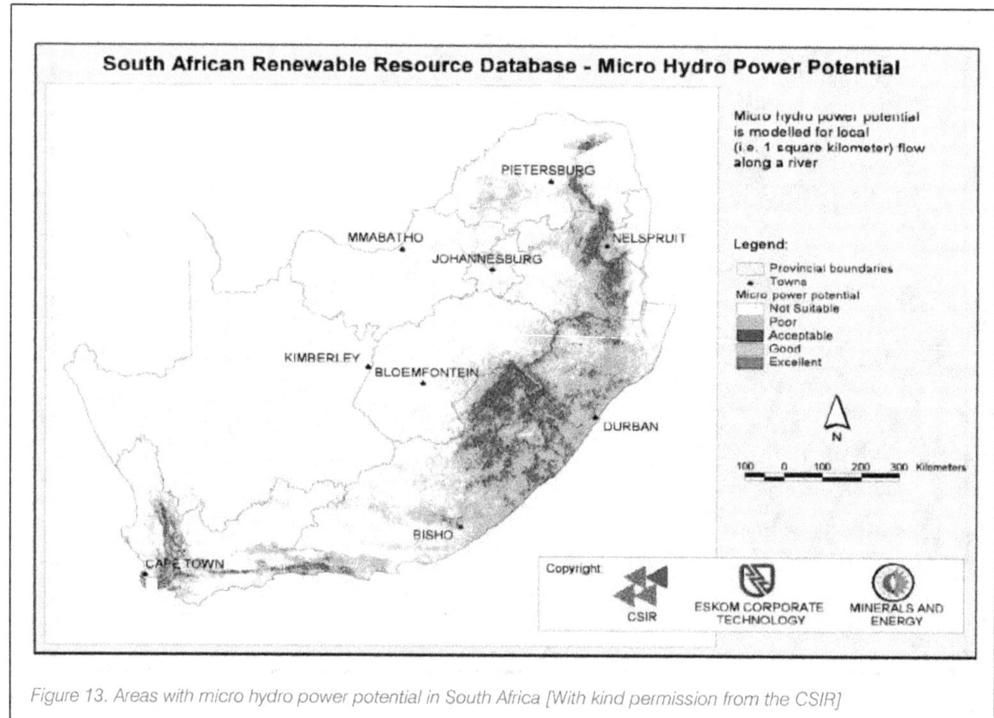

Figure 13. Areas with micro hydro power potential in South Africa [With kind permission from the CSIR]

Table 5: Potential job creation: Renewable energy vs. coal-fired power stations

Resource	Information Source			
	Danish Study[a]	New York State[b]	AWEA[c]	World Watch Institute[d]
	[man-years, same amount of energy]	[jobs per mil. US$ invest.]	[jobs per mil. US$ invest.]	[jobs per TWh]
Coal fired power plant	6 200	13.1	13	116
Photovoltaics		7.4		
Solar thermal electricity				248
Wind-generated electricity	14 200	10.0	14	542
Biomass-derived electricity		17.0-22.6		
Hydro-derived electricity		4.0	8	

As can be seen from all this data, there are opportunities in the Western Cape in all the Renewable Energy sectors to produce energy and employment.

The Western Cape government has set a target of 12% renewable power sources by 2014. In the recently released *A Renewable Energy Plan of Action for the Western Cape*,[19, 20] the following projections can be found:

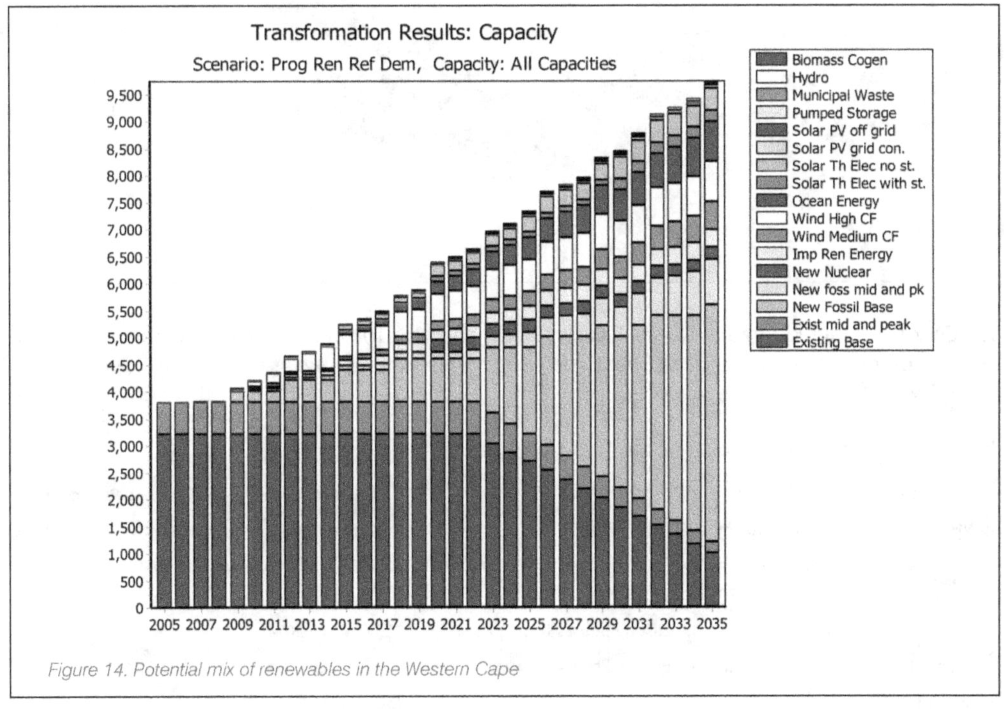

Figure 14. Potential mix of renewables in the Western Cape

---

19 Banks and Schaffler, *A Proposed Renewable Energy Plan of Action for the Western Cape*.

20 Ibid.

Wind	3,000MW
Ocean	1,000MW
Solar –PV	247MW
Hydro	15MW
Solar thermal	1,400MW
Pumped Storage	1,800MW
Total	**7,452MW**

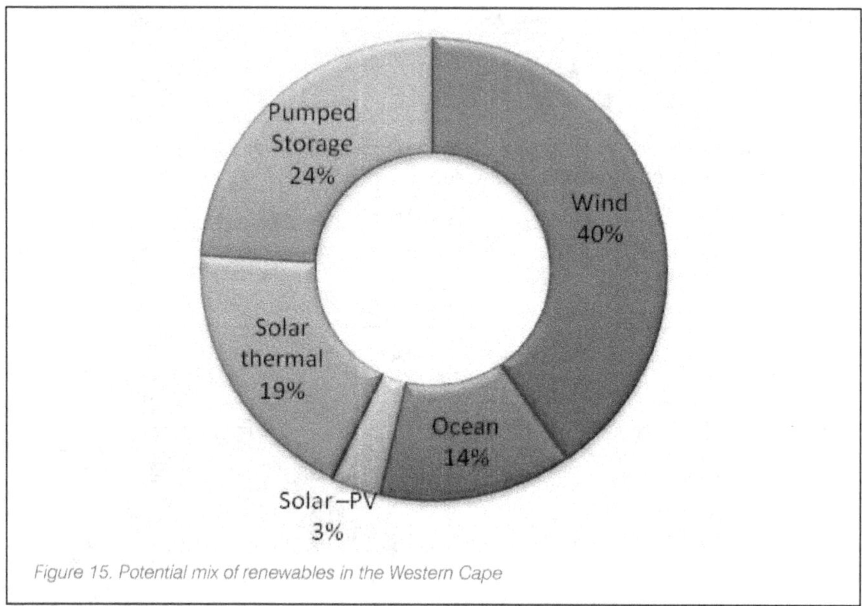

Figure 15. Potential mix of renewables in the Western Cape

The *Cape Town Draft Energy and Climate Change Strategy* calls for 10% of power to come from Renewable Energy Sources, in particular solar, wind and hydro. Thus, the following targets have been set for Cape Town:[21, 22]

Technology	Target	Energy Saving	Load Reduction
SWH	Houses: 10% by 2010; 50% by 2024	5,700GWh over 20 years	375MW
EE Lighting	Commercial and local authorities: 100% by 2010 Residential: 30% by 2010; 90% by 2020	7,700GWh over 20 years	290MW
Ceilings	CCT Target: Retrofit existing houses by 2020	365 GWh by 2024	
Efficient HVAC	20% reduction in energy used by HVAC by 2020	950 GWh by 2024	
Transport		150million PJ	

In light of a current total consumption of 150 PJ/year (42,000 GWh/year) and 12,000 GWh in electricity consumption, these targets are extremely low – let alone compared to the 'Business as

---

21 DEADP, *A climate change strategy and action plan for the Western Cape.*

22 Ibid.

usual' scenario for 2050's growth estimate of 783 PJ/year. Based on the large availability of Renewable Energy resources, we propose to do better!

5. Future Cape Town Energy Scenario

5.1 Key process

In times of war and crisis, governments have been able to roll out incredible expansions of infrastructure. We currently face a greater enemy than hostile nations, namely our own inability to care for our planet. As such, we really do need to respond as if it was a time of extreme crisis, and implement programmes that radically seize the window that we have to reduce carbon emissions in a manner that does not compromise people's happiness or the future well-being of our children.

In this section, a 'best case' scenario – where all interventions are pushed to the limit to produce an extremely sustainable city – is proposed.

As a first step, energy efficiency needs to be given the number one priority. With an aggressive approach, it may be possible to reduce the energy consumption of the 'Business as usual' scenario by up to 50% by 2050. However, this would require significant changes in the way that buildings are built and consume energy. Among others, energy efficiency interventions should include:

- Major industrial efficiency programmes;
- All new buildings will need to be energy efficient, with passive designs for heating and cooling;
- Extensive solar water heating, preferably in every house, by 2050;
- Correct orientation, ceilings, and insulation in all low-cost housing; and
- Efficient HVAC, with heat recovery for other processes.

The capacity for local renewable energy should be exploited as quickly as possible. The proposed 7.4GW power in the Western Cape should be fully developed by a date earlier than 2035. In fact, a goal of producing 50% of the City's energy from local sources should be set for 2050, which could include 100% of its electricity supply if this was sourced from Concentrated Solar Power (CSP) plants, which is currently the cheapest and most efficient renewable energy alternative.

A local carbon trading scheme, in the vein of the Chicago Climate Exchange, should be developed for the City. Companies should be strongly encouraged to participate.

Where possible, alternatives to cement and steel, based on local building materials, should be promoted where appropriate. Consumer behaviour change will need to be pushed hard.

We now move from the demand for energy to solutions around energy supply.

5.2 Proposed energy supply

The following table illustrates various targets to be achieved. Energy efficiency plays a major role in reducing the 'Business as usual' scenario from 784 PJ to 444 PJ. Electricity still plays a dominant role, but the source is shifted dramatically to renewable energy.

Table 8: Renewable energy source targets

User Group		Households	Industry & Commerce	Local Authority	Transport	Total	Total %
Electricty	PJ	37.32	79.27	5.44		122.04	27%
Petrol	PJ				87.85	87.85	20%
Diesel	PJ		34.00	1.22	35.00	70.22	15%
Heavy Furnace Oil	PJ					0.00	0%
Paraffin	PJ					0.00	0%

User Group		Households	Industry & Commerce	Local Authority	Transport	Total	Total %
Biofuels		5.00	20.00	0.50	35.00	60.50	14%
Jet Fuel	PJ				56.56	56.56	13%
LPG	PJ	2.84	14.11			16.96	4%
Coal	PJ	0.23	25.00			25.23	6%
Wood	PJ	1.86	2.91			4.78	1%
Total	PJ	47.25	175.30	7.16	214.41	444.13	100%
Total %	%	11%	39%	2%	48%	100%	

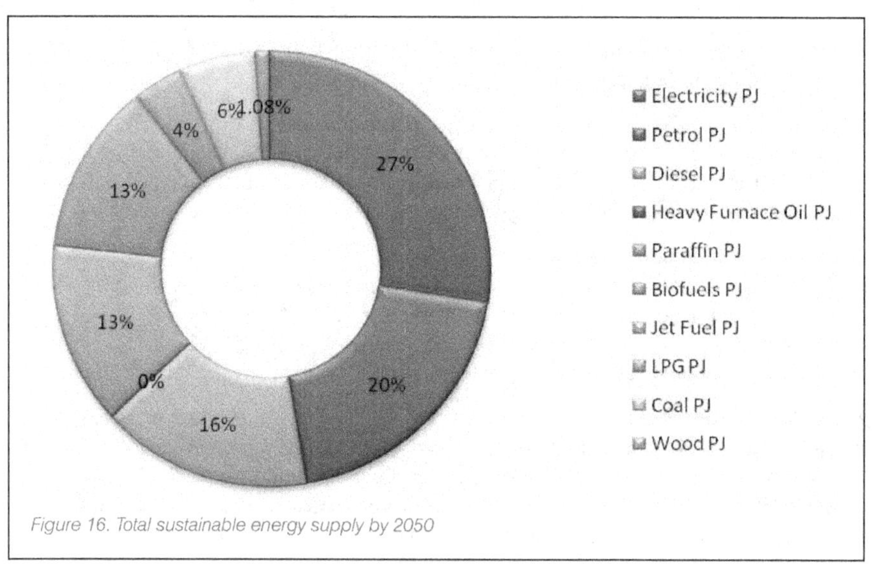

Figure 16. Total sustainable energy supply by 2050

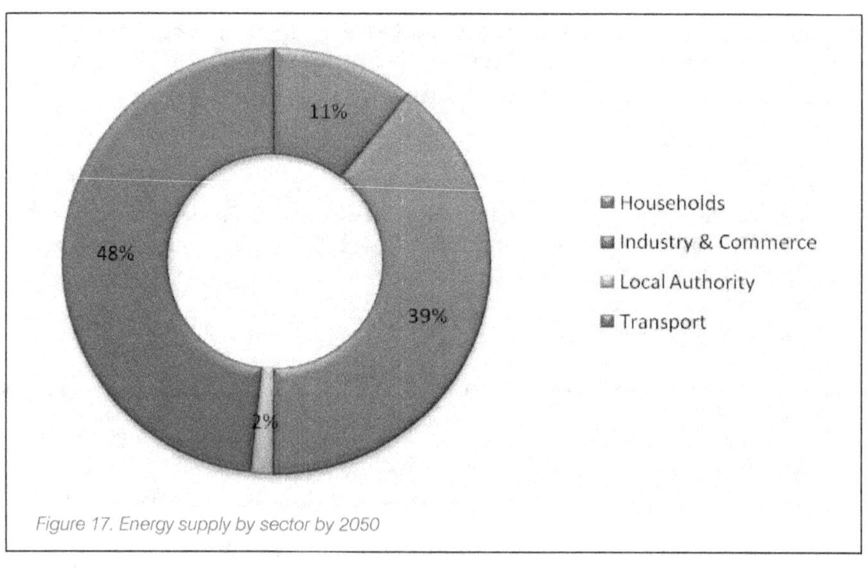

Figure 17. Energy supply by sector by 2050

The table below shows the proposed electricity breakdown – efficiency (as a 'virtual power station', or VPS) is a major contributor.

Table 9: Proposed electricity breakdown by source by 2050

Producer	Type	MW	%	GWh	%	Capacity
ESKOM	Transmission lines (Coal)	5,200	13%	11,388	14.1%	25%
ESKOM	Nuclear	0	0%	0	0.0%	
ESKOM	Palmiet Pumped Storage	400	1%	350.4	0.4%	10%
ESKOM	Gas Turbine	171	0%	150	0.2%	10%
CCT	Efficiency	3,523	9%	30,863	38.1%	100%
CCT	Pumped Storage	10,000	24%		0.0%	
CCT	Gas Turbine	2,000	5%	1,752	2.2%	10%
CCT	Renewables	20,000	48%	36,500	45.1%	21%
TOTAL:		41,294		81,003		

The graph shows the generation split between ESKOM and CCT (excluding pumped storage):

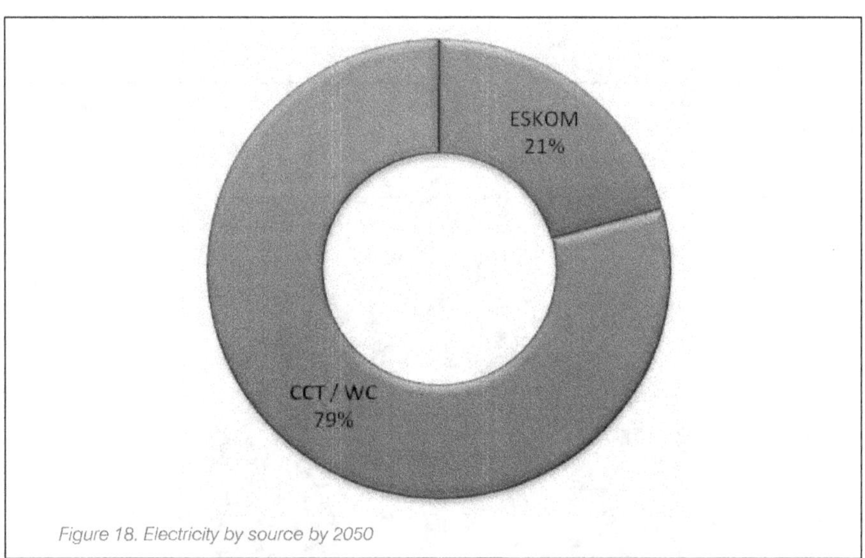

Figure 18. Electricity by source by 2050

Further break-down of renewables (excluding biofuels) is shown as follows:

Table 10: Renewable electricity sources by 2050

Renewable Source	MW	%
Wind	10,300	34.3%
Ocean	3,000	10.0%
Solar –PV	1,000	3.3%
Hydro	500	1.7%
Solar thermal	5,200	17.4%
Pumped Storage	10,000	33.3%
Total	30,000	

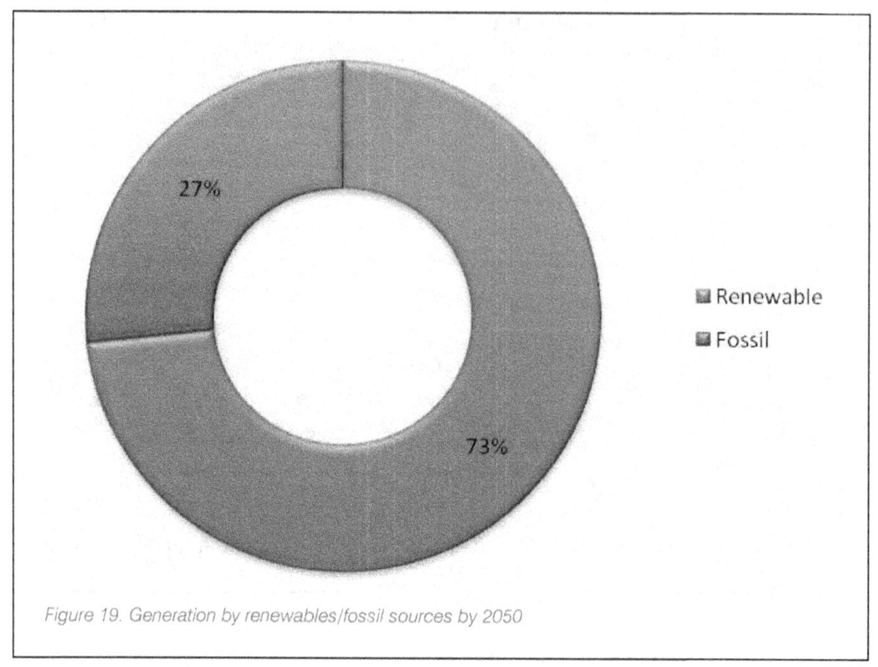

Figure 19. Generation by renewables/fossil sources by 2050

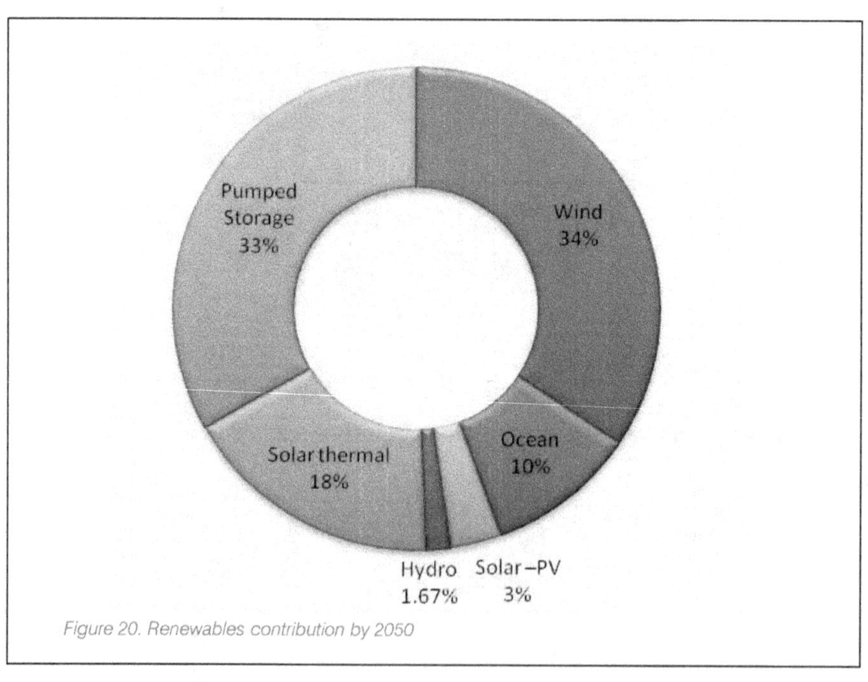

Figure 20. Renewables contribution by 2050

6. Technical Interventions

Technical Interventions can be classified into two main categories:

- **Production,** which relates to changes in technology choices to *produce* energy.
- **Consumption,** which refers to changes in technology choices that impact on how energy is *consumed* and how *efficiently* it is consumed.

6.1 Production interventions

This section suggests interventions related to local renewable production, and discusses what can be done with regard to 'virtual power stations' (VPS). When considering production of various energy sources, the end source energy service demand determines which needs are considered. For example, paraffin is produced to meet the energy service demand of lighting and cooking; thus the actual energy service required (lighting, cooking) can be met by an alternative, such as an ethanol gel fuel.

To mitigate the potentially disastrous impacts of climate change on energy security, electricity and biofuels should be produced locally. Suggestions for local renewable production include:

- Installing solar water heaters – they should be on every roof, and especially on government buildings, schools, hospitals, clinics, and hotels;
- Investing in small-scale renewable energy – the city could promote incentives, tax breaks, or feed-in-tariffs to those home owners who wish to generate their own power within the city;
- Investing in large-scale renewable projects built on the edge of the City and in the Western Cape (using concentrated solar power technologies); and
- Producing biogas wherever feasible.

Many of these have short paybacks – if there is adequate policy to support them. Environmental benefits and energy security concerns must also be factored into the evaluation of these technologies.

It will be necessary to promote the idea of a VPS, which is in fact the culmination of energy saving interventions – or what Lovins called 'nega-watts' (see.ccnr.org/amory.html).

6.2 Consumption interventions

Consumption interventions mainly impact on how energy is consumed, both in terms of energy efficiency (e.g. energy efficient lighting) and behaviour change (e.g. switching off lights in unoccupied rooms). The following recommendations are offered:

- Energy Saving Plan[23] – The City must develop an energy saving plan incorporating targets that align with the requirements of science, particularly the South African Long Term Mitigation Strategy (LTMS) for Climate Change.
- Energy efficiency – This should be the number one priority. Sector wide strategies for reducing energy consumption need to be planned. Targets should be 50% below 'Business as usual' by 2050.
- Variable tariffs – High energy users should pay a premium for their electricity usage.
- Smart meters, which allow users to regularly check their consumption, should be increased.
- Efficient cooking methods, such as microwaves and pressure cookers, making minimal use of traditional electricity. Also use LPG (liquid petroleum gas) where possible.
- Local building materials – Locally-sourced building materials, rather than cement and steel, should be used wherever possible.
- Transport – Public transport must be promoted extensively, especially the new proposed Bus Rapid Transport Systems.

If implemented correctly, the above items will help with decoupling growth from energy consumption.

---

23 *Cape Town City IDP.*

7. Key Goals for Sustainable Energy

Based on the discussion above, several key goals for Cape Town are recommended for the production of sustainable energy.

7.1 Policy

Policy should support the following outcomes:

- Energy security through local production of sustainable cost effective energy;
- Energy efficiency and reduced energy demand per capita;
- Extensive support for Independent Power Producers, particularly through the use of incentives, tax breaks, feed-in-tariffs, or the like;
- Increased support for skills in the energy sector;
- Promotion of job creation, where low energy consumption per job created takes preference over high energy consumption per job created (e.g. smelters);
- Support for local manufacturing of 'green' technologies;
- Incentives for houses and buildings that operate completely off-grid; and
- Marketing support for 'green buildings' or 'green floors'.[24]

7.2 Production

The following energy production outcomes are recommended:

- The re-use of waste energy sources for biogas or other energy production (e.g. waste biomass, sewerage, solid waste);
- Local production of electrical energy through the promotion of renewables, both small-scale and large;
- Local production of sustainable biofuels;
- Local job creation in energy production – renewables create more jobs than conventional energy systems; and
- All municipal buildings to produce 10% of their electricity or heat requirements on-site.

7.3 Residential consumption

The following outcomes are recommended for the residential sector:

- Gas or electricity for every household – no paraffin;
- Solar water heaters on every rooftop;
- Energy efficient lighting in every light fitting;
- Rated appliances in every kitchen;
- Insulation retrofitted in every ceiling;
- Passive heating and cooling design in every new house;
- Every new house to meet a minimum of SANS204 rating, and a 4-star Green Star building rating when it is launched for the residential sector;
- All buildings to have smart meters in a visible location, showing household consumption on a daily/monthly basis;
- Public awareness of possible energy savings in the residential sector; and
- An improvement of up to twice the usual energy efficiency (or 50% reduction in consumption) should be promoted.

---

24 A 'green floor' is a particular floor of a multi-story building, where the power is supplied exclusively from local sustainable renewable energy produced on, or in, the building.

7.4 Commercial and industrial consumption

The following outcomes are recommended for the commercial and industrial sector:

- Bulk solar water heating/heat pump systems for every large hot water consumer, e.g. hotels, guest houses, hospitals, schools, etc.;
- Heat recovery systems operating on all large generators and air-conditioning systems;
- Corrected electricity power factor throughout the city;
- Reduced peak electricity demand;
- Energy efficient lighting in every building;
- Efficient electrical motors and motor drive systems in all machines;
- All new buildings to meet a 4-star Green Star building rating;
- All traffic lights to be replaced with light-emitting diode (LED) systems; and
- An ultimate reduction in energy intensity of up to 80% should be targeted.

7.5 Transport

The following outcomes are recommended for energy usage in transport systems:

- Low energy consumption and carbon emissions per person transported through the extensive use of rail and mass transit systems (new busses) by most commuters;
- Reduced fossil fuel dependence through the use of sustainable second-generation biofuel as a blend in all diesel and petrol;
- Reduced energy and carbon emissions per person through significant movement to hybrid and electric vehicles;
- Energy for electric vehicles to be locally produced;
- Promotion of pedestrianisation throughout the city; and
- Energy intensity per capita should be reduced by at least 50%.

8. Conclusion

Cape Town is currently almost entirely dependent on energy resources that enter the City from beyond her control. As such, we are completely at the mercy of international markets and national policymakers. A significant portion of the City's revenue leaves its boundaries due to the sources of energy – this money could be put to better use within the City.

In addition, the behaviour of those consuming energy in both the City and throughout the rest of the world, is exacerbating climate change through the release of greenhouse gasses such as CO_2. This need not be the case. There are massive opportunities for improving the way the City produces and consumes energy and energy services. By changing the energy intensity and decoupling energy consumption from economic growth, it is possible that energy consumption could be halved, without compromising on service delivery, job creation, and social justice. In addition, the supply of energy could be redirected from fuels with high carbon emissions to a high proportion of renewables.

It is our hope that, as a City, we will dream big and act wisely to bring about a sustainable energy future for Cape Town.

References

Banks, D, and Schaffler, J. 2007. *A Proposed Renewable Energy Plan of Action for the Western Cape*. Cape Town: Western Cape Provincial Government, DEADP, May 22.

City of Cape Town. 2008. *Cape Town City IDP*. Cape Town: City of Cape Town.

Cape Town. 2005. *Cape Town Energy Futures Report*. Cape Town: City of Cape Town.

Cape Town. 2003. *Cape Town State of Energy Report*. Cape Town: City of Cape Town.

DEADP. 2008. *A climate change strategy and action plan for the Western Cape*. Cape Town: Western Cape Provincial Government, DEADP, March.

DME. White Paper on Renewable Energy. *Department of Minerals and Energy*, 2003. http://www.dme.gov.za/pdfs/energy/renewable/white_paper_renewable_energy.pdf.

Cape Town. 2006. *Household Numbers in Cape Town – Discussion Document*. Cape Town: City of Cape Town, unpublished mimeo.

Kenny, A. 2006. ESKOM CCT Electricity Dept Presentation, unpublished mimeo.

Lovins, A. and Imran, S. 2008. The Nuclear Illusion. *Ambio* Nov 08 preprint, dr 18 (May 27, 2008). http://www.rmi.org/images/PDFs/Energy/E08-01_AmbioNuclllusion.pdf.

PDG. *State of Energy Report for the City of Cape Town 2007*. Cape Town: Palmer Development Group, 2007. http://www.capetown.gov.za/en/EnvironmentalResourceManagement/publications/Pages/Reportsand.aspx#state%20of%20energy.

SEA. 2006. *SA State of Cities Report*. Cape Town: SEA.

Sustainable Energy Africa. 2007. *Energy Baseline Analysis*. Cape Town: SEA. Report commissioned by the UNDP and Sustainability Institute, unpublished report, 2007. (Available at www.sustainabilityinstitute.net)

9

Sustainable Integrated Solid Waste Management

Sally-Anne Engeldow

Figure 1. Landfill site in Cape Town [Source: Metelerkamp, 2009]

3.   Solid waste management in the City of Cape Town

The institutional and infrastructural capacity of the City of Cape Town to manage solid waste is significantly better than that of most developing countries elsewhere; however there are a range of constraints:

- ongoing challenges of integration with six formerly independent and self-administered municipalities;
- the legacy of Apartheid and unequal service distribution;
- decreasing landfill availability and rapid urbanisation; and
- associated increasing volumes of waste.

Solid waste generation has increased dramatically in the past 10 years with an interesting decline since 2007/2008. Figure 2 provides a graphic illustration of the amount of waste disposed of at the City's landfill sites from 1991 to 2008/2009. The data noted in the years from 1991 to 1997 are relatively stable, which could in part be due to poor data capturing of waste disposed at landfills. The figures do not take into account the private waste disposal facility (the Vissershok Waste Management Facility) and recycling or waste minimisation initiatives, therefore the actual waste generated could be higher. The declining trend in waste disposed was not projected in previous estimations and the City's Solid Waste Management Department generally calculated future planning requirements based on an 8% increase.

The rapid increase in waste generation prior to 2007/2008 could be attributed to factors including:

- urbanisation and population increase;
- increase in production and consumption (associated with an increase in affluence);
- formalised and equitable services; and
- more accurate record keeping.

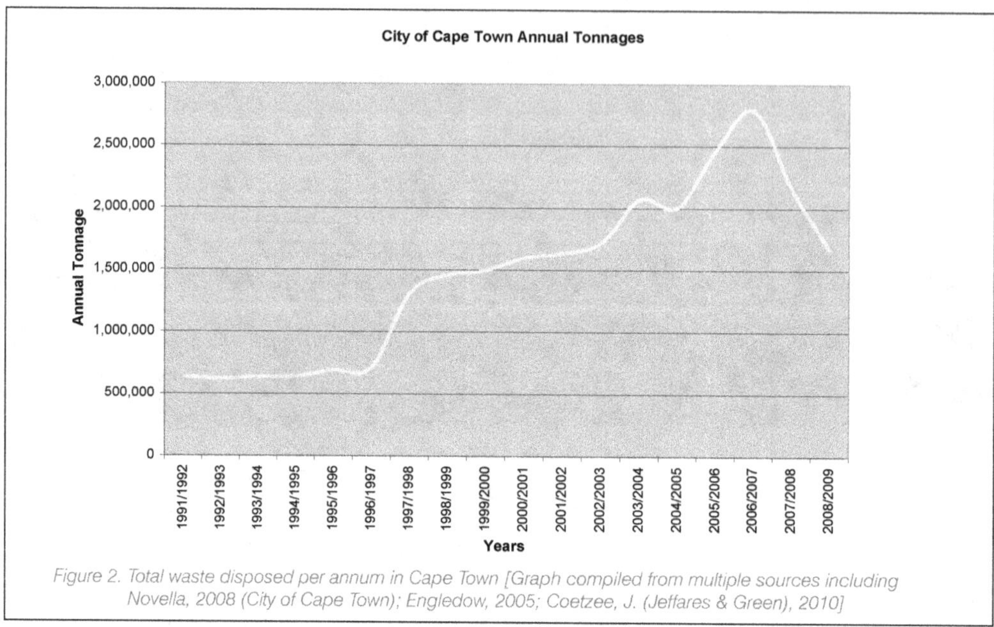

Figure 2. Total waste disposed per annum in Cape Town [Graph compiled from multiple sources including Novella, 2008 (City of Cape Town); Engledow, 2005; Coetzee, J. (Jeffares & Green), 2010]

The decrease in waste disposed in 2007/2008 has been attributed to the economic downturn in the economy and implementation of various waste minimisation interventions, with the latter probably the more significant factor.

3.1 Urbanization and population increase

The 1996 Census recorded the population to be 2,682,866, and in 2001 the recorded number was 2,893,251. The population in Cape Town has thus increased by 26% since the 1996 Census. The increase in waste generation for the same time period is 106%.[2] The increase in the waste generation therefore cannot solely be attributed to population increase, but could also be attributed to improvement in economic conditions and affluence.

3.2 Increase in production and consumption

International waste management literature has confirmed the relationship between economic affluence and increasing waste generation, as well as the change in the composition of waste generated. Figure 3 illustrates the relationship between GDP per capita (expressed in US Dollars) and per capita waste generation figures (expressed in kilograms). The numbers demonstrate that, as per capita GDP increases, there is generally an upward trend in the amount of waste generated. In South Africa's case, the graph illustrates an interesting trend: high waste generation rates per capita, but a low GDP per capita. This is a good illustration of the diversity within the country as a whole, in that there are large numbers of poor people and a smaller percentage of wealthy people. However, it is the more wealthy sectors of the population that tend to generate large volumes of waste.

The graph was compiled from information obtained from the Organisation for Economic Co-operation and Development (OECD) (2008). The information has been gathered for 2005 and 2006, and represents information that may not be entirely consistent. This is especially relevant in terms of waste generation data, as various waste streams are included or not included in the data. Therefore it is used here to present an indication of the trends and as a means of comparison.

---

2   Compiled from Novella, 2008; Mega-tech, 2004; and Engledow, 2005.

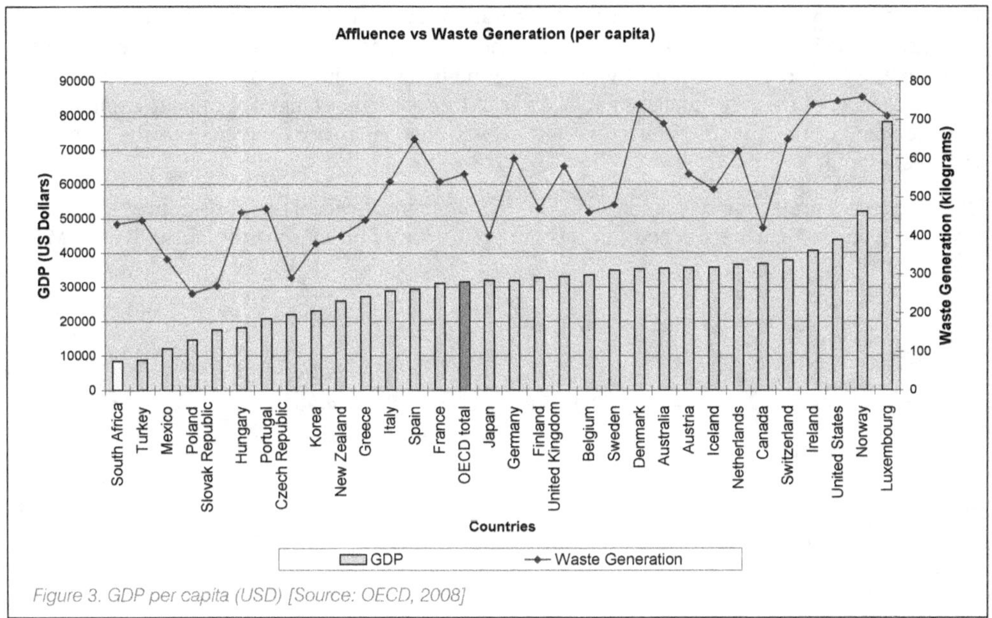

Figure 3. GDP per capita (USD) [Source: OECD, 2008]

An interesting observation can be seen in the numbers of the amounts of waste generated per person per day in the City of Cape Town from 1997 to 2007 in Table 1.

Table 1: Average waste generation per person per day[3]

Year	Kgs per person per day
1997	1.39
2001	1.55
2004	1.96
2007	2.23

These numbers are based on the total waste stream averaged for the whole population, and based on the population statistics from Census 1996 and 2001 and figures obtained for 2007. It provides a platform on which to assess the increasing generation of waste. Given that these figures are averaged over the population, it is important to remember that there are differences in the amount of waste generated per socio-economic group. Estimations from the City's Integrated Waste Management Plan (IWMP) (Mega-tech, 2004) suggest that low income households generate 0.16-0.50kgs per person per day, while the middle to high income group generates 1.50-2kgs per person per day.

There are two opposing arguments linked to the use of per person generation rates. The first argument is that using these rates is dangerous and inaccurate. The danger with using figures such as a 'per person' average rate of generation is that it seems to over-simplify the generation rates, and therefore they are not accurate for the following reasons:

- The way that the rates are calculated is often not clear, and it is often uncertain whether the rates are based on the total waste stream (which includes industrial and commercial rates);
- The rates are generally shown as an average per person; and
- There are no recent studies to either confirm or refute the calculations.

---

3   Numbers calculated using information from Engledow & Eichstadt, 2007; Mega-tech, 2004; Strategic Information, Strategic Development Information and GIS. These figures are estimates and should not be taken as completely accurate but rather an indication – refer to explanation in the text.

A contrasting view is that the 'per person' generation rate can capture the generation trend – it should therefore be viewed as an indicative tool and, as such, does need not be completely accurate. Individuals, while they are not directly generating industrial or commercial waste, are active in the market place, because they purchase goods and services and other tangible products manufactured by the commercial and industrial sectors. In this way, they indirectly contribute to the amount of waste generated.

Table 2 reflects a projection calculated for 2008. The numbers provide a categorisation of the amount of waste generated per person and per income group, based on information in the IWMP (Mega-tech, 2004).[4]

Table 2: Waste generation per person per day (2008)[5]

Income Group	Kilogram per person per day*		
	Household	Garden	Total
High	2.1	0.35	2.45
Middle	1.2	0.35	1.55
Low	0.5	0	0.5

* Numbers have been modified from original CCT IWMP (Mega-tech, 2004) to take into account the increase in generation – approximately 5%.

The main shift in South Africa in the last few years can be attributed to economic growth and the emergence of a distinctive 'Black' middle class, or so-called 'Black Diamonds' (Naidoo, 2007). This reference is an important one in that the middle class in South Africa as a whole is growing. Naidoo makes a comparison between the United States of America and the United Kingdom, where the middle class accounts for about 50% of the population, and South Africa, where it constitutes only about 20% of the population, but is growing. In the context of Cape Town, we can assume that the expanding emerging middle class (irrespective of whether racially defined or not) could be interpreted as both an expanding waste problem and a trend towards increasing packaging type wastes entering the waste stream.

In summary, the rapid increase in waste over the last ten years can be explained in part by the increase in affluence and the emergent middle class.

Cape Town has also seen a dramatic increase in demolition and construction activities. This sector has reportedly grown for the entire Western Cape Province by 6.6% on average from 1999 to 2003, as compared to the National average of 4% per annum for the same period.[6] The construction of new residential areas and commercial and retail sectors usually includes demolition activities that generate building rubble, which again increases the waste volumes.

3.3 Formalised and equitable service levels

The City is not only faced with increasing waste volumes from increased consumption and population growth factors, but also needs to manage past inequalities of service. Prior to 1994, the levels of service offered were dependent on area. Informal areas were not serviced, while middle to upper income areas were well-serviced (including a separate green waste collection service) (Engledow, 2005).

A more standardised level of service has been implemented across the City over the last few years. Most of the population of Cape Town have access to refuse removal services in the form

---

4   The calculations are based on a conservative increase of 5%, and the numbers only take into account the domestic waste stream and garden waste. Information such as this is very useful to determine where waste minimisation initiatives should be concentrated in terms of the socio-economic group.

5   Numbers calculated from the CCT IWMP (Mega-tech, 2004); Coetzee, J (J&G); CCT Discussions. They are estimates and should not be taken as completely accurate, but rather an indication – refer to explanation in the text.

6   Western Cape Provincial Government, 2005.

of containerised 240l bins to bagged[7] and skip container services.[8] The implementation of more formalised services has led to more appropriate waste management services, as well as more regular refuse removal. Although this may give the impression that more waste is being generated, the reality is that more (previously not collected) waste is now being accounted for.

3.4 Accurate record keeping

As technology improves, more accurate and up-to-date record keeping can be conducted. Weigh bridges at landfill sites have only been installed during the last ten to fifteen years. Previously, estimations based on the load capacity of the truck were used. This still occurs at some of the landfill and drop-off sites, especially for green waste and builders' rubble. This leads to inaccurate information being recorded due to over- or under-estimations being made.

Inaccurate information regarding waste generated within the City presents a challenge in terms of future planning. The NEM Waste Action (59 of 2008) (DEA, 2008) calls for the establishment of a waste information system (WIS) at the National level. This requires provincial and local government (municipalities) to ensure that information pertaining to generation, treatment, transport, and disposal is collected in such a manner as to be included into the National WIS. The City of Cape Town has promulgated an Integrated Waste Management By-Law (CCT, 2009),[9] which – in line with the National legislation – requires information to be forwarded to the City in the form of Integrated Waste Management Plans (IWMPs), and waste service providers to register with the City to obtain a license to operate.

This would assist in standardising a method of accurate record keeping in terms of the amount of waste generated, treated, and disposed. A WIS would require municipalities to ensure that accurate information is captured and then fed into the National WIS. In turn, data generated from this information system would not only assist the City in terms of future planning, but also improve the reliability of National statistics regarding waste generation and management.

We have reviewed some of the managerial elements of solid waste management, such as service levels and information management, and now shift to a consideration of the progress and challenges in practical delivery considerations, such as how waste is defined and categorised in various streams.

3.5 The economic downturn in the economy

The unexpected decrease in the amount of waste being disposed of at the City's landfill sites has been a puzzling anomaly in the waste figures. One of the main drivers of the decrease has been attributed to the economic downturn. This is as yet to be documented and therefore there is no academic or other formal literature to refer to in order to substantiate this claim further. Although, as waste and affluence has been linked in many references throughout waste literature, this assumption should be seen as a potential explanation of the decreasing amounts of waste disposed.

A further explanation or perhaps in conjunction with the downturn in the economy is that the waste minimisation interventions implemented by the City have been effective.

---

7   Bagged service refers to a formal service provided by the City where the household puts refuse out in black bin liners which the city collects.

8   A skip services is a formal service generally found in an informal area where movement between households (informal dwellings) is difficult. A skip container (varying in size) is placed at a central point and households would bring their refuse and place it within the skip. The skip would then be collected by the City and replaced with an empty container.

9   Accepted by Council, 30 March 2009.

The City implemented 3 main strategies since the end of 2007 including:

- Think Twice Campaign
- Builders Rubble Crushing
- Green Waste Chipping

The Think Twice Campaign was first launched in areas within the South Peninsula and Atlantic seaboard (i.e. Pinelands, Table View, Blouberg, Melkbos). The campaign requires households to separate recyclables from the waste stream. The recyclables are collected by a separate vehicle, taken to a material recovery facility for further sorting into the different categories and then sold on to the recycling industry for reprocessing. Further roll outs of the project have more recently[10] been implemented into the Helderberg (Somerset West, Strand, Gordons Bay) and the South Peninsula (Fish Hoek, Simonstown, Kommetjie, Noordhoek). These projects are still under evaluation and analysis in terms of volumes and tonnage diverted from disposal, but the disposal figures seem to reflect a decrease, therefore providing an insight into these projects having a positive impact thus far.

Since 2008, the City has implemented a program to crush builders rubble at their landfill facilities for re-sale into the market and used in various civil construction projects or fill material. The strategy to divert green waste from landfill sites was further strengthened by implementing additional chipping facilities at more of the drop off and landfill sites. The diversion of both builders rubble and green waste could also therefore be a contribution to the decreased tonnage disposed of at the City's Landfill sites.

4.   Waste Streams

Because of the decreasing landfill airspace[11] available, the City of Cape Town has made efforts to separate garden waste from the general waste stream in order to remove this waste type from landfills. Drop-off facilities have been established around the City, as well as at all of the operating landfill sites. The greens are then chipped and used as raw material in composting. Figure 3 presents a green waste stockpile at the City-owned Kommetjie drop-off facility, where the community is encouraged to drop off green waste.

For the same reason, the City is making an attempt to keep builders' rubble off landfill sites. In the past, landfill sites have used builders' rubble either as cover material or for on-site road maintenance. A study conducted by the City of Cape Town in 2005 estimated that builders' rubble accounted for about 15% of the total amount of solid waste (Ingerop Africa, 2005).[12] The study reported that about 900,000 tons of builders' rubble is generated per annum, of which 75% was recycled. Approximately 530,000 tons were privately recycled, and 130,000 tons were recycled at the municipal facilities of Coastal Park and Gordon's Bay. This is a very impressive rate of recycling and shows the value of builders' rubble.

However, the illegal dumping of builders' rubble is ongoing, which raises questions regarding the City's enforcement capacity, because fines can be issued only if offenders are caught or reported. An increase in awareness of these fines, as well as increased motivation for the public to report offenders, is required. The disposal of builders' rubble at the City's sites (landfill and drop-off facilities) is referred to as 'free waste' – i.e. there are no charges involved. As mentioned, the City has an IWM by-law that requires waste service providers to register with the City and individuals to use only registered providers. This is one attempt to regulate the industry and limit the amount of illegal dumping.

---

10 Since 2008.

11 Landfill airspace is a term commonly used within landfill management in South Africa as it refers to the amount of available volume that can be used to dispose of waste within a landfill site.

12 Builders' rubble consists of materials from demolished buildings and civil infrastructure, including rocks and soil arising from earthworks, civil works, and general foundations.

Figure 4. *Chipping of garden waste at a City drop-off site for mulch and composting [Source: Engledow, 2007]*

Electronic waste is likely to become another management challenge. Electronic waste (or 'e-waste') includes all computer hardware, cellular phones, and household appliances, and is rapidly increasing as electronic equipment swiftly becomes obsolete and is replaced by newer models. The Electronic Waste Association of South Africa (eWASA) has been formed in an attempt to increase awareness of e-waste and the potential toxicity of the substances contained within electronic equipment. Heavy metals contained within e-waste include lead, mercury and cadmium, and therefore management of this waste stream is imperative. In more developed countries, this problem has reached crisis levels, with e-waste being exported to less developed countries for processing (see Figure 5).

In Cape Town, e-waste has become more of a problem on landfill sites over the last couple of years. Previously, people would store obsolete electronic equipment in basements (i.e. government departments and commercial businesses) or in their homes, because they were unsure of where to take these items. Stored items are now making their way to landfill sites. Private industry and NGOs have led the way in terms of implementing initiatives to re-use and recycle old electronic equipment and to raise awareness of the importance of managing e-waste correctly.[13]

Figure 5. *E-waste at a Cape Town drop-off facility [Source: Engledow, 2007]*

13 eWASA; Footprints Environmental Center in Wynberg, Cape Town.

While this Chapter refers mostly to solid waste, the following provides a summary of hazardous waste.

Hazardous waste is collected, treated, and managed separately to general solid waste, and is defined in the Minimum Requirements of the Department of Water Affairs and Forestry (1998: 2-4) as:

"... an inorganic or organic element or compound that, because of its toxicological, physical, chemical or persistency properties, may exercise detrimental acute or chronic impacts on human health and the environment. It can be generated from a wide range of commercial, industrial, agricultural and domestic activities and may take the form of liquid, sludge or solid. These characteristics contribute not only to degree of hazard, but are also of great importance in the ultimate choice of a safe and environmentally acceptable method of disposal."*

The NEM Waste Action (59 of 2008: 14) (DEA, 2008) defines hazardous waste very broadly:

"Hazardous waste means any waste that contains organic or inorganic elements or compounds that may, owing to the inherent physical, chemical or toxicological characteristics of the waste have a detrimental impact on health and the environment."

Hazardous waste is determined using a hazard rating system and the South African Bureau of Standards classification of hazardous waste for transportation classification 1-9 (SANS 10228). The hazard rating system classifies waste according to extreme risk (hazard rating 1) to low risk (hazard rating 4). The hazard rating of a given type of waste will determine what type of landfill site that waste can be disposed at. Hazard ratings 1-4 can be disposed of at a H:H landfill site, while hazard rating 3 and 4 can only be disposed of at a landfill site licensed as H:h.[14] Therefore H:H landfill sites can receive all hazardous waste except for radioactive waste unless a special license is issued; and H:h landfill sites can only receive low risk hazardous waste.

There are many industries that generate hazardous waste during their manufacturing processes or after use, e.g. chemical and paint manufacturing plants, agricultural and forestry, industries, etc. Medical and infectious waste is included within the classification of hazardous waste and is classified as hazard rating 1, which is extremely hazardous. This type of waste must be incinerated or sterilised, or treated using nationally approved technologies (e.g. autoclaving or electro-thermal deactivation) prior to being land-filled at a licensed hazardous (H:H/H:h) landfill site. All anatomical waste (body parts) must be incinerated. Medical and infectious waste generally originates from hospitals and clinics, but also from biological research facilities and private homes (as a result of medical care). The collection, treatment, and disposal of hazardous waste must be in accordance with legislation, i.e. the Western Cape Health Care Risk Management Act, 2007 and the South African National Standards Authority (SANS), to mention two examples of the relevant legislation.

The following section reviews the principles of integrated waste management in terms of the integrated waste management hierarchy, including waste minimisation, recycling, treatment, and disposal.

5. Integrated waste management hierarchy

The integrated waste management hierarchy presents fundamental principles for the management of waste. It is a concept that promotes waste avoidance through prevention, and general minimisation ahead of any re-use, recycling, treatment, and disposal. Waste avoidance and reduction should be the first option; if waste cannot be avoided, then efforts should be made to minimise the quantities generated. Only once all avoidance and minimisation options have been explored, should on-site recovery, re-use, and recycling be considered. And treatment and disposal should only be considered and accepted as a last resort, in accordance with the ultimate objective of this policy.

Figure 6 provides a graphic illustration of the principles that should be used to promote more sustainable waste management systems.

---

14 The classification of landfill sites will be discussed later in the document.

While this hierarchy is what should be strived for, for the most part the City of Cape Town still landfills the waste generated within the City. However, there are policies and plans in place to move away from the traditional landfill approach, including expanded producer responsibility in both National and local legislation.

The discussion that follows will reflect the order of the approach currently in place in the City, i.e. treatment and disposal, followed by minimisation.

Figure 6. Integrated Waste Management Hierarchy [Source: Jeffares & Green, 2008]

5.1 Waste treatment and disposal

Landfill disposal is currently the predominant method used in the City of Cape Town to manage solid waste. However, initiatives have been started to compost solid waste and to minimise the amount of solid waste that requires landfill disposal.

From a sustainability perspective, the siting and management of landfill sites is important in order to minimise the impacts on the surrounding environment in the form of potential air, soil, and water contamination. Landfill sites situated on the Cape Flats (e.g. Bellville and Coastal Park) are situated directly above or within the recharge zones of the Cape Flats Aquifer System (Mega-tech, 2004). Sites constructed pre-1990 were not designed using engineered liners or leachate detection layers, but were quite rudimentary in design, thus there are real concerns about the potential for historical landfill sites polluting groundwater resources and causing soil contamination.[15]

The remaining municipally owned and operated landfill sites in the Cape Town area are: Bellville South (classified as a G:L:B+), Coastal Park (classified as a G:L:B+), and Vissershok (classified as H:h). A further site, referred to as the Vissershok Waste Management Facility, is a H:H landfill which is privately owned and managed by a joint venture between two waste service providers, EnviroServ and Wasteman.

---

15 Parsons (2002) presents evidence that historical waste cells of the Bellville South site (in operation since the 1960's) has had a significant impact on the Cape Flats Aquifer System. Newer 'cells' developed at Bellville South have been designed using engineered linings and leachate detection systems to mitigate potential groundwater pollution. This is also accompanied by groundwater monitoring procedures to assess whether these systems are still functional.

These facilities are the only remaining sites left to serve the City of Cape Town for the next ten to fifteen years. A regional landfill site has been planned, but can only be constructed pending the outcome of an appeal lodged against the Environmental Authorisation before the site can be constructed. It will be located in close proximity to Atlantis (about 50km outside of the Cape Town CBD) and will serve the City and potentially outlying areas, for example Stellenbosch.

Table 3 shows the limited number of landfill sites serving Cape Town, as well as the expected dates for their closure. Certain implications of the distances from proposed regional sites to serve Cape Town raise new challenges for transport.

Table 3: Remaining landfill sites in Cape Town

Municipal landfill sites	Classification	Permit status	Expected closure date	Soil type
Vissershok	H:h[1]	Permitted (1998)	2020	Clayey
Coastal Park	G:L:B+[2]	Permitted (2000)	2022	Sandy
Bellville	G:L:B+	Permitted (2003)	2013	Sandy
Privately operated landfill site				
Vissershok Waste Management Facility	H:H[3]	Permitted (1997)	2013[4]	Clayey

Notes:

1 **H:h**   Landfills that can accept Moderate (3) to Low Hazard (4) (as defined by the Minimum Requirements) and General Waste.

2 **GLB+**   A Large General Waste landfill which is likely to produce significant amounts of leachate.

3 **H:H**   Can accept all hazardous waste (hazard ratings 1-4).

4 There is potential for site expansion, if this is achieved then the lifespan can be extended to about 2020.

In order for waste to be transported to regional landfill sites in the future, a number of Refuse Transfer Stations (RTSs) need to be constructed. The Tygerberg Waste Management Facility, Helderberg, and the Oostenberg Integrated Waste Management Facility have been through an Environmental Impact Assessment Process and received the necessary Environmental Authorisation from the Department of Environmental Affairs and Development Planning (DEADP). All of the proposed RTSs have been designed to include 'clean' Material Recovery Facilities (MRFs) to sort commingled recyclables in a bid to reduce the amount of waste requiring landfill disposal.

5.2 Waste minimisation

Waste minimisation activities include reducing both the amount of waste generated and the amount of waste that requires landfill disposal. For recycling initiatives, major emphasis is placed on waste minimisation activities linked to disposal.

A number of recycling initiatives operate within Cape Town, and operators include NGOs, schools, churches, private companies, government drop-off facilities and informal collectors. These initiatives are conducted both as viable businesses and as strategies for fundraising attempts at schools and charities, as well as to raise environmental awareness. Homeless or formally unemployed people rely on informal salvaging to sustain a living. This activity includes picking from residential, street or business bins, and directly from landfill sites.

Recovered recyclables collected via this complex web of formal to informal collectors are taken – either directly or through a 'middle-man' – to the recycling companies that engage in processing for re-manufacture. The recovered recyclables are then either re-used or sold to buy-back centres for market-related prices per kilogram.

The current rate of recycling at the City of Cape Town's landfill sites by waste salvagers is 0.5-0.7%.[16] This very low recovery rate is partly due to the contamination of recyclables and the difficulty in recovering them. Salvaging directly from a landfill site is dangerous work, as the salvagers toil close to the working face[17] (see Figure 7). It is also unhygienic to dig through the mix of wastes that are disposed, because household waste often contains wet waste in the form of food wastes and recyclable packaging type wastes, and often a mixture of household hazardous waste, including old car oil, fluorescent tubes, paint and resin from household maintenance, and expired medicines. The City has attempted to formalise these salvaging activities by setting up contracts with the salvagers working on the landfill sites. This has improved the salvagers' health and safety, but the working conditions are still dangerous, and, in general, salvagers do not wear the personal protective equipment issued by the contracted company.

Waste salvagers and waste pickers not active on the landfill sites usually pick recyclables from bins in residential areas, or located outside businesses or other sources, on collection days. Waste salvagers are considered a nuisance, and there is a perceived association with littering and instances of increased crime in areas where they are active. There is no detailed information to either confirm or refute this perception. Many salvagers in Cape Town use trolleys from the big supermarket chains to transport their recyclables to buy-back centres or other recycling companies. This is evident in many areas around Cape Town, and salvagers are often referred to as 'trolley people'.

Figure 7. Waste salvagers on a waste disposal site (extreme example) [Source: Engledow, 2007]

Recycling has been recognised at both the National and local levels of government as an important strategy to divert valuable resources sent to landfill as waste. At the national level, recycling is highlighted within the National Waste Minimisation Strategy, the White Paper on Integrated Pollution and Waste Management for South Africa, and the NEM Waste Action. In Cape Town, it has also been recognised as an important method to divert waste, as evidenced by the many recycling initiatives previously implemented and the inclusion of recycling in the IWMP (Mega-tech, 2004.)

---

16 Resource Management Services, 2005-2007. External Audits of the City of Cape Town Waste disposal facilities.

17 Working face: term commonly used in waste management to refer to the active working area of a landfill site where the trucks are disposing waste and the landfill compactors compact the waste (Engledow, 2005).

Figure 8. Informal salvagers [Source: Engledow, 2008]

In many places, including the City, waste management has been regarded as a 'Cinderella' service, with the focus mostly being on the management of waste material. Fortunately, the waste industry is an evolving one and has recognised the value in viewing waste differently. This is, however, not without its own challenges.

6.   Key challenges in terms of sustainability

There are numerous challenges and barriers to making sustainable waste management in Cape Town an easy reality. The range of challenges includes:

- Legislation and other legal barriers that can delay proactive alternative solutions, e.g. the necessary Environmental Impact Assessments (EIAs), waste licensing procedures, and land use planning issues;
- Funding and capital expenditure – for the private and public sectors;
- The Municipal Finance Management Act and other procurement processes;
- Customers are generally driven by the price of the service, and not necessarily interested in alternative solutions (especially relevant for the private industry);
- Time constraints – roads are becoming more congested, which leads to transport delays;
- Lack of skilled people in the industry, both in the private and public sectors (including regulating authorities);
- Lack of inter-departmental and intergovernmental communication;
- Creating markets for recyclables to ensure a sustainable recycling economy, as well as ensuring that products are manufactured for recyclability;
- Public and private participation (what are the levers or incentives to change behaviour?); and
- Gaps in information (information from private and public sector).

6.1   Legislation and other legal barriers

There is a plethora of legislation pertaining to various aspects of waste management, from National to local government level. This can make waste management confusing, because it is difficult to discern which Acts and Bylaws are applicable. A related problem is the degree to which legislation is enforced. The lack of capacity within all the tiers of government tends to impede effective enforcement of legislation. The National Environmental Management: Waste Management Act that has been promulgated forms the overarching piece of waste legislation for the country, which should bridge the gaps in the array of legislation.

Almost all waste management activities (including small MRFs for the sorting of recyclables) trigger listed activities within the NEM Waste Action, which require waste licenses from the Provincial or National Department, depending on the nature of the application. The licensing process must follow the same process as the EIA regulations as per Chapter 5 of the National Environmental Management Act 107 of 1998 (NEMA), as amended by the NEMA Second Amendment, Act No. 62 of 2008. Confusion can creep in when there are activities listed in both the waste act and the EIA regulations.

The purpose of the EIA regulations and the need to apply for a waste license is to protect the environment, and is therefore necessary. However, it often leads to lengthy delays at the applicant's cost. In terms of the establishment of landfill sites, where environmental, social and economic impacts may well be high, the EIA process is essential. It is the smaller waste management companies that suffer and are deterred from establishing these low impact facilities. The smaller activities include composting and small MRFs, both of which are waste minimisation initiatives desperately required in the City, yet hindered by bureaucracy. This is not to say that the establishment of these sorts of facilities should not go through any environmental assessment process; however, the process used should be appropriate for the type of activity being applied for. The NEM Waste Action and revised EIA regulations are an attempt to streamline this process.

Another aspect of legislation that seems to complicate the establishment of waste management facilities is the outdated Offensive Trade Regulations (1944).[18] Waste management facilities generally need to be established in areas zoned for offensive trade, i.e. rezoning applications or departure use applications are necessary, despite the fact that the area may be zoned within an industrial area. Application for rezoning of a property can only be done once a decision on the EIA process has been granted. The EIA process can take anything from six months to three years, and the application for rezoning a further one to two years – and this time frame is often further affected if there are appeals.

6.2 Funding

The operating budget of the City of Cape Town is derived from tariffs and the capital budget is derived from a number of different funding mechanisms, including provincial and National government. The funding for both budgets is restricted, while the cost of waste management activities is increasing due to the rapidly increasing volumes of waste and transport costs. The available funding is not increasing in proportion to the rate at which waste is being generated.

6.3 Municipal Finance Management Act and Municipal Systems Act

As the name states, the Municipal Finance Management Act (MFMA) details how municipalities are required to manage their finances, while the Municipal Systems Act (MSA) governs the way municipalities must function in terms of service delivery and standards of service. In certain instances, the procedures required by the respective acts actually restrict the ability of municipalities to operate optimally in the waste sector.

Two examples to illustrate this point:

- In terms of the MFMA, contracts awarded to private contractors are limited to a three-year period. Many contractors argue that this time period is too short to justify major capital investment, and as a result, they tend to invest in the minimum equipment necessary. This unfortunately leads to substandard service at the expense of surrounding communities or the environment.

---

18 Government Notice 614 of 1944.

- In terms of the MSA, outsourcing of management functions in terms of waste management facilities is a lengthy process that is governed by Section 78. Municipalities must first investigate the need to outsource the function, followed by an approval process, and procurement and tender process to award the function to a suitable contractor. These processes can take months or even years, and result in lengthy delays for options that could offer more efficient service, e.g. the outsourcing of the management function of waste management facilities/services to private contractors.

While the MFMA and MSA have been drafted to protect municipalities and private contractors, the above examples show that they tend to have the opposite effect.

6.4 Price-driven market

The waste industry, like any other market, is driven by price and profitability. This applies especially to the private waste management industry. Customers are inclined to take the cheapest option available, even if this is not always the most sustainable one, e.g. land-filling as opposed to alternative treatment technologies. However, waste management companies are starting to diversify and offer their clients a wider selection of treatment technologies. But, until the legislative levers are more effectively in place, many companies will not feel compelled to take the alternative – and sometimes more expensive – option.

What is encouraging though, is the 'green revolution' that is currently underway. Companies are seeing the marketing benefits of choosing a more sustainable approach to waste management, e.g. in-house recycling programs.

6.5 Transport

Every year the amount of traffic on the roads increases. This is true for all vehicle types. More waste is also being generated every year, which requires additional long-haul collection vehicles on the roads and on the City's landfill sites. Due to the decreasing number of landfill sites around the City (now only three), the turnaround times of the trucks are currently longer than before, as further transfer stations must still be constructed. This may in turn affect the public and private sector in terms of lengthy delays and additional congestion of the City's roads.

6.6 Skills shortage

There is a general lack of managerial and operational skills within the waste management industry in both the private and public sectors. The situation has serious implications where landfill sites are not operated by qualified people. Actual numbers for the number of vacant posts within the City's Solid Waste Management Department are not available at present.

6.7 Communication

A common constraint remains the lack of interdepartmental communication and co-operation; however, this is not restricted to the management of waste. This aspect is clearly evident when applying for the various applications in terms of authorisation for the establishment of waste management facilities, e.g. Waste Licenses, EIAs and landuse.

6.8 Markets for recyclables

Reducing the volume of waste disposed on landfills is one of the City's key objectives for the future. Reduction initiatives include waste minimisation and the recovery of recyclables. In order to achieve this objective, it is important to ensure that markets are established for recyclables. An unstable recyclable paper market in 2004 drove the price of paper so low that the recovery rate dropped as a result (Engledow, 2005). The glass, metals, and plastics markets are good examples of strong, stable markets.

6.9 Participation

Implementation of new systems always present challenges in terms of the participation of various stakeholders. Yet, it is often the participation rate that will determine the success or failure of a new initiative.

Participation in waste minimisation and recycling initiatives is on a voluntary basis. It is important to implement structures where there are levers or incentives to change behaviour. Behaviour change is required from the individual, the manufacturing sector, and government.

6.10 Gaps in information

Like any other industry, the private waste industry is very competitive and generally does not provide consultants with any information. Unfortunately, this leads to a skewed picture of the waste volumes being generated, and of where and how they are treated. In fact, because of the gap in data from the private sector, it is assumed that the waste volumes generated are actually in excess of what has been presented.

While there appears to be numerous constraints and challenges, there are also many opportunities for both government and industry.

7. Opportunities for Sustainable Waste Management

Constraints can often inspire a range of opportunities. These include:

- Partnerships;
- Business re-engineering, including the establishment of new business and changing the business case from 'business as usual' to 'business as *unusual*' – Extended Producer Responsibility;
- New markets – promotion of alternative treatment methods to open up the recycling economy;
- Decreasing the reliance on raw materials if recycling initiatives are increased;
- Increased public awareness as to their role in waste minimisation and recycling;
- Increased landfill airspace[19] savings;
- Implementing instruments;
- Green Procurement Policies.

7.1 Partnerships

Partnerships, both public-private and business, are essential for the success of waste reduction and recycling objectives. It is also essential that the different tiers of government form partnerships and embark on waste minimisation initiatives together. This would enable programs to be rolled out from the national to the local levels. An example of such collaboration can be seen with the initiative that PETCO (PET Plastic Recycling South Africa) has launched at the City of Cape Town drop-off sites. PETCO has also sponsored bags for the collection of PET (Polyethylene teraphthalate), HDPE (High Density Polyethylene) and LDPE (Low Density Polyethylene) for recycling.

---

19 Landfill airspace is a term commonly used within landfill management in South Africa as it refers to the amount of available volume that can be used to dispose of waste within a landfill site.

Figure 9. PETCO initiative in collaboration with City of Cape Town drop-off sites [Source: Engeldow, 2008]

7.2 Business re-engineering

As environmental awareness becomes an area attracting more attention from larger corporations, business re-engineering gains more momentum. International pressures, as well as local constraints with regard to landfill airspace (see footnote 22), the increasing cost of transport, and the environmental degradation involved with traditional methods are demanding that corporations re-think their traditional approach and report on their performance. Waste management, and specifically waste minimisation and recycling initiatives, is finally receiving the necessary attention at the corporate level.

The marketing opportunities associated with business re-engineering in terms of sustainable waste management are great. For example, one of the major retail chains in South Africa has identified packaging wastes as a serious concern. The company has put pressure on the suppliers of various plastic packaging to include a stamp identifying the type of plastic on the base of the container as a minimum standard in an attempt to make identification – and therefore post-consumer recovery for recycling – easier. In addition, there are also ongoing investigations of alternative packaging (e.g. composting packaging) for food items.

Extended producer responsibility should also be considered under the banner of business re-engineering, as many businesses do not take the externality of the by-product (waste) into account when considering their financial costs and profitability. Take-back systems or deposit systems should be encouraged.

There are also opportunities to move away from the conventional business models and to explore new technologies and methodologies. This applies especially to the required move away from end-of-pipe technologies, such as land-filling, to a more continuous natural cycle, i.e. a move away from the 'cradle to grave' approach to a 'cradle to cradle' approach (Braungart & McDonough, 2002). It is time to re-think the way that we manufacture products to ensure their ongoing recyclability, thereby adopting a more natural cycle.

In addition, interdepartmental communication and co-operation should also be encouraged, because the field of waste crosses many boundaries, e.g. the Waste Water Department in terms of sewage sludge; the energy sector in terms of waste to energy projects; electricity regulators – as seen in the implementation of energy saving light bulbs (Compact Fluorescent Lamps (CFLs)) that contain mercury; waste oil recovery and recycling.

Even waste management companies are re-thinking their traditional end-of- pipe methodologies to waste management and the service they provide to their clients. Waste management companies in Cape Town are adopting and offering clients a broader range of alternatives, including on-site sorting and recycling, and composting.

The promulgation of the NEM Waste Action will hopefully speed these processes up, as the Minister has the authority to demand integrated waste management plans to be submitted by certain industries, as well as to require that waste management officers be appointed within organisations. Organisations will now be held accountable for the waste they generate.

7.3 New markets

Many opportunities are available as alternatives to land-filling. There is increased emphasis on the recovery and recycling of recyclables, green waste and sewage sludge for composting, and the crushing of builders' rubble for re-use in the construction industry. Building rubble crushing is taking place in and around Cape Town, and the recovered material is used in a variety of civil applications. This area has been identified at the municipal level as 'low-hanging fruit' and an accessible waste fraction to recover from landfill disposal.

7.4 Raw Materials

It is well known that the recycling of various recyclables reduces the need of raw material required for many applications. This is especially the case for metals, glass, paper and plastics. There are also energy, water savings and reduced emissions attached to the recycling of materials. For example, the large majority of plastic bottles that are thrown away or discarded in the environment are made from PET. South African Breweries produces 1 billion PET bottles per annum for the South African market and almost nothing to collect these bottles. And yet there is a shortage of low-cost PET material for manufacturing cheap plastic items such as plastic bags, coat hangers and water tanks. If PET recycling became a major industry, less virgin PET would need to be imported from overseas or bought from SASOL.

7.5 Implementing Instruments

A number of key instruments can be used to achieve policy initiatives, objectives, and targets of waste minimisation through push-and-pull type mechanisms. Push factors refer to legislation and law enforcement, licenses, and control in terms of direct regulations. Tariffs, levies, deposits, market creation, and financial support refer to indirect methods that can be applied. Pull factors rely more on self-regulation in terms of information, and the continuation of collective initiatives, like waste minimisation clubs.

7.6 Green Procurement Policies

Green Procurement Policies (GPP) or sustainable purchasing practices are effective tools for stimulating the demand-side of the economy for alternative markets, including the recycling economy. The basis of green procurement is the principle of pollution prevention and the reduction of risks to human health and the environment. Organisations need to evaluate providers of goods and services in terms of these principles. Government organisations should lead by example in terms of GPP. An example of this could be paper supply, where governmental departments (or any organisation) might stipulate that a certain percentage of the paper must include recycled content; others would be the adoption of policies to buy only rechargeable batteries, or to purchase cleaning products that are biodegradable, etc.

The Western Cape Provincial Government Department of Environmental Affairs and Development Planning has taken a proactive approach and is in the process of developing a Green Procurement Policy document. This document is not yet available to the public, but it shows the commitment of the provincial government.

8.   Future Plans

The City has planned well ahead in terms of solid waste management based on current waste generation volumes. Future planing has previously been based on an average 6% increase in waste volumes per annum, however, the decreasing volumes being received at landfills will require close monitoring to ensure that the plans being proposed are in line with the potential changing requirements of waste management within the City. Planning has also needed to take into account the 2010 Soccer World Cup and the potential impact of additional waste generated by preparations leading up to and after this three-week event.

As discussed, the City's primary objective is to minimise waste to landfills. In order to achieve this, a number of initiatives are planned or have already been initiated. The third phase of Waste Wise is to commence shortly to continue a broad education and awareness campaign about the importance of waste minimisation and recycling.

A dual collection system, as mentioned previously, called the 'Think Twice Campaign', has been rolled out in the Atlantic Seaboard of Cape Town, including Pinelands, Blaauwberg and Parklands. It involves about 11,000 households that separate their waste into dry and wet waste. The dry fraction is sorted at a MRF in Maitland, which has received an Environmental Authorisation from the DEADP. As of August 2008, the Helderberg area, including Somerset West, Gordon's Bay and Strand, was also included into the pilot campaign (Lourens, n.d.).

Melani Materials, a BEE subsidiary of Afrimat, has been contracted since February 2008 to crush and remove builders' rubble from Coastal Park and Bellville South. Mobile crushing units will be placed at the landfill sites to process up to 800 tons of rubble per day. Crushed material is being processed and sold in bulk to the construction industry for various civil applications.

The City's Integrated Waste Management Bylaw, which has been in effect since March 2009, will require greater collaboration between the various stakeholders involved in the network of waste management.

Behaviour change and waste minimisation starts with the individual, or, like charity, in the words of P. Novella (2008): "Good waste management starts at home."

References

Braungart, M. & McDonough, W. 2002. *Cradle to Cradle – Remaking the way we make things*. New York: North Point Press.

CCT, 2009. Integrated Waste Management By-law. Province of the Western Cape: Provincial Gazette 6651, 21 August 2009.

Department of Environmental Affairs and Tourism. 2000. *White Paper on Integrated Pollution and Waste Management for South Africa*. Government Notice 227, 17 March 2000.

Department of Environmental Affairs and Tourism. 1989. *Environment Conservation Act 73 of 1989. Juta Statutes*, Volume 6.

Department of Environmental Affairs. 2008. National Environmental Management: Waste Act, 59 of 2008. Government Gazette No 32000, 10 March 2009.

Department of Water Affairs and Forestry. 1998a. *National Water Act 36 of 1998. Juta Law Statutes*. 2001: Volume 6.

Department of Water Affairs and Forestry. 1998b. 2nd Edition. *Waste Management Series: Minimum Requirements for the Handling, Classification and Disposal of Hazardous Waste*. Cape Town: CTP Book Printers.

Department of Water Affairs and Forestry. 1998c. 2nd Edition. *Waste Management Series: Minimum Requirements for Waste Disposal by Landfill*. Cape Town: CTP Book Printers.

Engledow, S. 2005. *The Strategic Assessment of a curb-side recycling initiative in Cape Town as a tool for Integrated Waste Management*. Unpublished Masters thesis. Cape Town: University of Cape Town.

Engledow, S. & Eichstadt, L. 2007. *Integrated Resource Analysis – Solid Waste Management: Baseline Study*. Report prepared for the Sustainability Institute. Stellenbosch: Sustainability Institute.

Ingerop Africa. 2005. *Community waste drop-off centres: Investigation for the management of builders' rubble in the Metro*. Unpublished document.

Jeffares & Green, 2008. Integrated Waste Management Hierarchy Poster, unpublished.

Lourens, B. n.d. Waste Plan. Ongoing communication.

Mega-tech. 2004a. *Integrated Waste Management Plan for the City of Cape Town: Final Status Quo Report.* Cape Town: City of Cape Town.

Mega-tech. 2004b. *Integrated Waste Management Plan for the City of Cape Town: Draft Assessment Report.* Cape Town: City of Cape Town.

Naidoo, S. 2007. Black Power. *Business Times*, 20 May 2007. Article based on the findings of the 'Black Diamonds 2007 Survey' conducted in March 2007 by the University of Cape Town Unilever Institute, and TNS Research surveys.

Novella, P. 2000. Sustainability: Quo Vadis? Proceedings Volume 2: Wastecon 2000.Bienniel Conference, Cape Town.

Novella, P. 2008. Private telephone and email communication.

Parsons, R. 2002. The Geohydrology of the Bellville South Waste Site. Wastecon: International Waste Management Biennial Congress and Exhibition: Proceedings Volume 2, 342-348.

Resource Management Services, 2005-2007. External Audits of the City of Cape Town Waste disposal facilities, unpublished.

Vesiland, P.A., Worrell, W. & Reinhart, D. 2002. *Solid Waste Engineering.* California: Brooks/Cole.

White, P.R., Frank, M. & Hindle, P. 1995. *Integrated Solid Waste Management: A Lifecycle Inventory.* Scotland: Chapman & Hall.

Wolman, A. 1966. *A 'Scientific America' Book.* Harmondsworth: Penguin.

Sustainable Transport Options

Passenger Transport in Cape Town

Roger Behrens & Peter Wilkinson

ALL
BUSSES
MINI
BUS

"Our economic growth over the next decade and beyond cannot be built on the same principles and technologies, the same energy systems and the same transport modes, that we are familiar with today."

Minister of Finance, Parliament, 20 February, 2008

1. Introduction

This chapter calls for a radical transformation of Cape Town's transport sector. We begin with a brief overview of Cape Town's passenger transport system and outline the major environmental impacts associated with its current pattern of operation. These issues are viewed against the backdrop of rising concern about two key challenges in the contemporary world: the depletion of global oil and natural gas stocks, and the prospect of global climate change induced by greenhouse gas emissions. We examine the impact of the use of oil and fuel on the environment, and then draw out the possible implications for the sustainability of passenger transport operations in Cape Town.

The focus on the passenger transport system is a useful lens through which to view progress towards a sustainable transport system. A decrease in aggregate VKT (vehicle kilometres travelled) is a fundamental indicator of progress towards reduced petroleum-based liquid fuel consumption and reduced greenhouse gas emissions. This would be one of the critically important preconditions for achieving a more sustainable urban transport system where the routine movement of both people and goods would need to be dependent on 'localised' sources of renewable energy (e.g. solar and wind power provided via the grid into electric busses, cars and trains).[1] We explore the implications of moving the majority of commuters into high quality public and non-motorised transport systems and speak to the need for a sharp shift to a more sustainable pattern of resource use, arguing that it is an essential and urgent component for Cape Town's transport sector. The massive and rapid urban growth of Cape Town has not been coupled to investments in the kinds of urban infrastructure, energy, and transportation systems that are appropriate for a world that is running out of atmosphere, water, oil, and sinks for liquid, solid and airborne wastes.

Our explicit intention in this chapter is to put forward a strong position on a cluster of issues. We have written it in a style that is more polemical than is academically conventional. Some of our conclusions regarding the current situation, as well as some of the proposals we have put forward as to what might be done about it, are not substantiated by reference to the results of systematic and reliable empirical research. The relative paucity of such research in the transport sector is symptomatic, in our view, of the somewhat marginal status it has been accorded to date.

A suggested framework for a strategic response to these problems is offered. One of the keys to securing a sustainable future for Cape Town will be the design and establishment of a public transport system for the city that links the rail, bus, and minibus-taxi systems. The chapter closes with a blunt reminder of the urgency of the present situation. In our view transformative action simply cannot be deferred to some later date.

2. Cape Town's transport system: an overview

Metropolitan Cape Town has a diverse and well-established transport infrastructure that has evolved gradually over three centuries, with particularly rapid development during the second half of the twentieth century. Major sea and airport facilities connect the city externally into both national and international maritime and air transport systems, while comparatively extensive regional and national road and rail networks accommodate the movement of people and goods. Within the

---

1 While the intra-urban movement of freight is not inconsequential, it presents a set of issues which are even more intractable than those we wish to explore for passenger transport. The same proviso would apply to supra-local or inter-city passenger and freight movements by ground, air and sea transport.

city, a relatively dense road network serves road-based public transport. This includes scheduled bus services and minibus-taxi operations, as well as general traffic flows of private vehicles – both personal and freight. A unique feature of the contemporary South African context in Cape Town is the city's radially-configured passenger rail network, which carries the largest proportion of public transport users, particularly in peak periods.

2.1 Transport infrastructure[2]

The total length of the city's road network was estimated in 2002 to be 8,200 kilometres, of which some 2,200 kilometres (27%) could be described as being of metropolitan significance in terms of traffic functions, with the remaining 6,000 kilometres (73%) comprising local distributor and access roads – representing a then total asset replacement value in the region of R16 billion.[3] The total length of passenger rail track in the metropolitan area is currently of the order of 260 kilometres, served by 97 stations. The combined asset replacement value of rail track and rolling stock was estimated at R14 billion in 2002. The integrity of signalling systems has often been compromised by cable theft, leading to delays and cancellation of services, while vandalism of rolling stock remains a serious problem.

Provision for non-motorised transport (NMT) modes – predominantly walking and, very much less significantly, cycling – has been limited to the installation of conventional sidewalk and pedestrian crossing facilities in most areas, and the establishment of some limited and disconnected cycleway routes. There is a very high incidence of accidents involving pedestrians, including a significant proportion of fatalities. It is clear that the accommodation of pedestrian movements within the system remains woefully inadequate.

2.2 Vehicle fleets and rolling stock

The fleet of private vehicles of all types in Cape Town totalled 790,000 in 2000, including some 570,000 registered cars – an estimated ratio of 190 cars/thousand people. This is significantly higher than that found in other South African cities (Venter, 2007).[4] The mean annual growth rate of registered cars in the metropolitan area was 3% between 1995 and 2000, while the number of daily car commuters grew from around 285,000 in 1980 to 680,000 in 2001, at a mean annual rate of around 4.6%. The rate of growth in car usage exceeds population growth in general, as well as growth in vehicle registrations, and has resulted in relatively sharp increases in daily traffic volumes on the major road network with a concomitant intensification of peak period congestion levels.

The primary modes of public transport in the city are:

- passenger rail services operated by Metrorail,
- scheduled bus services operated primarily by Golden Arrow Bus Services (GABS), and
- unscheduled and only partially regulated minibus-taxi services provided by numerous individual and small-scale operators.

Together they serve a daily market of some 1.13 million passenger trips. In 2004, Metrorail was operating 66 train sets in peak periods, employing rolling stock comprising 231 motor coaches and 770 passenger coaches. Rail serves a 54% share of this market. Plans to upgrade and replace the ageing rolling stock through substantially higher levels of investment than were previously undertaken, were announced recently.

In 2004, Golden Arrow Bus Services (GABS) was operating a fleet of 852 single-deck buses with an average capacity of 90 passengers (60 seated) on 736 scheduled service routes during morning peak

---

2   Unless otherwise attributed, statistics included in this and the following sections are drawn from Behrens & Wilkinson (2003), which incorporates data from various official sources, or the 2004/5 Current Public Transport Record (City of Cape Town (CCT), 2005).

3   Regular scheduled maintenance of the road network was effectively suspended for a number of years in the face of budgetary constraints, but has since been resumed.

4   According to the National Household Travel Survey 2003, the National figure is 108 cars per 1,000 population

periods, using 50 terminals and 80 ranks, as well as 2 holding areas. Buses are responsible for a 17% share of daily passenger trips in the public transport market. In the same year, some 7,500 licensed and unlicensed 15-seater minibus-taxis – operating on 555 routes in morning peak periods from 112 terminals and 61 ranks, as well as 30 holding areas – provided unscheduled services to 29% of this market. Ongoing replacement and extension of the bus fleet has been negatively affected by uncertainty regarding plans to reform the road-based public transport system, while the much delayed government-funded taxi recapitalisation programme has yet to have a significant impact on the replacement of the minibus-taxi industry's generally aging vehicle fleet.

2.3 Modal split

Broadly indicative data on commuter travel patterns in Cape Town, obtained from household surveys conducted in 1991 and 2004, reveal an overall shift during this period from public to private modes of transport and, within the public transport sector, a shift from scheduled bus and rail services to minibus-taxi services (Table 1).[5]

Table 1: Changing modal split in commuter travel, 1991-2004

	Private transport	Public transport			Walk/other
		Rail	Buses	Minibus-taxis	
1991	44%	27%	16%	6%	7%
2004	48%	13%	7%	13%	13%

[Source: CCT, 2006]

The first of these trends accords with other evidence that suggests that those public transport users who are able to acquire private vehicles have done so as soon as their economic circumstances have permitted this, largely due to dissatisfaction with the levels of service offered by the public transport system. (This trend may be modified because of the recent sharp escalations in both fuel prices and finance charges, which will have substantially altered the parameters of such choices.) The modal shift within the public transport system towards minibus-taxis has long been evident. It can largely be attributed to their higher levels of convenience relative to the scheduled modes, despite widespread concerns about their higher (unsubsidised) fares, overcrowding, and compromised safety due to poor driving behaviour.

2.4 Systemic duality and differentiated mobilities

The dualistic structure of the passenger transport system is both derived from, and reinforced by, the legacy of socio-spatial segregation inherited from apartheid era urban planning practices. On the one hand, when they can afford it, members of lower income households, predominantly situated in what are often geographically peripheralised townships and informal settlements, constitute the majority of 'captive' users of public transport. Alternatively they choose to walk. On the other hand, members of middle and higher income households, situated in wealthier and sometimes more conveniently located suburbs, for the most part have acquired and routinely use private vehicles rather than public transport. The routine travel patterns and experiences of people within these two broadly defined segments of the passenger transport market therefore remain effectively discrete, reflecting significantly different mobilities.

Most immediately, such differentiated mobility is evident in the lack of convenience, comfort, and safety frequently endured by 'captive' users of the three primary public transport modes. Such gaps in opportunities for mobility are also apparent in lengthy travel times and transport expenditures, which tend to be relatively less affordable than those incurred by wealthier households using private

---

5   The substantial discrepancy in the public transport sector modal split for 2004 between this data set and that obtained from the 2004/5 Current Public Transport Record is not explained but the overall directions of the key trends are likely to remain generally correct.

transport. Lower income households – those earning less than R500/month – commit as much as 35% of their total income to meeting basic transport costs. This figure falls to 5% for wealthier households earning more than R3,000/month, who are likely to own and use private transport (Dept. of Transport, 2005).

2.5 Aggregate travel patterns

Although not uniformly so, the city's present spatial structure is dominated by sprawling, low-density development patterns. Together with the geographical eccentricity of its historical core area, where a large proportion of economic activities and employment opportunities continue to be concentrated, this imposes long average trip lengths on much of the population, particularly the lower income people living in the metropolitan south-east sector, in the more recently developed and heavily populated residential areas and informal settlements.

The overall pattern of significant geographical separation between zones of employment and major residential zones leads to the phenomenon of 'tidal flow' in commuter traffic, in which peak period congestion and overcrowding of public transport in one direction is combined with the presence of unused or underused road capacity and public transport passenger capacity in the reverse direction. This reflects the single most significant inefficiency imposed on the operation of the passenger transport system.

2.6 Institutional framework

The institutional framework governing the operation of Cape Town's passenger transport system is characterised by a high degree of fragmentation, resulting in poorly coordinated, occasionally incoherent, planning and regulation, as well as generally inefficient operational management (Wilkinson, 2008). The fragmentation cuts across both the parastatals and private agencies involved in operating the different modes of public transport, as well as the three spheres of government responsible for different aspects of infrastructure provision, and the planning, regulation, and disbursement of public subsidy funding within the system.

The establishment of local level transport authorities to integrate and coordinate the execution of most of these functions – excluding those associated with the provision of passenger rail services – has been legally possible since the passage of the National Land Transport Transition Act (NLTTA) in 2000. For complex reasons, however, to date there has been no effective progress in this regard in Cape Town, although there is some expectation that the recently passed National Land Transport Act may resolve some of the difficulties that have been seen to obstruct the installation of a suitably constituted metropolitan transport authority.

3. Cape Town's transport system and the environment

The impacts of transport system operation and associated patterns of travel behaviour described in the preceding section on resource use and the surrounding biophysical environment are considered here in terms of fuel consumption and gaseous emissions. The significance of the inefficiency of Cape Town's transport system, and its contribution to energy consumption and greenhouse gas (GHG) emissions, which are discussed here needs to be viewed within the context of global concerns about oil depletion (Box 1) and climate change (Box 2). Clearly, transport systems have a variety of aural and visual pollution effects, as well as 'barrier' or 'severance' affects, on local communities, but these are arguably less significant from the perspective of sustainable resource use, and are therefore excluded from more detailed discussion here.

With regard to fuel consumption, according to a 2008 report by the Sustainability Institute, *the transport sector is the largest consumer of energy, accounting for 47% of the city's total energy consumption,* followed by commerce and industry (38%) and households (14%) (*Table 2*, Sustainability

Institute, 2008).[6] Within the transport sector, 60% of liquid fuel consumption is in the form of petrol, and 20% in the form of diesel.[7] *Table 2* shows that the energy use profile for the city as a whole is dominated by petrol (28%), electricity (29%), and diesel (18%), with paraffin, liquefied petroleum gas, coal, heavy furnace oil, jet fuel, and wood accounting for the remaining 24% of energy use.

Table 2: Annual energy use by user group and fuel type (1,000 gigajoules)

	Electricity	Diesel	Petrol	Paraffin	LPG	Wood	Coal	Heavy furnace oil	Jet fuel	Total	Total %
Households	17,969	-	-	2,587	547	359	43	-	-	21,505	14%
Industry/commerce	24,755	13,160	-	444	2,718	561	10,788	4,696	-	57,123	38%
Local authority	1,747	234	119	-	-	-	-	-	-	2,100	1%
Transport	-	14,337	42,294	-	-	-	-	-	13,616	70,246	47%
Total	44,472	27,731	42,413	3,030	3,265	920	10,831	4,696	13,616	150,975	100%
Total (%)	29%	18%	28%	2%	2%	1%	7%	3%	9%	100%	

[Source: Sustainability Institute, 2008]

Box 1: Transport and global oil depletion

Key debates in the area of global oil depletion revolve around *when* and *how* the operation of transport systems – and economic systems more generally – will be impacted. As oil is a finite resource, the rising production of oil that has underpinned economic growth over the last 150 years cannot continue indefinitely. Currently, oil accounts for around 35% of global energy supply (*Table 3*). As an energy source, it is used for electricity generation, heating, and as a liquid fuel for transport systems. The world's transport systems (including land, air, and sea transport modes) depend on oil for around 90% of their energy requirements (Wakeford, 2007).

Table 3: Global energy supply by fuel type (2004) (million tons of oil equivalent)

	Oil	Coal	Gas	Combustible renewables and waste	Nuclear	Hydro	Geothermal/solar/wind
Million tons of oil equivalent	3,947	2,769	2,310	1,177	718	247	56
Percentage	35%	25%	21%	10%	6%	2%	0.5%

[Source: Wakeford, 2007, citing the International Energy Agency]

Hubbert (1956) assessed discovery rates, production rates, and cumulative production in the oil industry, and predicted that production in any given region would follow a bell-shaped curve (subsequently known as the 'Hubbert curve'), rising to a peak when approximately half of the total oil had been extracted, and thereafter gradually falling toward zero as extraction becomes progressively more difficult and costly.

---

6   An earlier study in 1997 estimated that the transport sector was responsible for 44% of energy consumption (Wicking-Baird *et al.*, 1997) – suggesting perhaps that the transport sector's relative energy consumption has grown.

7   A limitation of these data is their omission of electricity consumption by the city's extensive rail operations.

The Association for the Study of Peak Oil (ASPO) argues that there is growing evidence that global oil production is nearing – or is already at the at – the top of the 'Hubbert curve' (a point of production referred to as 'peak oil'). This evidence includes, among other things, a steady decline in new 'conventional oil' discoveries since the 1960s', oil consumption rates exceeding discovery rates since 1981, and 33 out of the 48 significant oil-producing nations passing their individual production peaks (Wakeford, 2007).

Predictions about the timing of the world peak vary among oil geologists and energy agencies (see Wakeford, 2007). Some predict 'peak oil' within the next five to ten years, while others predict a peak in about 25 years. The main sources of contention revolve around the accuracy of reported Middle Eastern reserves, and potential yields from 'unconventional' sources (e.g. tar sand and shale). On the current evidence, it appears unlikely that production will increase substantially beyond its current level (at around 86 million barrels/day), and that, following a 'bumpy plateau' of several years, it will begin to decline. The impact of stagnant or declining oil production – and the probable unavailability in the short term of alternative fuels distributable on a mass basis – will be a growing supply-demand gap, with inevitable consequences for fuel prices. As a heavy user of oil, Cape Town's transport system is likely to be hard hit.

With regard to gaseous emissions, important quantifiable emissions from the transport sector, which either directly or indirectly contribute to the 'greenhouse effect', include: carbon dioxide (CO_2), methane (CH_4), nitrous oxide (N_2O), mono-nitrogen oxides (NOx), carbon monoxide (CO), and sulphur dioxide (SO_2) (Energy Research Institute (ERI), 2002). A study of photochemical smog (or 'brown haze) in Cape Town in 1997 estimated that the transport sector was the largest contributor to nitrous oxide and the largest contributor to photochemical smog (52%) in the city (Wicking-Baird *et al.*, 1997). In Cape Town, 2,815,566 kg of CO_2 were estimated to have been emitted from petrol consumption in 2001, and 1,487,441 kg from diesel consumption (CCT, 2003).

An update of GHG emissions data in Cape Town by Kennedy *et al.* (2008) is illustrated in *Table 4*. These provisional data suggest that ground transport is responsible for some 12.7 megatons of CO_2 equivalent per year, representing around 33% to 45% of total direct GHG emissions.

Table 4: Provisional estimate of direct greenhouse gas emissions in Cape Town

	Annual GHG emissions (megaton CO_2 equivalent)	Percentage of total emissions included in estimate
Electricity	8.6	22.6%
Heating and industrial	3.7	10.6%
Ground transport [1]	12.7	33.5%
Air and marine transport [2]	12.6	33.3%
Direct industrial	not determined	
Waste	not determined	
Vegetation	not determined	
Total [3]	37.6	100%

[Source: Hansen & Gasson, 2008]

Notes:

1   The estimated split between petrol and diesel is 8.9 megaton and 3.8 megaton of CO_2 equivalent respectively.

2   This estimate allocates 100% of emissions generated from the consumption of fuels put on board in Cape Town, to the Cape Town area.

3   If it is assumed that only 25% of marine and air transport emissions are allocated to the Cape Town area, the total is reduced to 28 megaton CO_2 equivalent.

From an energy and emissions perspective, Cape Town's transport system will therefore be central to any attempt to place the city on a sustainable resource use path.

Box 2: Transport and global climate change

The Intergovernmental Panel on Climate Change (IPCC) has concluded that most of the observed increase in globally averaged temperatures since the mid-1900's is very likely due to observed increases in anthropogenic GHG concentrations via an enhanced 'greenhouse effect'. Paleoclimatic studies of ice cores, spanning the last 650,000 years, indicate that global atmospheric concentrations of CO_2, CH_4, and N_2O have increased markedly as a result of human activities since the mid-1700s, and now far exceed pre-industrial values (IPCC, 2007). The recent global increases in CO_2 equivalent concentration are argued to be due primarily to fossil fuel use and land use change. *Table 5* illustrates the contemporary relative contributions of key sectors to global GHG emission, with the transport sector estimated to be responsible for 14% (Stern, 2006).

Table 5: Global Greenhouse gas emissions by sector (2000) (gigatons of CO_2 equivalent)

	Power	Land use	Agriculture	Industry	Transport	Buildings	Other energy related	Waste
Gigatons CO_2 equivalent	10.08	7.56	5.88	5.88	5.88	3.36	2.10	1.26
Percentage	24%	18%	14%	14%	14%	8%	5%	3%

[Source: Stern, 2006]

Climate model projections summarised by the IPCC (2007) indicate that average global surface temperature will likely rise a further 1.1°C (to 6.4°C) over the next century if no significant action is taken. The Stern Review on the economics of climate change in 2006 argued that such a radical change in the physical geography of the world must lead to major changes in its human geography (Stern, 2006). Even at more moderate levels of warming, climate change will have serious impacts on economic systems and human life. The predicted impacts include, among other things, increased flood risk, reduced water supplies, decreased crop yields, rising sea levels, and reduced biodiversity (Stern, 2006).

4. Implications for the sustainability of transport operations in Cape Town

Imminent international political pressure to substantially decrease GHG emission through the introduction of carbon pricing (and trading) measures, or penalties for exceeding emission constraints, as well as sustained increases in the price of petroleum fuels, will have significant implications for Cape Town's transport system. We believe that the more important implications are likely to include the following:

- The accelerating 'automobilisation' – the acquisition and use of private cars – experienced in the city's passenger transport sector over the last 50 years is likely to be gradually halted, and then permanently reversed. In the likely event that affordable alternative fuels are not readily available at least in the short term, a substantial number of 'choice' passengers will be unable to bear the growing cost of extensive car use. A significant level of private car (and perhaps motorcycle) usage is, nevertheless, likely to remain in the medium-term future, unless additional constraints are introduced and the levels of service offered by the public transport system improved significantly.
- Generally, within the public transport passenger market, a shift from minibus-taxi services back to the commonly cheaper, publicly subsidised rail and bus services is likely where such services are reasonably accessible. More specifically, among the poorest sections of the 'captive' public transport passenger market, reliance on non-motorised travel (NMT) modes, particularly walking, is likely to deepen as public transport fares rise to incorporate fuel price increases. It is likely that for the poorest of the poor any form of motorised travel will become increasingly unaffordable.

- The likely effects of possible future carbon pricing measures and associated energy price escalations would be felt across both electrified and petroleum fuel-based transport systems. In the inter-city land freight transport sector, such price escalations, in conjunction with those induced by oil depletion, would probably lead to a shift from road to electrified rail services, provided the reliability and competitiveness of the latter can be improved, and the long-term decline in its 'reach' or penetration across the national space-economy can be reversed. Within the intra-urban land freight transport sector, however, a similar shift from road back to rail is less likely, due to inherent inertia in the current locational patterns of economic activities in the city's land use system.
- The viability of air transport for both freight and passenger movements – particularly those of a discretionary (e.g. tourism) and short-haul nature – is likely to decline fairly rapidly over the next 20 years in the face of aviation fuel price increases. Given the substantial investment currently being made in the expansion of the city's facilities to service anticipated growth in air transport, there must be some possibility that such new infrastructure could become an underutilised 'white elephant'.

5. Towards more sustainable resource use in the transport sector

Ideally, to achieve a more sustainable pattern of resource use in Cape Town's transport sector, the routine movement of both people and goods would need to become as 'localised' and non-dependent on carbon energy sources as possible. The key indicator of progress in this regard would be a reduction in the total amount of vehicle-kilometres travelled (VKT) – both within the metropolitan area and in connecting it externally to other cities and regions, specifically in travel using the conventional motorised modes. Petrol and diesel-powered vehicles obviously fall into this category, but, given the current level of reliance of the national electricity grid on coal-based generation, the use of electrically-powered vehicles on either the road or the rail networks would also have to be regarded as problematic, at least for the foreseeable future. In the absence of practical alternatives, the growing economic significance of the city's external connections – particularly those made by air transport – presents what is probably an even more pressing problem from this perspective.

Focusing most immediately here on the issue of intra-metropolitan passenger movement, however, the feasibility of bringing about a significant reduction of VKT can be seen to be determined primarily by the realistic prospects of reducing the need to travel in the first place and, where that is not possible, by promoting the use of more fuel-efficient and less polluting modes of motorised transport, particularly public transport, as well as the key NMT modes of walking and cycling. The first of these prospects – reducing aggregate travel demand – is shaped directly by the pattern of accessibility of workplaces, shops, schools, health care, and recreational and other public facilities embodied in the spatial fabric of the city, which has evolved historically over long spans of time and which therefore has a substantial degree of inertia built – literally – into it. Transforming the current pattern of accessibility into one that is less structured to accommodate, and consequently is dependent on, motorised transport is therefore by no means an option likely to be realised in the short or even medium term.

The second prospect involves promoting a shift towards the use of more sustainable modes of transport, including NMT. This can only be realised through the emergence of a very significant shift in current patterns of travel behaviour. In the context of contemporary Cape Town, as in other South African cities, the key target here would be the relatively wealthier segment of the dualistically structured passenger transport market that relies largely (or exclusively) on petrol- or diesel-driven private transport (cars and two-wheelers) to meet its mobility needs. The absolutely essential precondition to effect appropriate change in such people's travel behaviour would be to establish a well-integrated, reliable and safe public transport system. Criteria for such a system would include the need –

- to remain affordable to its (current) mainly lower-income users;
- to provide much more systematically and comprehensively than at present for the needs of pedestrians and cyclists; and
- to enable pedestrians and cyclists to access public transport facilities safely and conveniently.

Building on this – still provisional and undoubtedly incomplete – understanding of the issues involved, we would advocate that a strategically framed response to the complex and difficult task of inducing movement towards more sustainable resource use in Cape Town's passenger transport sector should incorporate, and appropriately elaborate, at least the following set of key policies or actions:

- Introduction of an appropriately structured and phased programme of travel demand and road space management measures, including – but not limited to – the prioritising of available road space for public transport operations over any accommodation of general traffic flows (through the provision of dedicated public transport lanes, intersection signalling priority, etc.), instituting direct or proxy road use pricing for private vehicles, encouraging the formation of lift clubs, firm-based travel planning and other 'mobility management' measures, and the promotion of compressed working week schedules or telecommuting options, or both in combination, among local employers.
- The establishment of a systematically planned public transport network that operates efficiently and effectively across appropriately and comprehensively integrated road and rail-based modes to facilitate the easy ('seamless'), reliable, safe, and affordable passage of its users throughout the metropolitan area. Very substantial amounts of capital expenditure, as well as public funding for operating subsidies, are likely to be involved.
- Initiation of a programme of significant investment in the extension and upgrading of pedestrian and cycling infrastructure, systematically integrated with current and planned public transport facilities, but also offering opportunities for safe non-motorised travel within and between local areas.
- As the corollary of prioritising provision for public transport and NMT modes, investment in infrastructure or facilities that primarily or exclusively serve the least sustainable modes of transport should be discontinued, particularly the use of private cars and air travel. This may exclude certain cases – justified on the basis of careful and comprehensive assessment of the full range of social and environmental costs that may be involved to realise any claimed benefits of such investment.
- The planning and regulation of integrated public transport operations should explicitly acknowledge and build on the significant physical and human capital assets represented by key components of the current public transport system – local passenger rail services, in particular, but also the privately-operated and long-established scheduled bus services, as well as, perhaps more problematically, the minibus-taxi industry. This implies the modification of any proposed *tabula rasa* or 'clean sheet' approaches to the necessary far-reaching reform of the city's public transport system in such a way that present contextual realities are appropriately accommodated.
- The formulation of robust, well-grounded, and widely canvassed plans ('spatial development frameworks') that seek to promote, through appropriate land use management measures and careful planning of the installation of urban infrastructure, the evolution over time of less travel-intensive patterns of urban development. These would generally include, but not be limited to, the emergence of polycentric spatial structures at the city scale, and the facilitation of 'transit-oriented' mixed use and higher density development associated with public transport interchange, terminal or station precincts at the local area or neighbourhood scale. This obviously requires the abandonment of any planning or regulatory practices that underpin the extension of current patterns of low-density and spatially fragmented 'urban sprawl' and 'automobile dependent development', at either city-wide or local area scales.
- Spatial development plans should, to the degree possible in the face of uncertainties and imponderables in this regard, allow for and support the increased 'localisation' of appropriately reconfigured economic activities as transport costs become increasingly burdensome – including, perhaps most importantly, the production and marketing of foodstuffs.

It seems clear that at both the global and the immediately local scale, we are rapidly approaching what may prove to be a critical 'tipping point' in the way that our present transport systems operate. This will have fundamental implications for the manner in which our cities themselves are structured and function. When liquid fuel prices exhibit a sustained rise above some threshold level – which we cannot readily predict, but may be imminent – the now unaffordable travel behaviour of many people will unavoidably have to change. In the wake of the changes we envisage, the lifestyle choices of all or most people will be challenged. New considerations will affect where to live and work, where to send children to school, and where to shop and recreate, among others. The viability, not only of our transport systems in particular, but also of our urban systems more generally, will then certainly be thrown dramatically into question.

In addition, while this remains the object of ongoing controversy, it is possible that another critical tipping point, which will have equally significant – if not even more fundamental – impacts on our behaviour and lifestyles may already have been passed. There is now a weight of informed opinion that anthropogenic climate change as a result of increased GHG emissions, to which transport systems have been a major contributor, is already a reality. Even if we were to manage to reduce the absolute volume of GHG emissions immediately – requiring a massive and fundamental transformation of the way our social and economic systems work – we may already have set processes of irreversible climate change in motion on a planetary scale. At the very least then, acceptance of the so-called 'precautionary principle' would imply that we need to act <u>now</u> to contain and reduce the level of such emissions in all sectors, but certainly, from the perspective of our concerns here, in the transport sector.

There are expectations some may hold that technological 'fixes' will inevitably emerge in the form of greatly more fuel-efficient and less polluting or non-petroleum-based vehicle propulsion systems. It is argued that these will obviate the need for any radical systemic transformation. In our view, such expectations are likely to prove self-deluding. As in other arenas of contemporary existence, the issue is primarily not one of simply substituting one technology for another, but rather of reforming or transforming the social structures and practices in which the use of <u>any</u> technology is embedded. Efforts to suppress recognition of the urgency of the current situation are unwise. And delaying intervention to secure transition towards a more sustainable urban transport system in the interests of continuing 'business as usual', we believe can no longer be seen as tenable.

References

Behrens, R. & Wilkinson, P. 2003. *Metropolitan transport planning in Cape Town, South Africa: a critical assessment of key difficulties*. Working Paper No. 5, Urban Transport Research Group. Cape Town: University of Cape Town.

City of Cape Town. 2003. *State of Energy Report for Cape Town 2003*. Cape Town: City of Cape Town.

City of Cape Town. 2005. *Summary of 2004/5 Current Public Transport Record*. Cape Town: City of Cape Town.

City of Cape Town. 2006. (Draft) *Public Transport Plan*. Cape Town: City of Cape Town.

Department of Transport. 2005. *Key Results of the National Household Travel Survey, South Africa*. Pretoria: Department of Transport.

Energy Research Institution. 2002. *Energy Outlook for South Africa: 2002*. Cape Town: Energy Research Institute, University of Cape Town.

Kennedy, C., Gasson, B., Pataki, D., Phdungsilp, A., Ramaswami, A., Steinburger, J. & Mendez, G. 2008. *Greenhouse gas emissions from global cities*. ConAccount 2008 Urban metabolism: measuring the ecological city, Prague.

Hubbert, M. 1956. *Nuclear energy and the fossil fuels*. Paper presented before the Spring Meeting of the Southern District, Division of Production, American Petroleum Institute, San Antonio.

Intergovernmental Panel on Climate Change. 2007. Summary for policymakers. In Solomon, S., Qin, D., Manning, M., Chen, Z., Marquis, M., Avery, K., Tignor, M. & Miller, H. (Eds.). *Climate change 2007: The physical science basis*. Contribution of Working Group I to the Fourth Assessment Report of the Intergovernmental Panel on Climate Change. Cambridge: Cambridge University Press.

Republic of South Africa. 2000. *National Land Transport Transition Act*, No. 22 of 2000.

Republic of South Africa. 2009. *National Land Transport Act*, No. 5 of 2009.

Stern, N. 2006. *Stern Review on the economics of climate change*. HM Treasury, Cambridge: Cambridge University Press.

Sustainability Institute. 2008. *Integrated analysis energy baseline report*. (UNDP Project No. 00038512). Stellenbosch: Sustainability Institute.

Venter, C. 2007. *Some observations on car availability and car use, and implications for TDM policy*. Proceedings of the 26th Southern African Transport Conference, 9-12 July 2007. Pretoria, South Africa. (Note: TDM means 'Travel Demand Management'.)

Wakeford, J. 2007. *The nature and timing of peak oil and its possible impacts on South Africa*. Cape Town: Association for the Study of Peak Oil – South Africa.

Wicking-Baird, M., De Villiers, M. & Dutkiewicz, R. 1997. *Cape Town brown haze study*. Cape Town: Energy Research Institute, University of Cape Town.

Wilkinson, P. 2008. Reframing urban passenger transport provision as a strategic priority for developmental local government. In Van Donk, M., Swilling, M., Pieterse, E. & Parnell, S. (Eds.). *Consolidating Developmental Local Government: Lessons from the South African Experience*. Cape Town: University of Cape Town Press.

Natural Space and City Growth

Matthew Cullinan

"Rondevlei is valuable to me because it's where you can forget about your problems, sitting there watching the birds..."

Bonga Mboyiya (City of Cape Town b, 2008)

1. Introduction

What would Cape Town be without Table Mountain?

Table Mountain means different things, and plays various roles, depending on who you are and what is important to you. It forms the visual backdrop of the city. A botanist might see it as the last refuge of an endangered plant (the Cape Floral Kingdom is the smallest and most diverse of all the world's floral kingdoms, meaning that rare and endangered species can be found across the city). A youth leader in Mitchell's Plain may see it as a place to take groups on weekend outings. For the tourism marketing agency it is a fantastic icon around which various economic opportunities are promoted. A newly arrived rural migrant living in Khayelitsha might see it as symbolic of what she has come to Cape Town for – work and a better life (and thus also the hardship, challenges and possible ennui). In any place, natural and green open spaces play all these roles and more – they are at once symbolic, social and recreational, economic and a contributor to the services we get from ecosystems, such as storm water detention, CO_2 absorption, and food production via urban agriculture.

When imagining a sustainable Cape Town, it is not only important to think about economic recovery, but also the transition to a more sustainable socio-ecological regime of accumulation and well-being. This chapter begins by reflecting on how Natural and Green Open Space mirrors the spatial and income distribution of the city, and makes the case for investing more in natural space. We define the different elements of Natural Space – Public Open Space, Natural Green Space, Metropolitan Open Space – as well as the scope and roles played by each, and then examine the social, economic and ecological functions of the different types of spaces. The chapter considers an approach to the sustainable management of Cape Town's natural space and raises questions of equity and value (and how it is valued), as much as questions of its quantity and accessibility. Suggestions are provided for a more strategic approach towards natural space in Cape Town at the metropolitan, municipal, neighbourhood and site levels.

2. Natural Space and Sustainable Urban Development

However it is seen and used, natural and green open space forms a critical element of Cape Town's character and the protection of unique habitats. How the Natural and Green Open Space is used and looked after also reflects the spatial patterns and income distribution across the city. Cape Town was (and remains) a spatially divided city. The high quality of open spaces and natural space is often (although not exclusively) associated with the former white (and currently wealthy) parts of the city. Every year the City spends millions on maintaining and looking after these spaces, while in poorer parts of the city they often remain unkempt and unused. For this reason, environmental protection is often viewed as the preserve of the rich. This is sometimes objectively true. But at the same time this is an unfortunate perception. What does Cape Town lose by this perception and the complementary actions or non-investment it may choose as a result? This chapter illustrates how and why investment in natural and green open space is a fundamental part of maintaining the health of the city as well as species diversity.

There is no doubt that ecosystems, and the natural spaces and systems within which they exist, are under threat worldwide – not just in cities, but wherever human activity, such as industrial agriculture, takes place. As these natural spaces and systems are eroded, so too is the ability of our planet to renew and sustain itself.

In exploring the role of natural space and city growth it is useful to look at the broader context of sustainable urban development. The Natural and Green Open Space is one element of a sustainable city. As with any of the other critical elements, its mismanagement has widespread implications.

In Chapter 2, it was demonstrated that in 2006-7 Cape Town spent R9.3 billion of its total annual budget of R17 billion on energy, water, sanitation, and waste services alone. This is 10% of the Gross Geographic Product (GGP) of the metropolitan economy, even though it excludes the full environmental costs arising from the 'free' ecosystem services these systems depend on (such as CO_2 sequestration, catchment areas regulating water supplies, traditional medicines, polination, stormwater control, landfills to absorb waste, air quality, and negative effects on health). These excluded costs provided by the ecosystem are often located within the Natural and Green Open Spaces of the city. The health and effective functioning of the city's natural spaces and green systems thus forms a critical component of a sustainable city. From a sustainability perspective, the 'health' of a city can be determined as much by the pollution of its rivers, beaches and harbours as by anything else.

Natural Open Space is a resource. Some of it needs protection, but all of it needs good management. In Cape Town we have more than enough open space, but it is often under-utilised and inaccessible. Paradoxically, we also do not have enough of the right natural space to ensure effective operation of key ecosystems (specifically lowland fynbos). How much open space we have – and how accessible it is – depends on how you look at it, and what you want to use it for. The discussion around Natural and Green Open Space – how much of it is required and its level of accessibility – hinges on key concepts which require definition.

3.   Definitions and concepts

3.1   Public Open Space (POS)

Public Open Space (POS) is publically owned land that is open to all citizens and may be used by them in accordance with whatever rules or restrictions are applicable. It includes parks and playgrounds, as well as larger natural systems – such as river corridors and forestry areas. It thus exists to –

– enable the functioning of the natural systems around (and within which) the city has developed, and
– accommodate community needs for various forms of physical recreation and play, and psychological development and health (play and games are recognised as a key form of learning among children, and accessible open space provides scope for creative play).

Its size and distribution is shaped both by human need – as in the case of parks – and the demands of urban settlement planning and logic, such as flood-lines and high water marks. Cynics typically remark that POS is the space left over after planning. While not strictly true, this perception is based on the fact that many pieces of our POS are derelict, neglected pieces of weed-infested land that are sometimes a source of threat to the communities surrounding them. (On maps, these pieces of land are coloured green. Hence POS is also sometimes referred to as Green Open Space. Although in reality this is not always true, there is a strong and positive connotation to the word 'green' from a sustainability perspective. Therefore, in this chapter, the term Green Open Space refers to the potential role of POS in sustainable urban management.)

Yet another term defined as Green Open Space is used in *The Red Book* (Council for Scientific and Industrial Research (CSIR), 2000). This is soft open space, which is predominately vegetated, or has a porous surface, rather than hard open space, which is paved. POS is then a broad zoning category within which these different roles may be captured. It is thus also applicable to the open spaces within urban settlements (i.e. within the urban edge).

Provision of POS is a requirement when any new settlement is planned. Set standards and criteria are used to determine how many parks, playing fields and such should be supplied per new household. In this regard, *The Red Book* (*ibid*.) has clearly defined standards that are as good as any when it comes to guidelines. However, many cities and provinces have applied their own standards to reflect the peculiarities of their context. As we shall discuss later in the chapter, the problem of access is not an issue of proximity, lack of guidelines, or even lack of appreciation of its importance and role.

3.2 Natural Space

The concept of Natural Space is not one that is commonly used in South Africa. However, it is becoming an increasingly important concept in urbanised countries, where there is a greater scarcity of natural spaces.

For the purposes of this discussion, *Natural Space is defined as all places within the city that are managed and run in order to maintain and preserve their natural state or the natural functioning of ecological systems.*

Natural Space also has a number other roles:

- It has a critical absorptive function, assisting in the natural processing and recycling of liquid and solid wastes, filtering atmospheric pollutants, and trapping CO_2.
- It plays an important productive role when it comes to urban agriculture. For example, the nutrients we put into the system as waste could represent an important input into urban agriculture.
- It mitigates natural or extreme environmental hazards, such as wetlands regulating run-off from heavy storms.
- It enables ecological processes to continue to occur sustainably and safely within environments significantly altered by human action.
- It has a psycho-social or spiritual dimension.

Examples of large Natural Spaces include nature reserves (such as Table Mountain National Park and Cape Peninsula), protected critical habitats and wetlands (e.g. the Edith Stephens Nature Reserve), river corridors (e.g. Kuils River), and dune systems (e.g. False Bay Coastal Park and Macassar Dunes). These areas may be part of the official, zoned POS. Their role, however, is more about enabling ecological processes, and they would incorporate sensitive environments, like wetlands, rivers, coastlines, and remnant patches of indigenous flora, which are necessary to maintain the diversity of indigenous flora and fauna habitats. Natural space thus often extends beyond POS to capture the full extent of a dynamic natural system.

Their size is therefore not only a function of direct human need, but also of the natural processes they are expected to support. Sometimes their protection is due simply to the fact that they represent something unique.

Natural Space plays a critical spiritual role. It enables us to make the connection between ourselves and the natural system upon which we depend. Being in, engaging with, and observing the complexity of a natural system helps build the understanding that humans are but one piece in a complex interconnected ecological chain of existence:

"I have never been to the wild before and seen nature so close and beautiful. I believe animals also need a place to be safe and feel comfortable about their nature."
Khanya Moni (quoted in Environworks, City of Cape Town b, 2008)

This psycho-social or spiritual dimension is becoming increasingly important as we seek to build awareness among urban dwellers of the fragility of this chain.

Ideally, Natural Spaces should be created and managed to ensure sustainability in terms of:

- *Location* – incorporating wetlands, rivers, coastlines and key areas of indigenous flora and fauna;
- *Size* – large enough to maintain the natural populations viably and with sufficient/healthy genetic variety and/or large enough to absorb and manage the waste the system is expected to absorb;

- *Dimension* – shaped so as to provide sufficient protection to 'core' areas (round and fat is generally better that long and thin); and
- *Landscape connectivity* – relationships to other pieces of natural space are important factors in ensuring the effective and sustainable operation of the system, genetic interchange, and migration between one part and another.

The concepts linked to these features have been explored in theories relating to 'Island Biogeography' and has led to various debates about appropriate approaches to conservation in cities (conservation biology).[1] This is also an underlying principle in the City's own *Biodiversity Strategy* (Primary Biodiversity Conservation) (City of Cape Town, 2003).

The ICLEI – Local Governments for Sustainability (ICLEI) "Local Action for Biodiversity" (LAB) initiative recognises, that, while cities cover just 2% of the earth, they consume 75% of its natural resources. Managing biodiversity in Cape Town (and especially critical habitats, such as lowland fynbos) is a key challenge. Urbanisation, invasive vegetation, climate change, poor communication, and lack of capacity are some of the challenges faced by the City in achieving this (City of Cape Town b, 2008).

3.3 Natural Green Open Space

The term Natural Green Open Space is used here to capture the combination of both Natural Space and POS. The scaled application of this generally means that at the city and district level the emphasis is on systems – and thus natural space system – while at the neighbourhood and site level the focus is primarily on public open space.

3.4 Accessibility

For the purposes of this chapter a broad definition of accessibility is used. In general terms, accessibility refers to the ease with which a place or activity is physically reached. In an urban context physical accessibility is a function of:

- *Distance*
- *Location* (top of a mountain vs. near a major road)
- *Travel mode by which it is/can be reached* (on foot, cycle, public transport, car, etc.)
- *The quality of the infrastructure* (quality of roads, public transport frequency)

But mobility (and thus accessibility) is also a function of income. Higher income households generally have a greater range and choice when it comes to mobility, while lower income households will tend to have greater reliance on walking and public transport. Walking distance to a facility and/or public transport stop thus becomes a critical factor in determining accessibility.

However, there are in fact many other factors that determine whether an urban space (such as natural space or public open space) is accessible, and it is these factors which pose the greatest challenge. They are effectively management challenges, and include:

- *Cost of access* – which may include the cost of travel or the fees applicable to gain entry;
- *Opportunity cost of access* – linked to the cost of access, e.g. going to Kirstenbosch vs. making the payments on the lounge suite;

---

1 Conservation biology, or conservation ecology, is the science of analysing and protecting the Earth's biological diversity. It is an interdisciplinary field, drawing on biological, physical and social sciences, economics, and the practice of natural-resource management. The rapid decline of biological systems around the world means that conservation biology is often referred to as a 'Discipline with a deadline'. Conservation ecology addresses population dynamics issues associated with the small population sizes of rare species (e.g. minimum viable populations). The term 'conservation biology' is the scientific study of the phenomena that affect the maintenance, loss, and restoration of biological diversity, and the application of science to the conservation of genes, populations, species, and ecosystems. (http://en.wikipedia.org/wiki/Conservation_biology#Threats_to_biological_diversity)

- *Awareness, knowledge and perception* – if people are not aware of a facility or opportunity, or have no knowledge of its benefits, or perceive themselves to be excluded, then it effectively becomes inaccessible;
- *Security and safety* – perhaps one of the biggest challenges to natural and green open space access in Cape Town. Table Mountain and the local play park are both potentially easily accessible from a physical perspective, but remain unusable because of the threat of mugging, rape, or other physical violence. Living next to a park that is the domain of a local gang means that it is inaccessible to the parent wishing to take the children out to play;
- *Quality/Maintenance and Management* – the quality of a space constitutes a critical component of its use. If the local park is overgrown or barren, the swings and play equipment broken, and the area littered and used as a dump, then it is also effectively inaccessible. This is a function of management and maintenance. Next to safety and security, this is the second greatest problem facing natural and green open space accessibility in Cape Town. (Vandalism in poorer communities is often blamed for the state of disrepair, but this needs to be explored in more detail.)

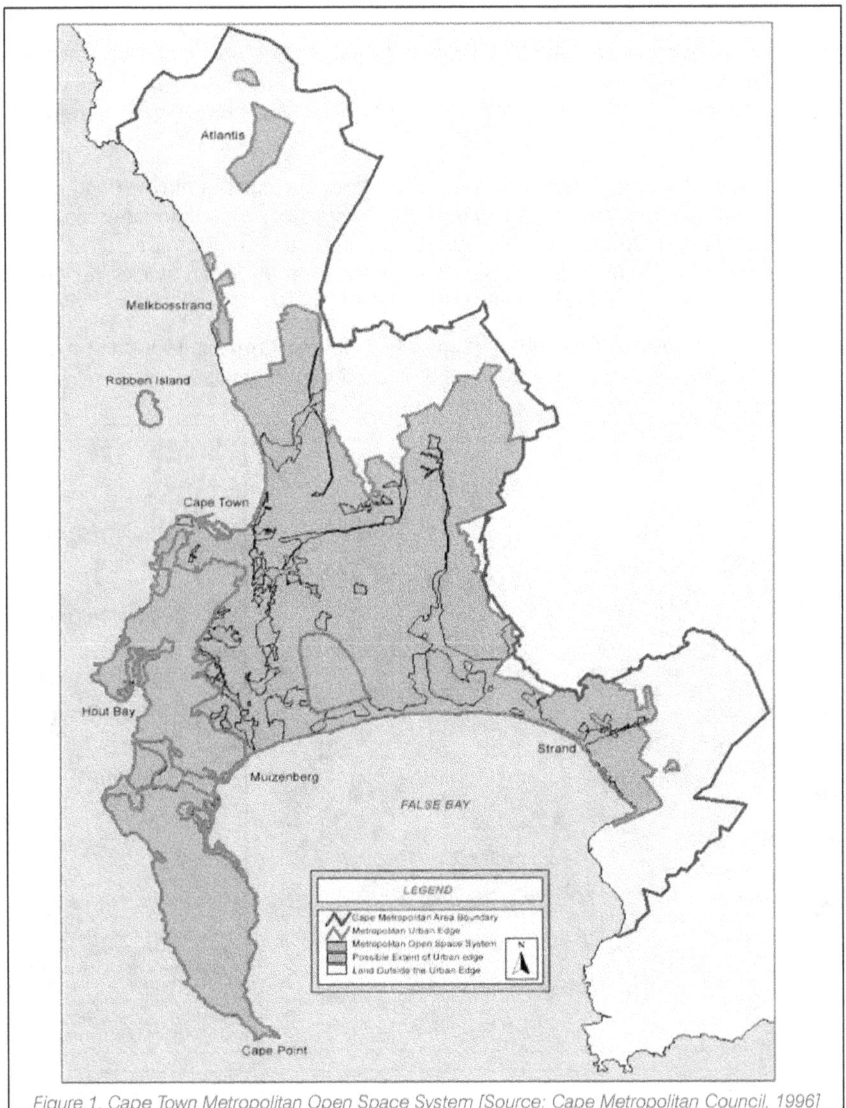

Figure 1. Cape Town Metropolitan Open Space System [Source: Cape Metropolitan Council, 1996]

Ideally every Cape Town resident should have equitable access to a range of open space facilities, resources, and opportunities. The lack of safety and poor management or maintenance of POS in poorer communities in particular, means that these are not generally accessible.

3.5 Metropolitan Open Space System (MOSS)

Drawing together both the Natural Space and the Green Open Space across the metropolitan area is the Metropolitan Open Space System (MOSS). MOSS is defined in the *Metropolitan Spatial Development Framework* (MSDF, 1996), as follows:

"A Metropolitan Open Space System (MOSS) is an inter-connected and managed network of open space, which supports interactions between social, economic and ecological activities, sustaining and enhancing both ecological processes and human settlements. MOSS comprises public and private spaces, human-made or delineated spaces, undeveloped spaces, disturbed 'natural' spaces, and undisturbed or pristine natural spaces."

The MOSS strategy aims to identify the MOSS elements and their associated management strategies. Thus it is an attempt to both draw together the physical delineation of the system and to address the management challenges posed by managing and maintaining this system. Figure 1 shows the MOSS system at a general level (Cape Metropolitan Council, 1996).

The MOSS generally focuses on large scale natural systems, such as riverine/wetland environments and mountain chains, but incorporates a range of aspects such as scenic landscapes, protected natural areas, sensitive environments, and formal and informal recreation areas. The MOSS offers an integrated system of elements linked into a network of green open spaces for conservation, recreation and farming throughout the metropolitan region.

Figure 2 is extracted from the 1996 MSDF (Cape Metropolitan Council, 1996) and it conceptually draws together the various elements that would be included in the MOSS.

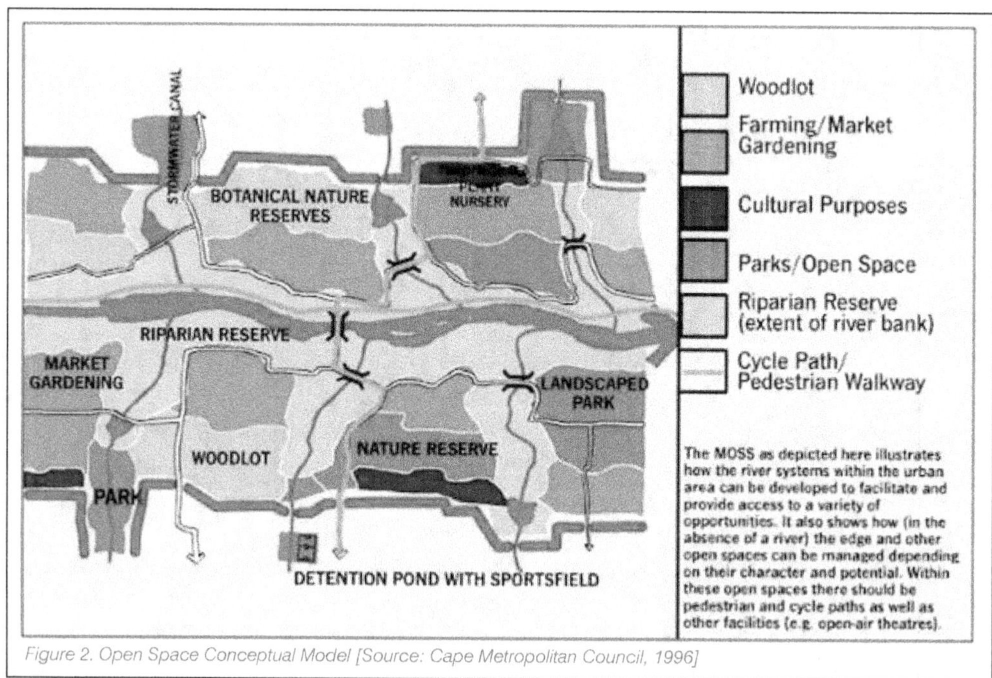

Figure 2. Open Space Conceptual Model [Source: Cape Metropolitan Council, 1996]

4.   Functions of natural and green open space

Natural and Green Open Space provides a range of functions within the city, all of which are necessary for the healthy and sustainable operation of the city.

4.1   Social functions

Open spaces provide for critical social functions – especially in the context of higher densities, and where there are large households on small properties. It provides space to socialise, relax, play, and escape the close confines of small homes and spaces.

Quotes from Grade 11 learners involved in a geographic information system (GIS) mapping exercise in Rondevlei Nature Reserve make the point:

"It is good to have nature in a city, and not just buildings and roads."

David Baak (City of Cape Town b, 2008)

"I just realised that the life of animals and other creatures are much more important in our daily lives, so we should not destroy our environment."

Bonga Mboyiya (ibid.)

However, the current state of most of our Public Open Spaces (community parks and sport fields) is poor. This is a reflection of lack of safety and poor management. The spaces are available, but not usable. This is the key challenge.

4.2   Economic functions

A recent study undertaken by the City's Environmental Management Department (and quoted in City of Cape Town b, 2008) illustrates that Open Space in the metropolitan area is highly valuable due to the ecological and social functions that it performs. The City assessed the value of ecosystem services by estimating the costs to the city of having to artificially provide the services that are delivered by the environment. The Open Spaces within the Cape Metropolitan Area provides hundreds of millions of Rands worth of services every year to the City, yet the maintenance budget for these 'service areas' is general extremely low, and they are often thought of as a luxury by administrations, instead of viewing them as essential for ensuring future services.

Moreover, well-managed Open Space can increase the value of adjacent properties, leading to a healthier rates base for local administrations. The property values were found to have a 10% discount rate or premium depending on the management of the adjacent Open Space. In neighbourhoods integrated into well-managed open space and conservation areas, the property values of adjacent houses were on average 40% higher than other comparable properties.

The greatest asset and competitive advantage of Cape Town lies in its unique environment. The city has all the qualities of a great tourist destination, and therefore tourism plays an important role in generating income for the city. Eco-tourism also adds economic value to recreational areas, with the Cape Metropolitan Area being a popular destination for both local and international tourists. Despite limited investment in green spaces, there are significant returns. With increased attention given to properly maintaining and managing the natural open spaces, land could be made even more attractive. Not only would neighbouring developments be significantly enhanced, but property values in general would increase as well.

4.3   Ecological functions

Ranked as one of only three cities in the world that are 'hotspots', Cape Town's biodiversity merits the highest priority. The City of Cape Town's biodiversity strategy defines biodiversity (biological diversity) as "the totality of the variety of living organisms, the genetic differences among them, and the communities and ecosystems in which they occur" (City of Cape Town, 2003).

In the context of Cape Town, the value of biodiversity can be measured by –

- the economic value of functioning ecosystems, e.g. clean water and clean air;
- its intrinsic value by virtue of its mere existence;
- its contribution to tourism;
- its consumptive use value, e.g. harvesting;
- its educational value;
- its social value (through recreation and open space);
- its aesthetic value (through beauty and scenic drives);
- its spiritual value;
- its bequest value (the value of retaining biodiversity for future generations); and
- its option value (the value of retaining biodiversity for future use).

In these terms, biodiversity is then the 'natural wealth' of the earth, which supplies all our food, and much of our shelter and raw materials. The mountain, coast, unique habitats and rivers then become key areas for ecological conservation. These areas have been defined in the city's biodiversity network, and are essential to the future survival of Cape Town's rich – but threatened – biodiversity. This network includes:

- *Core conservation areas* (where public access is strictly controlled),
- *Buffer areas* (where limited, short-stay public access is allowed), and
- *Transition areas* (where active public access and a wide range of activities is permitted).

The remnants of the unique biodiversity of an area consist of valuable plant communities, wetlands, and dunes, which are critical to the ecological sustainability of the open space system, as well as having economic potential. The open space system is both a means to conserve indigenous flora and fauna, and an important step in maintaining ecological balance within the city. How this balance is protected depends on factors such as quality and accessibility, the attention given to security, and the management of those spaces.

5.   Quality and Accessibility – how much is enough?

The MOSS is contained within the urban edge. The urban edge is designed to prevent sprawl and protect abutting natural resources, and generally demarcates the maximum limit of urban development for the next twenty years. It is calculated that some 14 000ha of land is available within the edge to accommodate the projected population growth of Cape Town. According to the *City of Cape Town Sustainability Report* (www.capetown.gov.za)of 2006, there is approximately 160 sq meters of green space (nature reserves, parks and public open space) per person in Cape Town. There is also 300km of coastline. This means that Cape Town has more green space than most other large cities in the world.

Once again, we see that the issue is not one of quantity or physical accessibility, it is about accessibility in terms of safety, security, and the proper management of the spaces that are available.

Research shows that even in developed countries such as the United Kingdom (UK), access to and use of natural open space was less about physical proximity and more about management. The example below is drawn from research conducted in the District of Rother in the UK:

"Of respondents who use a natural/semi-natural site as their primary open space, 41% walk less than five minutes to reach it, and a further 29% walk between five and ten minutes. The most significant problems were vandalism/graffiti and dog fouling. The consultation showed that the majority of respondents are content with the current quantity of natural/semi-natural sites in the District." (www.rother.gov.uk/)

Interestingly, there are 15ha of natural or semi-natural open space per 1000 people in Rother (or 150m² per person – equivalent to that available to residents of the Cape Metropolitan Area). The study recommended that for new large scale developments (only) 2ha of natural or semi-natural open space should be set aside per 1000 population. It was further recommended that this should be accessible within 1.2km (a 15-minute walk).

While it is possible to review local standards (such as those set out in *The Red Book*, CSIR, 2000) and apply it back to Cape Town, this would reveal little about the actual use, value and accessibility of the available natural open space. Indeed, there is more than enough open space – public and natural – for all residents of Cape Town. The central issues are management, maintenance, and control.

For example, we have already referred to the spatial distortions in the quality and accessibility of open space inherited from apartheid. Monwabisi Resort and the Wolfgat Nature Reserve are located in close proximity to Khayelitsha, and are accessible by foot and public transport. However, the area is unsafe and is hardly ever used – for example, Monwabisi is used on Boxing Day and New Year's Day, when it is massively overused and overcrowded, and people frequently drown. The challenge for Cape Town becomes one of management, and ensuring continuous, effective security and safety to create the possibility that people can enjoy it the whole year round.

The same arguments can be applied to the open space system in Khayelitsha and Mitchell's Plain. In both areas, green belts run through the settlements. However, these are unsafe, and in the case of Khayelitsha, neither managed nor maintained. The bottom line is that the City has an ample supply of Natural Green Open Space – but this is often poorly managed and/or maintained, and is often unsafe.

6.   Strategic Direction

In looking at the way forward for the use, management, conservation, and health of the Natural Green Open Space in the City of Cape Town, our approach is presented on various levels of action, namely:

- Metropolitan or City-wide level
- District level
- Neighbourhood level
- Erf/site development level

At each one of these levels different issues and systems come to the fore.

6.1   Metropolitan and City-wide

Contextual informants

Cape Town will need to accommodate an estimated population of 5 million people by 2025, therefore appropriate areas need to be identified for development. If there is an unanticipated population growth, Cape Town will need to be able to manage it in a way that does not destroy the city. With growing energy costs and an increasing shortage of water, steps must also be taken to localise food production and adopt water efficient strategies (primarily through water demand management).

Institutionally, there is still too much fragmentation in the management of Natural Space. As a result, problem areas fall between the cracks and are left unmanaged. A good example of this is the Open Space south of Khayelitsha. Nobody has taken responsibility for the management of this land, and it is being invaded by homeless people desperate for land.

The Civic Centre and Station Precinct

The Cape Town Civic Centre, City Hall, Station and Artscape precinct represent important public institutions and huge investment. These are however located in a very hostile and unwelcoming environment. The lack of amenities in the city centre for ordinary citizens, and those dependent on public transport, is a major concern. The opportunity exists to build a new public square at the most accessible place for ordinary people in the city, i.e. the Central Station. It could become the heart of the CBD and a hub of 24-hour activity and safety. The square could be used as an important lever to build confidence in the precinct and attract other appropriate investments and facilities, including an inner-city crèche, a medical facility, apartments for young people, and places of entertainment.

A city-wide strategy would need to address the following issues:

- The critical challenge is to ensure that the institutional mechanisms are in place to enable the implementation of what are already good existing strategies and plans (although some review may be necessary, as some of these are five years old).
- Areas with high biodiversity or agricultural value must be identified and their protection and management enhanced.
- Areas exposed to natural or man-made hazards must be identified and incorporated into the MOSS.
- The City's Biodiversity Strategy, which covers key areas for biodiversity management, must be reviewed and the implementation mechanisms strengthened. In reviewing the biodiversity strategy, attention must be paid to the institutional mechanisms and capacity challenges faced in managing Cape Town's Natural Space. This function will only increase in importance as energy and water constraints become more apparent.
- Opportunities for well managed, productive, and integrated urban agriculture (such as at the Philippi Agricultural Area) should be encouraged.
- Subsistence urban agriculture is important, but highly problematic, and resources should rather be directed to where they have most benefit in terms of poor livelihoods.

6.2 District level

Contextual informants

The City has embarked on a process of preparing District Level Spatial Development Frameworks (SDFs) and Environmental Management Frameworks (EMFs). From a Natural Space perspective, the City has also identified opportunities for new urban green spaces, including urban agricultural complexes, multi-purpose city parks, active recreational parks, and collective sport facilities. Considerable work has also been undertaken to consolidate coastal nature areas on the Cape Flats.

A district strategy would need to aim to achieve the following goals:

- Create a structured green and conservation web for the protection of biodiversity and to ensure access to green and urban conservation areas;
- Identify where public planting can reinforce this web, which includes rivers, canals, and stormwater and detention systems;
- Focus on the management and control of powerlines and major road servitudes, because these present opportunities to conserve and connect key lowland fynbos areas; and,
- Identify where the need exists for regional sporting and recreational facilities of a higher order. For the most part, these facilities exist, but in the underprivileged areas they suffer from neglect and poor maintenance. Therefore, programmes for the proper maintenance and operation of sports facilities, in poor areas in particular, should be put in place.

6.3 Neighbourhood level

It is at a neighbourhood level that the demand for Open Space to satisfy recreational, play, sporting, and psychological needs comes to the fore. Different age groups, genders, and cultural groups have different needs. Many of these can be met through single facilities, if properly designed and, very importantly, managed.

Over the past five years, the City of Cape Town's Dignified Places Programme has also been investing in creating open spaces in neglected parts of the city, but it needs to be scaled up in order to have a greater impact. One example – which could be considered and replicated as part of the Dignified Places Programme – is a memorial park and cemetery. There is a great shortage of dignified, accessible burial space in Cape Town. This project envisages creating a 'memorial park' somewhere on the Cape Flats (possibly on the edge of the Philippi Agricultural area). Functionally, the park would serve as a metropolitan cemetery and garden of remembrance, elevated well above the water table. It would create a major place-making element which would contribute significantly to an otherwise featureless landscape. The foundation of the park could be constructed with builders' rubble, covered with soil and sand. The project could be implemented as a public works initiative over a prolonged period, and could be linked to the general goals set out below.

A neighbourhood strategy should address the following issues:

- Community awareness of biodiversity issues should be raised and the safety and security of local neighbourhoods should be improved in order to make them safe places for children to grow, play, and learn. To achieve these goals –
  - joint initiatives with the Department of Education, and Safety Forums should be considered (a 'Parks for the People' type initiative);
  - community involvement in local greening, with an emphasis on fynbos (and thus a water-wise approach) should be encouraged (the BOSSIES project, supported by the City, is an example of this; NGOs, such as Abalimi Bezekhaya, are also key role-players that the City is already working with on other programmes, e.g. the Aachen Partnership); and
  - the general education and awareness programmes of the City (which are excellent) should be extended, but stronger links with community safety and education are encouraged.
- New neighbourhood development should be designed in such a way as to ensure that Open Space, and especially local parks, have good visual surveillance (i.e. it should create defensible spaces), and are safe from major traffic hazards (e.g. fast roads).
- Existing drainage patterns must be identified, along with wetlands, streams, and aquifers.
- Open Spaces should form a network to accommodate natural water flows.

6.4 Site development level

It is at the level of individual erven that the greatest opportunity to introduce a whole host of initiatives that ultimately impact positively on Natural Space, biodiversity, and the Open Spaces of the city, presents itself. The principle here is that all new development – and any re-development – should be low impact. This implies a host of measures and strategies that should be considered when approving buildings or site development plans.

The Cape Flats, for example, is devoid of quality recreation areas. A significant opportunity exists to provide a major urban park in the vicinity of Manenberg and Hanover Park. The area is currently used for sand mining, but a park could be created as part of the ongoing restoration of areas already mined. Apart from active and passive recreational opportunities, the park can also provide for productive uses (including urban agriculture and nurseries) and cultural activities, as well as attracting high yielding uses to its edges (because of the potentially high amenity value of the land). The site-level goals would need to include the following objectives:

- Promoting and encouraging development which makes minimal environmental impact through careful choice of materials, surfacing and siting to ensure that run-off is reduced and maximum water absorption takes place on or near-site. This includes initiatives such as porous paving and creative design of open space to enable local absorption. Techniques can be used for a particular site, with the aim of mimicking the watershed's natural hydrologic functions or the water balance between runoff, infiltration, storage, groundwater recharge, and evapotranspiration. With this approach, receiving waters may experience fewer negative impacts in the volume, frequency, and quality of runoff, so as to maintain base flows and more closely approximate pre-development runoff conditions. If done well, sensitive development which aligns with and even mimics the functioning of the natural systems can –

- help maintain drinking water supplies;
  - reduce the maintenance costs of storm water facilities;
  - improve the functioning of natural water systems by desiging drainage channels and stormwater retention areas as part of the natural water systems;
  - lower the maintence costs of streets, kerbs, gutters and other infrastructure;
  - increase property and community appearance and aesthetics;
  - increase property resale values through the appeal of natural landscaping;
  - reduce the possibility of contamination of sediments in bays (and ensure blue flag status is retained!); and,
  - increase opportunities for public/private partnerships and public education.
- Encouraging the use of low-flow water technology in order to reduce water wastage and improve water quality.
- Promoting renewable energy usage and reducing energy demand (e.g. through the installation of solar geysers and insulation of hot-water pipes).
- Encouraging water-wise and indigenous landscaping.
- Situating buildings in such a way as to preserve sensitive areas.
- Establishing and enforcing appropriate 'green building codes' in terms of features such as passive heating and cooling.

7. Conclusion

There is sufficient Natural and Green Open Space for all residents of Cape Town. The issue is not the quantity (and thus physical accessibility) of Open Space. Neither is the problem a lack of understanding or appreciation of the role and importance of Natural Green Open Space among the City Departments. The key challenge is bringing together the various role-players to ensure that plans and projects are properly implemented, and that these places are properly managed and maintained. Natural Space in the City of Cape Town is about establishing effective institutional structures (management) and effective maintenance (budgets and human capacity) regimes, coupled with greater community awareness and ownership, and control of their own public spaces. It is often unsafe for people to use these assets and resources. Accessibility is thus fundamentally an urban management issue – and not an urban development issue.

In managing and developing Natural Green Open Space, two clear imperatives emerge:

- The ecological systems – and assets they represent – must be acknowledged and catered for, and
- The human development role it plays must also be addressed, especially at a neighbourhood level.

References

Cape Metropolitan Council. 1996. *MSDF: The Metropolitan Spatial Development Framework*. Cape Town: Cape Metropolitan Council.

City of Cape Town. 2003. *Biodiversity Strategy*. Cape Town: City of Cape Town.

City of Cape Town a. 2007. *Envirowork*. Volume 2/07, August 2007. Cape Town: City of Cape Town.

City of Cape Town b. *Enviroworks*. Volume 1/08, May 2008. Cape Town: City of Cape Town.

Council for Scientific and Industrial Research. 2000. *The Red Book: Guidelines for Human Settlement Planning and Design*. Pretoria: CSIR.

Prince George's County. 1999. *Low-Impact Development Design Strategies – An Integrated Design Approach*. Maryland: Maryland Department of Environmental Resources and Planning Division.

Snohomish County Tomorrow. 1992. *Residential Development Handbook for Snohomish County Communities – Techniques to increase liveability, affordability and community viability*. Washington DC: MAKERS Architecture and Urban Design.

Wackernagel, M. & Rees, W. 1962. *Our Ecological Footprint*. New Society Publishers, Gabriola Island, Canada.

ww.capegateway.gov.za

www.capetown.gov.za

www.environment.gov.za

www.info.gov.za

www.rother.gov.uk

www.sacities.co.za

12

Urban Agriculture in the City of Cape Town

Gareth Haysom

Prospects for a food secure world in 2020 look bleak if the global community continues with 'business as usual'.

International Food Policy Research Institute, quoted in Scherr, 1999: 32

1. Introduction

This chapter contextualises food security issues in the larger global setting, outlining the flaws relating to global food security, the impact of unfair trade policies, and the negative effects of agribusiness on communities in developing countries. Not only is the world food situation rapidly being redefined by new driving forces, but changes in food availability, rising commodity prices, and new producer-consumer linkages all wreak havoc on the lives of poor and food-insecure people (Von Braun, 2007). We explore the implications of these changes and the scope of the challenges for Cape Town. Worldwide, as long distance transportation becomes increasingly expensive, cities will have to source local supplies. In essence, these cumulative and interacting driving forces require that food production and agricultural approaches will increasingly need to support sustainability.

Particular focus is given to the relationship between food and other elements of a sustainable Cape Town. What will it take to give priority to nutritional security for vulnerable households? What are the key challenges for food security within the urban context, and what do City decision makers need to do in order to respond effectively?

The chapter defines urban agriculture, describes its broad range of impacts, and examines how it fits into the larger landscape of sustainable development. We advocate for the formal recognition of urban agriculture as an industry, and a social development process that supports Cape Town residents in their ability to access healthy, nutritious local food, as well as the broader sustainability objectives of the City.

We then examine how this can be done in a manner that integrates ecological services and enhances social justice, while still supporting the efficient functioning of the City. During a time of great insecurity, when approaches must consider a far deeper and longer term view, we outline strategies most likely to actively build capital, rather than degrade it. The chapter suggests specific pathways and possible solutions. It also reviews the opportunities in Cape Town for increasing supply from local urban and peri-urban sources, and makes a case for the complementary strategic interventions required to support this in the long run, calling for integration of agriculture in urban environmental and health policies. While these pathways may not always provide direct solutions, the chapter takes a different approach – one that requires far deeper evaluation and analysis of the state of agriculture in the City, as well as offering unconsidered opportunities for a range of stakeholders. Finally, we propose a way forward that builds on innovative approaches.

2. Global context: the call of sustainable agriculture

A recent review of the current industrialised approach to food production by the United Nations IAASTD[1] (International Assessment of Agriculotural Knowledge, Science and Technology for Development) highlighted significant flaws in global policies relating to food security. The IAASTD report argues for fundamental changes in the world's agricultural systems (cited in McIntyre, *et al.*, 2009). It documents the inequitable distribution of costs and benefits of the present global agricultural systems, and focuses attention on how the undue influence of agribusiness and unfair trade policies negatively affect communities in the developing world. According to the Washington-based International

---

1   The *IAASTD Synthesis Report* captures the complexity and diversity of agriculture and AKST (agricultural knowledge, science and technology) across world regions. It is built upon the global and five sub-global reports that provide evidence for the integrated analysis of the main concerns necessary to achieve development and sustainability goals. It is organised in two parts that address the primary animating question: how can AKST be used to reduce hunger and poverty, improve rural livelihoods, and facilitate equitable environmentally, socially, and economically sustainable development?

Food Policy Research Institute (IFPRI), low growth rates in food production "will be insufficient to meet the expected increase in demand. ... IFPRI research suggests that prospects for a food secure world in 2020 look bleak if the global community continues with 'business as usual'." (quoted in Scherr, 1999: 32).

In addition, the IFPRI report states: "In 2006, global cereal stocks – especially wheat – were at their lowest levels since the early 1980's. Stocks in China, which constitute about 40 per cent of total stocks, declined significantly from 2000 to 2004, and have not recovered in recent years. Year end cereal stocks in 2007 are expected to remain at 2006 levels" (Von Braun, 2007: 2). The IAASTD report concluded that farming methods will have to change drastically. This expert group from around the world concluded that declining agricultural yields are the result of the practice that farmers work against, rather than with, nature. The IAASTD has made a resounding call for a global conversion to sustainable agriculture. The IAASTD report proposed that smallholder agro-ecological farming will be more effective at meeting today's food production challenges than the old energy and chemical-intensive paradigm of industrial agriculture if societal inequalities are to be reversed (Watson *et al.*, 2008).

While many factors contribute to the advancement or decline of sustainability, four key focus areas can be identified. These include:

- Shelter – where we live and work;
- Energy – the fuel that 'powers' our way of life;
- Transport – how we use transport and how the goods we consume are moved around; and
- Agriculture – what we eat and how the food we eat is produced.

A sustainable city is one that recognises the interrelationship between these networks, and the decisions and processes that connect them. These factors, as well as many local and site specific needs, become critical in the formulation of strategies and approaches pertinent to the functioning of the urban environment. The nexus between development, ecological services preservation, social justice, and the economy comprises interconnected sets of relationships and socio-economic dynamics, and impacts directly on all strategies and policies of the city.

It is through this web of interconnected relationships that we now consider the specific food security challenges of Cape Town.

3.   Cape Town's Food Security Challenge

In South Africa, 35% of the land surface receives sufficient rain for dry land crop production, but only 13% (or 14 million hectares) is suitable arable land. Most of the available 14 million hectares is marginal land, with only 3% of the land considered high potential land. If we use the international norm of 0.4 hectares of arable land to feed a person, then South Africa's 14 million hectares would feed at most 35 million people. If we only followed the international norm, the net result would be over-exploitation, as we exceed the carrying capacity of our soils (DEAT, 2007: 41).

Estimating Cape Town's food needs, based on this norm, means that Cape Town's population of 3,240,000 would require 1,3 million hectares to sustain its population, or 9.2% of arable South African land.[2] Considering the anticipated growth of Cape Town's population, this figure highlights the potential scale of the challenges. Taken together, they speak to the claim for an alternative approach to agriculture and food security. Food supply within the City needs to be robust in the face of increasing diversity, deepening complexity, drastic shocks, and rapid change. A sustainable approach needs to be locally and community adaptive, resource building in terms of all the capitals – human, social, economic, and biological – and should promote social and ecological justice (Frayne *et al.*, 2009).

---

2   2006 population figures (PGWC, 2006). This figure for arable land considers only food production and not food production for export, fibre, and non-food items or animal feed.

4. Why urban agriculture?

Smit and Nasr (1999) argue that cities require processes to close the open loop system and that the throughput of resources in cities needs to be reduced. Typically consumables are imported into the urban areas, and what remains, including the packaging, is dumped as waste into the bioregion and biosphere. This chapter makes the case for a set of strategies that support the continued and aggressive reduction of external inputs and wastes over time.

Food is listed as being one of the main contributors to the Ecological Footprint[3] of the City of Cape Town (Gasson, 2002). As Gasson (2002) demonstrates, the total ecological footprint for Cape Town is 128,264 square kilometres – and that **41% of this footprint (52,780[4] square kilometres) can be attributed to food**,[5] which is significant if compared to the 44% contribution of energy to the footprint. The most significant consequence of this input-output model is that it demonstrates how resource-intensive the Cape Town urban system really is. Every oil price rise corresponds to net increases in the amounts of cash transferred from the Cape Town economy to National and global financial circuits.

Urban agriculture (UA) is core to the development of a sustainable city – an inclusive, food secure, productive, and environmentally healthy city (Van Veenhuizen, quoted in RUAF, 2006). Because urban agriculture links cities and their environments, it is an increasingly acceptable, affordable, and effective tool for sustainable urbanisation (Girardet *et al.*, 1999). Urban agriculture is thus seen as a key component of urban planning processes. If correctly implemented, it has the potential to address a variety of the urban planning and developmental challenges because of its integrative nature.

UA is defined as an industry located either within intra-urban or on the peri-urban fringes of a city. It grows, raises, processes, and distributes a diversity of food and non-food products and (re)uses both human and material resources, products, and services found in and around that urban area. Urban agriculture initiatives also supply human and material resources, and products and services largely to that urban area (Mougeot, 2005). In addition, it is characterised by cross-cutting interventions, in which a variety of services and stakeholders contribute in some way to productive agriculture. This could include the provision of labour by urban dwellers and the recycling of wastes.

The real potential of urban agriculture lies in satisfying one of the most basic needs – i.e. food – through improved production and distribution systems, income, employment, and environmental protection. UA also plays a constructive role in the wider context, including a contribution to savings in transport costs and even saving in foreign currency costs for developing countries (Egziabher, 1994). Our challenge is to view urban agriculture as a strategic approach to addressing a number of the city's challenges. The potential range of benefits of urban agriculture extends to and includes:

- food security;
- food sovereignty and economic challenges;
- employment creation;[6]
- ecological restoration;
- urban greening;

---

3   Wackernagel *et al.* (2006) assert that ecological footprinting estimates, for a given population, the area of biologically productive land and sea required to produce the resources that population consumes and to assimilate the waste it generates. Ecological footprinting employs a standardised measurement unit to make results for different areas comparable, that is, global hectares.

4   This figure has been revised from the figure used in Gasson, B. (2002). This revision was communicated to the author by Dr Yvonne Hansen, a senior researcher on the Gasson (2002) work. The revision, completed in 2009, was informed by a review of data used to assess the food consumption of the City in the 2002 report.

5   This figure excludes the 2,5 billion tons/year of seawater used to cool the Koeberg Nuclear Power Station (input), and the return to the sea of heated seawater (output).

6   A recent review of the Philippi Horticultural Area found that a large proportion of the workers were recent unskilled female rural migrants and work in UA was one of the only employment opportunities available to them

- water recharge and cleaning; and
- social cohesion and general urban renewal .

An appropriate strategy would not limit urban agriculture to the poor areas within the city, but rather, would need to be diverse enough to consider the needs of all city dwellers throughout the city. A sound approach would include those living in various urban settlement modalities from differing economic backgrounds. Urban agriculture traverses economic groups, and if approached in a holistic and comprehensive manner, would serve all residents of Cape Town (anecdotal evidence indicates that middle income areas are experiencing significant growth in home-based food production). This perspective is currently lacking in the approach to urban agriculture within the CCT and will be addressed later in the chapter.

For Smit and Nasr (1999), urban agriculture is a large and growing industry that uses and adapts urban waste as inputs to close ecological loops. It also uses idle land more effectively. The potential positive impact of this neglected industry as described by Smit and Nasr are wide-ranging including:

- improved nutrition and health;
- an improved environment for living;
- increased entrepreneurship; and
- improved equity.

Cape Town has witnessed some expansion in urban agriculture, though this has been largely a response to past and current food crises, rather than the proactive development of a formal industry. Some benefits have accrued to those proactively adopting urban agriculture as a livelihood strategy within the City. Some examples include groups such as Abalimi Bezekhaya, Ikamva Labantu, and the urban farmers of Philippi, Gugulethu, and Khayelitsha. Other UA activities and projects include the SEED food gardens in schools, independent community groups establishing their own gardens, or the livestock farmers on the side of the N2, all of whom play a vital role in supporting specific communities. While these groups are all delivering on the needs of the communities they represent, their integration in the broader urban agriculture activities within the City remains limited and disconnected.

In this chapter, we are advocating for the formal recognition of urban agriculture as an industry and a social development process that supports the sustainability objectives of the City.

In practice, urban agriculture farming typologies range from small-scale backyard gardens to larger-scale 'urban farms', and from fish farming to horticulture. To implement urban farming would require a management team that is able to respond to all these options and support these interventions in a proactive manner. The approach in this chapter calls for a far broader view of the benefits and advantages to the City in respect of urban agriculture. The integrated nature of urban agriculture invites policy makers and practitioners to take a holistic view regarding its potential benefit to the City. Optimising the potential of such interventions will require that it earns greater prominence in future planning of the functioning of the City. In the next section, we discuss these issues from an urban planning and policy perspective.

5.   Planning for sustainable urban agriculture

What role can cities play in creating an environment that provides a response to the structural and policy challenge of 'business as usual'? What new perspectives are required to successfully incorporate these issues into the planning process? Cape Town's UA challenges are best articulated by Swilling:

"Between 40 and 60 per cent of the domestic waste stream is organic waste … this is a rich source of nutrients that could be composted and ploughed back into urban agriculture. Instead, it is combined with all other wastes and dumped into toxic landfills. In the meantime, 1.3 million tonnes of food are imported … middle- and high-income households may be able to afford prices that include the cost of transporting all this food (fuel, cold storage, packaging, energy, etc.), but this is certainly not the case for poor households … [urban agriculture] reduces prices for the

consumer and increases the returns for farmers. It also stimulates the growth of local small-scale growers who tend to be much less dependent on oil and are more efficient users of water."

(Swilling, 2006: 37)

Urban planners commonly used to consider urban gardening and livestock keeping as merely 'hangovers' of rural habits, a marginal activity of little economic importance, or as a health risk and a source of pollution. Such biases, sustained by the limited exposure of policy makers and planners to grounded information on urban agriculture, have resulted in important legal restrictions on urban agriculture. Nevertheless, urban agriculture has continued to grow in most cities in the South (De Zeeuw, 2003). However, few authorities recognise urban farming as an urban form of land use, despite its prevalence (Gabel, 2005). This attitude is reflected in the City of Cape Town (CCT), where urban agriculture is viewed primarily as an economic activity, and is not a critical function within the urban form. This challenge is not unique to Cape Town and can be noticed in the planning approaches in most cities. It is for this reason that Halweil and Nierenberg (2007) argue that planners interested in making room for farming in cities must look beyond farmers' markets and community gardens to much broader issues in overall city design.

The need to elevate the status of UA in the CCT has been articulated in the City's *Urban Agriculture Policy*:

"In order to improve and make urban agriculture more sustainable, it is necessary to give it a formal status. This will be done through the inclusion of urban agriculture as a multifunctional component in municipal land planning and standard development processes concerning land use and environmental protection, i.e. land use plans, zoning schemes and site development plans should provide for urban agricultural activities."

(CCTa, 2007)

The questions are: To what extent has policy been implemented and embedded within the City's functions? And to what extent does the Urban Agriculture Unit have the necessary authority to play the role defined in the policy? We explore these questions further in the following section.

6. The City of Cape Town's Urban Agriculture Policy

The Urban Agriculture Policy for the City of Cape Town (2007), was approved by Council in December 2006. The stated purpose of the policy is to develop "an integrated and holistic approach for the effective and meaningful development of urban agriculture in the City of Cape Town … to create an enabling environment wherein public, private and civil society agents can work collectively to create more real and sustainable opportunities for local area economic development" (CCT, 2007: 5). The City's vision for a prosperous and growing urban agricultural sector is framed by the following strategic goals:

- To enable the poorest of the poor to utilise urban agriculture as an element of their survival strategy (household food security);
- To enable people to create commercially sustainable economic opportunities through urban agriculture (jobs and income);
- To enable previously disadvantaged people to participate in the land redistribution for agricultural development programme (redress imbalances);
- To facilitate human resources development (technical, business and social skills training).

The City follows a dual approach to urban agriculture, on the one hand focusing on achieving household food security (poverty alleviation and improved nutrition), and on the other the creation of income (economic development) (*ibid.*).

The policy defines urban agriculture as being –

"... (t)he production, processing, marketing and distribution of crops and animals and products from these in an urban environment using resources available in that urban area for the benefit largely of residents from that area."

(CCT, 2007: 3)

This definition differs from the definition suggested by Mougeot earlier. It ignores non-food products and does not sufficiently emphasise the use and reuse of human and material resources, products, and services found in and around that urban area. Neither does it suitably consider the supply of human and material resources, products, and services. These omissions form the basis for the critique of the CCT's UA policy. To address its many challenges, we need interventions that go beyond urban greening, second economy economic interventions, and food security. It is essential that UA becomes a core thrust of the planning and development of the City.

UA needs to be multi-sectoral, diverse, innovative, and relevant, and at the same time, span economic sectors within the City. The policy's primary focus is to address household food security and support economic activity. It lists a number of strategic imperatives, including:

- UA in land use management and physical planning;
- creating linkages with other strategies;
- the establishment of urban consultative forums;
- building strategic partnerships; and, most importantly,
- the release of municipal land for UA purposes (*ibid.: 5-9*).

The key questions for future research are: how will the UA Unit be suitably empowered to be able to achieve these strategic objectives, and how will it be able to intervene in critical areas, such as planning, while located within the Economic and Human Development Department?

7.   Urban agriculture and the City of Cape Town: where to from here?

Achieving sustainability goals will involve creating space for diverse voices, perspectives, and a multiplicity of options. Alternative solutions to agriculture and food supply within the City need to be identified. These solutions need to be packaged into an overarching UA strategy for the City. The current policy falls significantly short of this need.

A viable approach to UA needs to embody the principle that communities identify their own, specific needs. Identification of specific needs ideally evolves as part of a process that maps the food status of the various regions of the City. Such a process should pinpoint potential solutions – specific to the various regions – and select the community structures best suited to support the development process. The City and the identified structures then need to work collaboratively to map out a path that is agreed, supported, and sustainable. Only once these needs have been identified will the CCT be in a position to respond to the realities of each situation. Examples from existing groups that have succeeded in building social capital over time should be drawn on to support this process and to provide much needed insight into the strategies required. There are good examples of small community-based interventions in many cities around the world, such as Havana (Funes *et al.*, 2002), Addis Ababa, and Harare (Mougeot, 2005) – demonstrating that it is possible to address livelihood and nutritional needs while also providing communities with the necessary resilience needed to sustain themselves and contribute in a positive manner to the City.

This contrasts with large macro projects, typically, through the top down approach, where the development of social capital and the ability to meet food and nutritional security needs are often removed from community ownership and transferred to officials and political figures. Usually this approach ultimately requires large capital funds to ensure sustainability and often results in ownership of the process being removed from the community.

8. The call for an integrated approach

Girardet *et al.* (1999) argue that the interdependence of food, agriculture, health, and ecology calls for a more integrated approach to UA. They propose the formation of a municipal working group that can deal with food issues from a total system perspective, allowing for interventions that cross specific functions and needs within the City. While the CCT articulates this in its policy, it is unclear how it would be achieved, particularly given where the UA Unit is located. This structure is seen as a critical component of the UA policy and needs to be activated with urgency.

Further analysis of the experiences of several cities (although not listed directly) regarding the integration of UA in urban planning and programmes (Dubbeling *et al.*, 2001, in Bruinsma and Hertog, 2003) leads to the conclusion that, although developed separately, a similar logic and methodological process is generally followed. This includes the creation of an enabling institutional policy framework.

Other areas speak to critical connections where UA and urban policies intersect and are intrinsically linked:[7]

Integration in urban land use planning

- Revising urban zoning by-laws and indicating in which zones specified modalities of UA are allowed or promoted;
- Enhancing access to land by offering vacant urban open spaces and semi-public spaces (schools grounds, hospitals, prisons, etc.) with medium-term leases;
- Promoting multifunctional land use[8] and the promotion of community participation in the management of urban open spaces; and
- Including space for individual or community gardens in new public housing projects, and requiring the inclusion of such spaces in private building schemes.

Inclusion of agriculture in urban food security policies can be supported through the following:

- Providing budget and expertise to catalyse the implementation of broader UA programmes;
- Stimulating participatory adapted research, oriented towards development of technologies suitable for farming in confined spaces and with low risks for health and the urban environment;
- Organising farmers' study clubs and the provision of training and technical advice to urban farmers;
- Improving urban farmers' access to credit schemes for investments in production infrastructure and innovation of production technologies;
- Facilitating the local marketing of fresh urban produce; and finally,
- Promoting small-scale enterprises linked with UA.

Integration of agriculture in the urban environmental policies is enhanced by:

- Establishing low-cost facilities for the sorting of organic wastes and production of compost and animal feed, or biogas, and stimulation of practical research to develop adequate composting and digesting technologies;
- Promoting investments in systems for rainwater collection and storage;
- Establishing localised water-efficient irrigation systems in order to reduce the demand for expensive municipal water;
- Implementing projects with decentralised collection and treatment of household waste water for use in agricultural production; and
- Promoting the supply of natural fertilisers, bio pesticides, soil amendments, and quality seeds to urban farmers.

---

7   Adapted from De Zeeuw, 2003.

8   This refers to land that is used for a variety of functions and would not be exclusively for agriculture. It could include commonage areas where there is a mix of uses, or could be areas that are used at specific times of the year only.

Integration of agriculture in urban health policies can be strengthened by:

- Farmer education on the health risks associated with urban farming;
- Promotion of ecological farming practices, such as integrated pest and disease management, ecological soil fertility management, and soil and water conservation;
- Organising joint agriculture/health programmes on prevention of vector-borne diseases. (The emphasis could be on adequate environmental management and the placement of restrictions on the production of certain types of crops or animals, or certain farming practices in specific parts of the City, where such crops, animals, or practices may cause unacceptable health risks.)

9. Creating an enabling environment

In addition to the above-mentioned integrative approaches, there are a number of aspects that should inform the overall UA strategy of the City. These are covered below.

- The UA strategy should strive to facilitate the creation of the required enabling environment, linking other role players and facilitating the roll-out of UA in the City. Calls for an enabling environment are often vague, but it is critical that the UA Unit (UAU) be given the mandate to make this a reality. By elevating the status of the UAU, it would be put in a position to play the necessary facilitation role that would complement and catalyse other aspects, such as waste recycling and sewage re-use. Facilitating distribution and coordination of materials, inputs, and services would need to be included in this programme as well.
- The creation of a body – effectively used in other countries – could facilitate this enabling environment to support urban food issues. Urban food policy councils are useful structures to help guide government decisions. Such councils are informal coalitions of local politicians, hunger activists, environmentalists, sustainable agriculture advocates, and community development groups. Inclusion of diverse stakeholders would allow food policy decisions to reflect a broad range of interests and tap possible synergies (Hamilton, 2002; Pothukuchi & Kaufman, 1999). Food policy councils are critical as they bring public and non-government agencies into the debate. Often, cities make the mistake of addressing food security issues independently. Although partnerships can support the process, these are often considered to be secondary to the overall strategy. Collaborative partnerships could be a primary catalyst to an effective implementation process. In Cape Town, as is the case in most cities, the strategy is formulated prior to the formation of the citizens groups, who are well placed to liaise with government and officials on the matter.
- The creation of such urban food councils would be a critical component of UA in the City, stimulating the called-for partnerships and integrating a variety of perspectives and skills into the process – knowledge is of critical importance. While the CCT may have some idea of what land may be available, this needs to be reviewed from a far wider perspective.
- Any approach to an integrated UA strategy would begin with a thorough land audit, as well as an audit of work already completed for the City with regard to UA.
- Key questions need to be asked in respect of the role of planners and planning departments, and their relationship to UA. The UA policy should not focus on specific groups within the City, but rather respond to needs across the City, ensuring that in all cases it responds to the current global challenges in a manner that insulates the entire City from future shock.

10. Complementary strategies for integrated implementation

While it is noted that the implementation of programmes to support UA in the City would be complex, the current food crisis and sustainability challenges highlight the need for alternative approaches and greater support for those currently involved in the process. Some of the on-the-ground interventions could include – but should not be limited to – the following options:

- *A review of animal husbandry practices and proactive engagement with communities and other departments* (such as Health and Protection Services). The intention would be to find cross-cutting solutions.

Animal husbandry (although not necessarily cattle) takes place in various parts of the City, in areas such as Greenpoint (prior to the construction of the stadium), and in Pinelands, most townships, as well as in areas such as Hout Bay and Constantia. While the issues vary in the different areas of the City, strategies need to be able to respond to overall City needs, while at the same time addressing challenges at the local neighbourhood level.

- *The effective resourcing of the UA team within the CCT* is essential. This mobilises human resources, as well as the necessary equipment and utilities to respond to the needs of the different communities, ideally with a strategic escalation planned into the future.

- *Public-private partnerships with major businesses in the City could also be considered.* Programmes such as the Organic Freedom Project,[9] for example, do not need to be limited to distant rural areas. Private sector assistance, although advantageous, should be designed to benefit the broader development objectives of the programme and not specific interest groups only.

- *Cooperation between public bodies, residents and civil society is critical* for a robust, equitable and beneficial UA strategy.

- *An area-based plan to identify which product would suit which areas best.* For example, certain SEED (Schools Environmental Education and Development) schools are having success farming herbs where the soils and climate are not suitable for vegetable production. This knowledge is essential in allocating the correct resources to specific activities.

- *The development of areas that become drop-off areas for organic waste* and other waste streams that could support the programme need to be considered. Such drop-off areas should ideally be located in areas where processing facilities have been established to support the UA growers within the City, and specifically located near the sites where UA activities are in place, thus eliminating excess transport and other related costs. The City of Curitiba[10] example – where the collection of waste is free of charge – is a further item for consideration in the context of a broader UA policy.

- *Waste water run-off* is a challenge for the CCT, particularly in times of excess rain. If managing this problem considered UA more fully, it could have a significant impact on how waste water is managed and planned for. This would provide significant support to UA farmers and reduce costs, while also supporting the CCT in certain aspects of disaster management and general civil engineering plans. It could be argued that effective UA could also serve to reduce rain water run-off, partially eliminating some of the challenges associated with this. One of the main inputs that could limit the viability of UA is when costly municipal drinking water is used for irrigation. Making use of waste water and run-off could assist the overall UA programme greatly in the reduction of these costs.

- *Urban greening:* While urban greening is an important aspect of a sustainable City, the cost of such programmes to the CCT do at times mean that some areas are not suitably maintained, or at the very least place a burden on the City's ability to manage a developmental agenda. In addition, some urban green spaces can become unsafe places to be. Placing members of the community as 'farmers' in these areas is an opportunity that can be used to resolve these challenges. This option gives communities a chance to work the land, even upgrading it. Access to the land would need to be strictly controlled via either an allotment system or other municipal land usage agreements, similar to those currently in use in Stellenbosch through the Stellenbosch Small Farmers Trust. Alternatively, international examples of allotments, such as in the town of Ely in Cambridgeshire, UK, or even the allotments in the Kent area, UK, can be used to ascertain governance and leasehold arrangements.

- *Alternative economic models are also required.* The establishment of cooperatives have proved effective in the stimulation of UA in other regions (Mougeot, 2005). The connection between UA and alternative localised economies, such as seed saving groups and seed banks, provide

---

9   The Organic Freedom Project (OFP) is a programme supported primarily by PicknPay, and is a Section 21 (not for profit) membership organisation incorporated in South Africa, with the aim of promoting job creation and sustainable trade in the region through the facilitation of organic farming, processing, and marketing of organic products, including food, textiles, and bio-fuel (see http://www.urbansprout.co.za/sa_goes_organic).

10   In Curitiba, collection of unwanted garden waste is done free of charge as it has been found that once a fee is levied, this waste is often dumped, exacerbating waste management challenges, rather than being able to be productively used.

opportunities that are often not considered as benefits associated with UA. Seed saving is an old tradition that was practiced by many farming communities in the past. Seed saving, exchange, and sharing is social capital that has been lost, but which is central to the growing community of urban farmers (Saruchera, 2008). These alternative economies become a critical component of UA, and UA strategies should support these alternative economies, rather than undermine them.

11. Conclusion

When considering UA and the critical components of development in which the City is engaged, food is the ultimate 'cross-cutting' issue. Food and nutritional security are at the centre of all sustainability challenges. While activities supporting UA are certainly taking place within the CCT, there is a need to move the interventions from individual projects to an all-inclusive strategy. There needs to be a sustained and coherent integration of the responses to existing challenges faced by the City. For example, energy, water, and waste must be linked to a broader set of urban development objectives that integrate UA into the broader policy and planning regime of the CCT. By adopting a proactive and integrative approach to UA firmly embedded in the planning processes, the City will unlock significant potential for the communities of Cape Town. However, any strategy would need to be supported by proactive steps to reduce the footprint of the City. Without strategies to effectively reduce the ecological footprint of the City, efforts at sustainability will remain tokenistic and inconsequential.

For UA to effectively address the developmental requirements of a given city, that city must incorporate the integration in urban land use planning and include UA in overall food security policies. UA needs to be a core component of urban environmental policies as well as urban health policies. This integrated nature of UA makes it an ideal development opportunity and one that needs to come to the forefront of urban planning and management.

UA has the potential to provide palpable ecological benefits to the City, and it presents fewer costs than the more 'business as usual' approaches. However, UA is not a simple intervention and requires technical and human resource interventions, although not at the scale of other processes. The key challenge is that the benefits emerge over time, often taking far longer periods of time than is advantageous to communities who legitimately seek more immediate solutions. However, UA provides significant opportunities in addressing the issues of poverty, inequality, and resource consumption.

If food and nutritional security are to become the focus of how we provide food in society, we will have to change our relationship with food in its entirety, including all elements of the food system. UA allows for this to take place. By viewing the environment through the lens of food and nutritional security – rather than that of industrial agriculture, where food is a market commodity only – food gives us a promising pathway towards sustainable agriculture and broader sustainability practices.

References

Bruinsma, W. & Hertog, W. 2003. *Annotated Bibliography on Urban Agriculture*. Prepared for the Swedish International Development Agency (Sida) by ETC – Urban Agriculture Programme, in cooperation with TUAN. Leusden, the Netherlands.

City of Cape Town. 2007. *Urban Agriculture Policy for the City of Cape Town*. Available from www.capetown. gov.za/en/ehd/Documents/EHD_Urban_Agricultural_ Policy_2007_8102007113120.pdf. (Accessed 21 September 2008)

City of Cape Town. 2007. *City of Cape Town Socio-Economic Profile Report* (PGWC). Available from http://www. capetown.gov.za/en/stats/CityReports/Documents/ Population%20Profiles/ City_of_Cape_Town_Socio-Economic_Profile_ Report(PGWC)_181220069025_359.pdf. (Accessed 21 September 2008.)

DEAT. 2007. *People, Plant Prosperity: A National Framework for Sustainable Development in South Africa*. Pretoria: Department of Environmental Affairs and Tourism.

De Zeeuw, H. 2003. Introduction to Urban Agriculture. In Bruinsma, W. & Hertog, W. *Annotated Bibliography on Urban Agriculture*. Prepared for the Swedish International Development Agency (Sida) by ETC – Urban Agriculture Programme, in cooperation with TUAN. Leusden, the Netherlands.

Dubbeling, M., Prain, G., Warnaars, M. & Zschocke, T. (Eds.). 2001. *Feeding Cities in Anglophone Africa with urban agriculture. Concepts, tools and case studies for practitioners, planners and policy makers*. CD-ROM. International Potato Center – Urban Harvest, Lima, Peru.

Egziabher, A. 1994. Urban Farming: Cooperatives, and the Urban Poor in Addis Ababa. In Egziabher, A. *et al. Cities Feeding People: An Examination or Urban Agriculture in East Africa*. Ottawa: International Development Research Centre (IDRC).

Frayne, B., Battersby-Lennard, J., Fincham, R., Haysom, G., 2009. *Urban Food Security in South Africa: Case study of Cape Town, Msunduzi and Johannesburg*. Development Planning Division Working Paper Series No.15. Midrand: DBSA.

Funes, F., Garcia, L., Bourque, M., Perez , N. & Rosset, P. 2002. *Sustainable Agriculture and Resistance: Transforming Food Production in Cuba*. Oakland, CA: Food First Books.

Gabel, S. 2005. Exploring the Gender Dimensions of Urban Open-space Cultivation in Harare, Zimbabwe. In Mougeot, L. (Ed.). *Agropolis: The Social, Political and Environmental Dimensions of Urban Agriculture*. London: Earthscan.

Gasson, B. 2002. *The ecological footprint of Cape Town: Unsustainable resource use and planning implications*. Paper presented at the National Conference of the South African Planning Institution, 18-20 September, Durban.

Girardet *et al.* 1999. Urban agriculture and sustainable cities. In Bakker, N., Dubbeling, M., Guendel, S., Sabel-Koschella, U. & De Zeeuw, H.(Eds.). *Growing Cities, Growing Food: Urban Agriculture on the Policy Agenda: A Reader on Urban Agriculture*. Leusden, The Netherlands: ETC.

Halweil, B. & Nierenberg, D. 2007. Farming the Cities. In *State of the World Report 2007*. Worldwatch Institute. Washington DC: Norton & Company.

Hamilton, N. 2002. Putting a Face on Our Food: How State and Local Food Policies Can Promote the New Agriculture, in Halweil, B. & Nierenberg, D. 2007. Farming the Cities, in *State of the World Report 2007*. Washington DC: Norton & Company.

McIntyre, B., Herren, H., Wakhungu, J & Watson, R. (Eds.). 2009. Agriculture at a Crossroads, *International assessment of agricultural knowledge, science and technology for development* (IAASTD), Global Report. Washington DC: Island Press.

Mougeot, L. (Ed.). 2005. *Agropolis: The Social, Political and Environmental Dimensions of Urban Agriculture*. (IDRC). London: Earthscan.

Pothukuchi, K. & Kaufman, J. 1999. Placing the Food System on the Urban Agenda: The Role of Municipal Institutions in Food Systems Planning. *Agriculture and Human Values*, 16: 213-224.

PGWC. 2006. Provincial Government of the Western Cape, *Socio Economic Profile: City of Cape Town*. Research online at: www.capegateway.gov.za/.../city_of_cape_town_se_profile_optimised.pdf (Accessed 23 May 2009)

RUAF. 2006. *Cities Farming for the Future: Urban Agriculture for Green and Productive Cities*. Available from http://www.ruaf.org/node/1135. (Accessed 30 June 2008.)

Saruchera, M. 2008. *Seed savers and seed banks for development*. Stellenbosch: Sustainability Institute. Unpublished discussion document.

Scherr, S. 1999. *Soil Degradation: A Threat to Developing-Country Food Security by 2020?* International Food Policy Research Institute, Washington, D.C. Food, Agriculture and the Environment Discussion Paper 27. Available from www.ifpri.org. (Accessed on 12 April 2008.)

Smit, J. & Nasr, J. 1999. Agriculture: Urban Agriculture for Sustainable Cities: Using Wastes and Idle Land and Water Bodies as Resources. In Satterthwaite, D. (Ed.). 2001. *Sustainable Cities*. London: Earthscan.

Swilling, M & Fischer-Kowalski, M. 2010. *Decoupling and Sustainable Resource Management: Scoping the challenges*. Decoupling Working Group, International Panel for Sustainable Resource Management: United Nations Environmental Programme, Paris.

Swilling, M. 2006. Sustainability and infrastructure planning in South Africa: a Cape Town case study. In *Environment & Urbanization*, 18(1): 23-50.

Von Braun, J. 2007. *The World Food Situation: New Driving Forces and Required Actions*. International Food Policy Research Institute, Washington DC. Food Policy Report. Available from www.ifpri.org. (Accessed 11 April 2008.)

Wackernagel, M., Kitzes, J., Moran, D., Goldfinger, S. & Thomas, M. 2006. The Ecological Footprint of cities and regions: comparing resource availability with resource demand. *Environment & Urbanization: Ecological Urbanization*, 18(1): 3-8.

Watson, R. T., Wakhungu, J. & Herren, H. R. 2008. *International Assessment of Agricultural Science and Technology for Development (IAASTD)*. Available from www.agassessment.org. (Accessed 21 April 2008.)

Action	Detail	Scale of action in CCT
Integration in Urban Land Use Planning	The revision of actual urban zoning by-laws and indication of where specified modalities of urban agriculture are allowed or even promoted, and other zones where certain farming systems will be prohibited due to special conditions.	
	Access to land can be enhanced by offering vacant urban open spaces and semi-public spaces (school grounds, hospitals, prisons, etc.) with a medium-term lease for gardening and other agricultural purposes to community groups, farmer cooperatives, and/or unemployed people (purpose-specific leaseholds).	
	Promotion of multifunctional land use and community participation in the management of urban open spaces. Under certain conditions, urban farming can be combined with other compatible land uses; farmers can be used as co-managers of parks, recreational areas, water storage areas, nature reserves, fire break zones, etc. – by doing so, the management costs of such areas may be reduced, and protection against unofficial uses and informal rezoning may be enhanced. Agriculture can be used to make degenerated 'green zones' green and keep reserve areas free from being built upon. It can also act to form a buffer zone between competing land uses (e.g. residential and industrial areas).	
	The inclusion of space for individual or community gardens in new public housing projects, and requiring the inclusion of such spaces in private building schemes. In case of planned conversion of agricultural areas for other land uses, the urban farmers could be supplied with alternative land.	

Action	Detail	Scale of action in CCT
Inclusion of agriculture in urban food security policies	Provision of budget and expertise to boost the preparation of broader urban agriculture programmes. Stimulation of participatory adapted research, oriented towards development of technologies suitable for farming in confined spaces and with low risks for health and the urban environment. Organisation of farmers' study clubs that actively engage in the technology development and adaptation process. Provision of training and technical advice to urban farmers, with a strong emphasis on ecological farming practices, and the organisation of low cost and participatory systems for animal health services. Improvement of the access of urban farmers (with an emphasis on women producers and the resource-poor) to credit schemes for investments in the production of infrastructure and innovation of production technologies; revision of loan conditions and/or establishing micro-credit schemes for urban farmers. Facilitating the local marketing of fresh urban produced food, by – Authorising local farmer markets, food box schemes and other forms of direct selling of fresh agricultural produce from urban producers to local consumers Creation of the minimum infrastructure required for local farmers markets. Promotion of small-scale enterprises linked with urban agriculture, i.e. input suppliers and enterprises for processing and marketing locally produced food.	
Integration of agriculture in urban environmental policies	Establishment of low-cost facilities for sorting of organic wastes (households, vegetable markets, agro-industry) and production of compost and animal feed or biogas; stimulation of practical research to develop adequate composting and digesting technologies. Promotion of investments in systems for rainwater collection and storage, construction of wells, and the establishment of localised water-efficient irrigation systems (e.g. drip irrigation) in order to reduce the demand for expensive piped (drinking) water. Implementation of pilot projects with decentralised collection and treatment of household wastewater (preferably with biological methods) with a view to its re-use in agricultural production. Promotion of use of untreated or partially treated (household) waste water for the irrigation of woodlands and parks, orchards, pastures, root crops and grains, nurseries for tree seedlings and ornamental plants, etc., in order to reduce the demand for expensive piped (drinking) water and to make productive use of waste water and included nutrients. Promotion of the supply of natural fertilisers, bio-pesticides, soil amendments, and quality seeds to urban farmers, e.g. by providing incentives (such as reduced taxes) for enterprises that produce ecological friendly agricultural inputs.	

Action	Detail	Scale of action in CCT
Integration of agriculture in urban health policies	Farmer education on the health risks associated with urban farming and their causes, and practical ways to prevent such problems can be highly effective. Examples of preventive measures that can be taken by farmers themselves include proper choice of crops in relation to the location of production and the quality of the soils and water, proper choice of irrigation methods, proper handling of the products, adequate siting of animal housing, hygienic handling of feed, manure handling, and the proper handling of waste products and waste water. Promotion of ecological farming practices, such as integrated pest and disease management, ecological soil fertility management, soil and water conservation, etc. through: Farmer training and practical demonstrations; Promotion of the production and supply of natural fertilisers, bio-pesticides, soil amendments, and quality seeds to urban farmers, by providing incentives for enterprises that produce environment-friendly agricultural inputs and meet certain quality standards (nutrients, health standards, etc.); and Support to local initiatives for marketing of ecologically grown food and the establishment of 'green labels' for organically grown and safe urban produced food. Consumer education on preventive measures, safe food labels, and locations where these can be obtained, etc. Organisation of joint agriculture/health programmes for the prevention of vector-borne diseases with an emphasis on adequate environmental management. Restrictions on production of certain types of crops or animals, or certain farming practices, in specific parts of the City, where such crops, animals, or practices may pose unacceptable health risks. Education of food processing and marketing micro-enterprises on health risks and the hygienic standards to be maintained, and strict control of slaughterhouses.	

[Adapted from De Zeeuw, 2003]

13

Sustainable Quality

Architecture for a Sustainable Cape Town

Mokena Makeka & Peta Brom

1. Introduction

The chapter begins by contextualising the compelling challenges for sustainable architecture globally. Local features that constrain proactive attention to the design of sustainable buildings by the architectural profession in South Africa are highlighted. Drawing on our practice as architects in Cape Town, we then explore some of the factors which have limited innovation in sustainable architecture – both within the architectural profession itself, and in relation to other stakeholders such as the construction industry, the insurance industry, cost control professionals and the public. We also highlight the creative leadership vacuum that constrains the progress of sustainable architecture. The chapter speaks to what will be required to change the design paradigm so that, in future, buildings in Cape Town will require far fewer energy-intensive materials, and over the life cycle they can even become positive producers of strategic resources.

This chapter paints a picture of how architecture could become more sustainable by deepening and enriching the network of relationships between designers, architects, engineers and manufacturers in a way that supports innovation and respect for the environment. Two case studies illustrate what practical barriers exist on the ground, and what can be learned about the potential for sustainable architecture in Cape Town, as well as the role of politics and the reasons for hope.

1.1 Motivation and Background

In 1972, a publication called *The Limits to Growth* was published by the Club of Rome, a group of about fifty self-appointed wise men and women. The publication, although highly criticized, brought the world's attention to the notion that, if population, technology and human growth were to continue along the then current trajectory, there would be a shortage of natural resources that would potentially lead to the decrease of population and economic growth. The 1970's and 1980's saw a number of large-scale disasters, including, among others, the first fuel crisis; the release of dioxin in an industrial accident at a pesticide plant in Seveso, Italy; the discovery of the hole in the ozone layer; a chemical accident in Bhopal, India, that killed thousands and maimed many more; and a fire in Basal, Switzerland, that released chemicals into the Rhine and killed fish as far north as the Netherlands (United Nations Environment Programme (UNEP, 2002).

By 1983, it had become clear that development and the environment were inextricably linked, but communicating about it needed a source which was both credible and had authority. The Brundtland Commission (formally known as the World Commission on Environment and Development (WCED)) was formed to hold global hearings and to put together a formal report of its findings. In 1987, it produced the Brundtland Report, entitled *Our Common Future*, that coined the term 'Sustainable Development':

"It expressed concern that the rate of 'change is outstripping the ability of scientific disciplines and our current capabilities to assess and advise'. The Commission concluded that existing decision-making structures and institutional arrangements, both national and international, simply could not cope with the demands of sustainable development."

(UNEP, 2002)

The 1992 World Summit is still the largest conference ever held to discuss the environment (UNEP, 2002). During the following year, the Green Globe benchmarking standard was established to set up 'green' standards for tourism. It included a certification system, still widely used, for 'green' hotel management and construction (Green Globe International, 2010). But it was the UK Building Research Establishment Environmental Assessment Method (BREEAM), established in 1990, that was the earliest formal response by the construction industry to the environmental crises facing the world. Australia responded in 1993 with the National Housing Energy Rating System (NatHers), and America in 2000, with the Leadership in Energy and Environmental Design (LEED) (Beck, 2005). In 2002, Australia took a second step with the establishment of the Green Building Council of Australia (GBCA) and its 'Green Star' rating system (GBCA, 2008).

When UNEP (n.d.) wrote about the history of sustainable development, they had this to say about the relationship between Northern industrial activities and the South:

"In the wake of the industrial accidents of the 1980's, the pressure on corporations grew. … By the end of the decade, the concept of eco-efficiency was being introduced into industry as a means of simultaneously reducing environmental impact while increasing profitability. Few if any of these interests were shared by corporations based in developing countries, but there were already debates on the implications of industries migrating to 'pollution havens' in the South."

In 2002, the second World Summit on Sustainable Development was held in Johannesburg; yet the South African building industry's response to sustainability challenges has been relatively slow. At the time of writing, the Green Building Council of South Africa (GBCSA) had been established, with its own rating system adapted from the Australian Green Star one.

Against the background of a history of apathy and exploitation, slow movement, and the perception that the South African building industry is ill-equipped to meet international standards for sustainability, what challenges specifically rest with architects, and how do they respond?

2.   Challenges to Innovation for Green Architecture

The announcement of the GBCSA's new green certification standards has inspired a number of manufacturers to alter their practices. Cape Brick and Thermocoustex, for example, began to utilise recycled material and introduced lower toxicity levels, while Dulux lowered the VOC (volatile organic compounds) content in one of their paint ranges to meet GBCSA standards.

However, currently there isn't yet a culture of product development among architects, or the construction industry as a whole. This is in sharp contrast to other economies, where innovation and material exploration is often initiated by inventive demands on the part of the architect. In such economies, the construction sector is highly supportive of the approach to develop variations and new products. They not only adapt well to the construction sector, but often become lead materials for re-interpretation by other sectors of the industrial base.

This creates the conditions for innovation and increased building performance, and diversifies the economy, as has been the case in many parts of Europe and the East, where architectural technology has developed a research ethic of its own. In South Africa, the prevailing habit is one of selection and categorisation of existing, and often outdated, products. The situation is exacerbated by the poor nature of architectural training in product development and industrial design. The fact that local industries also have a poor research and development base, preferring instead to buy manufacturing licenses from the international community, doesn't help either, because it allows them little flexibility for adaptation. Although the relative size of the economy limits the potential for innovation, South Africa also lacks a strong sense of creative enterprise in this sector.

As primary decision makers, architects usually exert influence through their performance specifications or aesthetic design criteria. Unfortunately, they are often unaware of the potential for new products in their designs. Largely passive, architects tend to wait for product innovation to take place *before* they explore possibilities of their own. Moreover, they are often directly influenced by risk-averse clientele, who are unimaginative and suspicious of innovation, and who insist on previewing the application in other built works. The resulting scenario is a fruitless cycle based on client fear and mistrust of innovation, and, in the interests of maintaining client relationships, the professional team becomes a complicit contributor to this cycle.

Insights offered by Damanpour and Wichnevsky (2006) distinguish between organisations that *generate* innovation, and those that *adopt* innovation. The former relies on their technical knowledge and market capabilities, while the latter rely on their managerial and organisational capabilities to select and assimilate innovations.

The culture of 'cherry picking' in the latter approach is based on a methodological paradigm of observation and analysis, rather than exploration and experimentation. Architects tend to wait for technology to become available, rather than actively engaging in the developmental phase of innovation. Such passive behaviour corresponds to the notion of innovation adopters as described by Damanpour & Wichnevsky (ibid.). The reason why this is particularly problematic is expressed especially well by Holden *et al.* (2008:315)

"The very freedom of pursuing an ideal like sustainable development is that prototypes of the finished product do not exist."

Since architecture is the practice of generating new, custom designs, it is paradoxical that architects tend to be innovation adopters, rather than innovation generators. The call for sustainability in architecture will require architects to shift to active engagement, so that they become innovation generators. The next section paints a picture of how this can be interpreted.

2.1 Architectural Innovation: Constraints and Opportunities

Sustainable architecture involves deepening and enriching the whole network of relationships between designers, architects, engineers and manufacturers in a way that supports and shares the risks and the benefits associated with the shifts and the innovation that is required. This is particularly relevant to South Africa, where there is protectionism on the part of design professionals, and an inherent conservatism in the flow of the process from the design of a building to its eventual construction. The sustainable design of a building is normally seen as something that happens *after* the concept has been developed, rather than as the point of departure for the entire design team.

The first phase of the traditional architectural design process comprises site appraisal and definition of the brief. Site appraisal involves the analysis and study of the geographic elements of the site – access, zoning ordinances, topography, site survey, constraints (e.g. position of a servitude, building setbacks, etc.) – and opportunities such as views, local climatic conditions, and materials that are available on site. The development of a brief occurs in consultation with the client. A precedent study of the surrounding/neighbouring buildings and the typology of the building required by the client will follow. The architect then responds. While contextual analysis and response is essential, there is a tendency to be over-reliant on precedent, which hinders experimentation. In essence, the architectural concept is generated from a methodology that looks backwards and requires that 'someone else' first proves that it can work. The exploration of a new design intervention, or a new use of a material, will also rest on identifying a precedent. If no precedent exists, then the intervention could be labelled 'too risky' by the client.

Innovation tends to fall within the scope of the engineer's work, because the engineering methodological paradigm is one of test, scientific experiment, and modelling based on physical laws and principles. In other words, the architect's client will first ask the question: "Has this been done before?", while an engineer will ask the question: "How can we ensure that this will be practical?" Then again, for the engineer, the issue of professional indemnity often reinforces the tendency towards great caution. Insurance bodies do not cater for the creative process of research and development. Since the insurance industry's integral role in the professional execution of large works does not support innovation, this hinders many professionals from fully exploring their body of work, thus hampering the creative capacity across a whole network of business and professional relationships. Innovation also calls for new ways of manufacturing products and interpreting technical details, and is catalysed by dialogue between designers and manufacturers/crafters.

There are, however, some encouraging shifts that support a more integrated and creative approach, linking stakeholders together through opportunities for dialogue and collaboration. The annual Design Indaba expo, which began in 2003, has provided a platform that encourages dialogue and exchange between craft and design among South African industries. This has resulted in the development of a number of products that are uniquely South African, such as the work produced by the designer-manufacturer collaboration Nucleus of Kraft (NOK), and Michaele van Jansen's nylon designs, which, through technology advances, has made the process of prototype manufacture for her designs more affordable (Design Indaba, 2009). Such conversations between designers and manufacturers result in an exchange of innovative ideas that find the balance between aesthetic value and technical expertise.

Through communicating with manufacturers and building contractors, managing the risk associated with an innovative design can also be improved and even shifted. A new design only finds expression if someone is willing to carry the liability, should it fail. In the case of an engineered product, the engineer will sign off on a design, while most manufacturers will carry the liability in the form of a guarantee of the product for a period of years. By engaging in dialogue between suppliers and manufacturers about applications and designs, it then becomes possible to understand where the risks lie. Designs can then ensure durability, and performance criteria are met. However, there are other pressures which militate against sustainable architecture, and these relate to the demands and preferences of consumers, developers, and financiers, who often require expedited contracts.

2.2 Accelerated Construction

Accelerated construction often becomes the norm at a time when all stakeholders – clients as well as architects and engineers – could make more informed decisions by slowing down. What would a slower society, recognising the value of slowness in sustainability, look like in the realm of architecture? The Slow Society – "a survival strategy, a model for sustainability, a movement of people who are trying to live slowly, in harmony with the complex web of life" (http://www.slowsociety. org) – offers an idea: such a society will be "inspired by all philosophies that maintain that wise decision-making presupposes time and space for reflection. If we do not devote ourselves to the important issues every day, every month, every year, how can we then change the world in a more sustainable direction?" (Mission of the Slow Society cited in http://www.sustainableneighbourhoods. co.za/resources/Documents/architecture_presentation.pdf.)

In South Africa, we tend more towards a quickening revolution of client demands for fast delivery. Architects are expected to design and build homes in less than a year. Residential clients expect the design process to take between three and six months, compared to the American and Australian norms of approximately two years (Worby, 2006). The situation was worsened by the economic and construction boom of the early and mid-2000's. The 2010 Soccer World Cup Finals also expedited the construction of (much needed) infrastructure and public buildings. Between 2004 and the third quarter of 2006, it can be assumed that land values doubled in Cape Town, and construction costs increased by 50% (Napier, 2007). Accelerated building cycles simply don't allow latitude for exploration and development, nor does it allow time for reflection on sensitive and wise interventions that will inform adaptable and durable architecture.

There is an additional cost associated with design and product prototype development in South Africa, a cost which few clients are prepared to carry. The construction industry is characterised by under-regulation of the artisan trades and lack of consistency between available market choices. Quality, risk and price vary significantly in any given tender, but through slowing down the processes, risk could be minimised. For example, the risk of doing business with a fly-by-night contractor could be ameliorated by the use of quantity surveyor reports. These assess the most likely cost of a line item for budgeting and quality control purposes. When final budget decisions are made, the quantity surveyor's assessment could help to determine what is included, what is not, and what is considered a reasonable amount of money to spend on a line item. A tremendous amount of power to influence the final outcome of the design process rests with the quantity surveyors. Too frequently, quantity surveyors revert to default costs and specifications to the detriment of innovation and sustainability. Since finances are allocated according to their cost estimates, the net result is that the building costs are not fine-tuned to evaluate the best options for sustainability.

On government projects in particular, the quantity surveyor's role expands to include recommendations on responsible or appropriate allocation of funds. In the early phases of the project they can assist in motivating the need for additional funds, but if sustainable line items are not included in the bill of quantities, there is seldom opportunity to add it at a later point in time.

The primary role of the quantity surveyor is to calculate upfront capital expenditure. The challenge with this approach is that the focus is on the short-term costs of a building, as opposed to the life cycle cost. More rigorous cost-benefit analysis of sustainable alternatives is therefore required. This role is currently taken on by design and sustainability consultants, rather than the quantity surveyor. It requires an extension of services that is not easily accommodated in accelerated design cycles and falls outside of the standard scope of services.

Some headway is being made into making environmental products more affordable. Increased awareness in the market place for sustainability will lead to an increase in demand for environmentally responsible products. Pressure from the industry makes it viable for manufacturers to invest in greener product development and to upscale supply as the market place proves lucrative. In turn, this will make environmentally responsible products more widely available and affordable.

Our experiences have been confirmed by PricewaterhouseCoopers (2008):

"Consumer demand for sustainable alternatives exists, but PwC research shows that price and availability are large barriers to sustainable products becoming mainstream. The downside of failing to manage costs and reputational risks associated with sustainability is growing increasingly material in this sector."

There are some encouraging signs which speak to the emerging importance of sustainability in the field of architecture. The Commission for Architecture and the Built Environment (CABE) has published a number of documents on creating excellent buildings and improving quality in buildings. Some themes focus on sustainability as resource management and awareness of the value versus cost across the whole-life of the building. One of these documents, compiled by the (UK) National Audit Office (2004), provides some tools for evaluating the costs associated with maintenance vs. the costs aspects of construction.

"Auditors should consider the whole-life value generated to the business and taxpayer and not simply focus on minimising initial capital costs. This impact on service delivery can be seen clearly if we break down a building's costs. Over the lifetime of a building, the construction costs are unlikely to be more than 2-3% of total costs, but the costs of running a public service will often constitute 85% of the total. On the same scale, the design costs are likely to be 0.3-0.5% of the whole-life costs, and yet it is through the design process that the largest impact can be made on the 85%."

(NAO, 2004: 6)

While there is increasing awareness of resource management aspects, there is still very little consideration of the benefits of full whole-life assessments. The next section focuses on long term considerations.

2.3 Consideration of the Long Term

The choices of multiple stakeholders have an impact on the long term sustainability of buildings. They include:

- Government departments
- Entrepreneurs
- Construction companies
- Cost control professionals: quantity surveyors and accountants
- Architects

Buildings and cities are typically designed for a fifty-year regenerative life span. This is reinforced by the imposition of a heritage protection clause on buildings older than sixty years, which are protected as having 'light heritage significance' in terms of the National Heritage Resource Act, 25 of 1999 (SA Government, 1999). Furthermore, cheaper materials of much shorter life spans (often twelve to twenty-four years) have flooded the construction market. These products have built-in redundancy in order to encourage consumerism and promote economic growth through increased demand. Various industries have emerged to take advantage of business opportunities afforded by construction sector economics, with no intention of continuing business in the long term. Typically, products are sold by companies more interested in short-term profits, and by entrepreneurs who have no real intention to honour long-term warranties and liability conditions. Recent observations of industry shifts have seen those with a poorer focus on service and quality going under. Our opinion is that companies that were relying on the property boom for their success, rather than the strength of their product or services, were the first to fold.

There has been a complementary increase in professionals committed to cost control, such as quantity surveyors. However, there is usually scant regard and little training given to entrepreneurs to advise the client. Whole-life costing has often placed them at odds with the architect, whose chief concerns for long-term durability and quality are sometimes at odds with short-term financial interests and planning. As a result, clients increasingly tend to prefer materials that will need to be replaced within two to five years. The preference for quick, short-term solutions are all too prevalent in certain government departments, such as the Department of Public Works, who know full well that other government departments or spheres of government will have to shoulder the burden of the long-term implications of the project in terms of up-keep and maintenance. Competition between departments for additional funds also means that projects are distorted in their execution to the benefit and detriment of different departments. Performance criteria phrased as completing 'works on time, and on budget' supersedes the ideal of completing 'quality works for future generations.'

At the same time, one should acknowledge that company branding/image can change as frequently as every five years, so it might be necessary for front offices to be cyclically refinished according to updated company image. This changes the way specifications are approached. Instead of asking the question "Which finishes are the most sustainable over twenty to fifty years?", it might be necessary to shift to an analysis of which finishes will be most sustainable over five or ten years, depending on the commercial client's facility management cycles. It may therefore be necessary to encourage facility managers to consider longer-term alternatives, but understand the need for branding to stay contemporary. This is a challenging paradox for architects and specifiers to balance. For public buildings, the separation of construction and maintenance budgets continues to compromise the long-term integrity of the buildings and consequentially waste taxpayers' money.

The (UK) National Audit Office (2004) has prepared a document entitled *"Getting Value for Money through Construction Projects, How Auditors Can Help"*, which puts forward a set of guidelines for assessing the value for money of a building across its entire life cycle. It also discusses a set of value drivers for good design – each proposes a set of key questions which are exemplified in case studies – and a value assessment tool is presented as an appendix. The value drivers include:

1. *Maximise business effectiveness*
2. *Ensure effective project management and delivery*
3. *Achieve the required financial performance*
4. *Impact positively on locality*
5. *Minimise operation and maintenance costs and environmental impact*
6. *Comply with third party requirements.*

The idea of good design pervades all design drivers as a cross-cutting issue. In summary:

"Delivering value means maximising the benefits delivered by a project by satisfying or exceeding the needs of the various stakeholders whilst simultaneously minimising the use of resources. Value can be described as the function of the relationship between the 'satisfaction of needs' (business benefits and requirements) and the resources needed to deliver them, i.e. Value = Benefits/Resources....

…Good design also contributes to staff recruitment, retention and motivation, issues that increase value for money across the life of the asset."

NAO, 2004: 6

The private sector remains more dedicated to design and the use of quality materials – because of the personal capital investment involved – and there is merit in developing a system whereby long-term impacts are fed back not only into the design process, as is the case, but also into the cost control analysis. Facility managers routinely examine the life cycle costs of interior finishes over the tenant's branding cycles. Again, in cases where the developer is building for the purposes of sales, buildings will be cheaply finished in order to achieve practical completion certificates for sales purposes, intentionally under-specifying quality – in the full knowledge that once a tenant (who isn't necessarily the purchaser/intended building owner) takes occupation, they will retrofit according to their specific needs.

The cost control sector in South Africa has increasingly demonstrated two critical weaknesses:

- an inability to accommodate true costs into their financial planning and budgetary models, and
- the periodic unrealistic representation of building costs based on the rapidly changing economic environments.

To maintain face and client relationships in the light of poorly articulated initial budgets, cost control professionals often advise the client that cost savings can be achieved by reducing the architectural specifications. The quality of the final product and true sustainable design thence-forward rapidly spirals into the abyss of unsustainable products, processes, and project implementation methodologies.

Sustainable architecture therefore involves a responsible approach to creating buildings which offer one of two alternatives, where either –

- maintenance is of the lowest possible environmental cost, or
- maintenance and upkeep has a robust regimen in its execution.

Common to both ideals is the very clear notion that architecture responds to the needs and dictates of its brief and contextual condition, including climate, aspiration, joy, and quality convivial environments. Given the growth in population in the last century and a variety of associated – but unanticipated – social impacts, it is not realistic to suggest that a timeless architecture is either achievable or desirable in the context of today's rapidly evolving society.

Sustainable architecture will be sustainable due to the ability of buildings to adapt to conditions and evolve with seamless and minimal disruption, including responding to increasingly unpredictable weather phenomena, such as earthquakes, typhoons, and net heat gain. The notion of an architecture that will be of significance for future generations is seldom honoured, and instead there is a plethora of pseudo-heritage, modern copy-cat interpretations that find their way into architectural guidelines for gated communities (examples include the Welgevonden Estate, De Wijnlanden, De Zalze, Boschendal and Zevenwacht developments in the Cape Winelands). It is our contention that architecture that produces sustainable buildings and neighbourhoods, is also architecture that is durable – not only through quality and sound construction methods, but also through an aesthetic that will be cherished and a flexibility of design that can be adapted to changing social needs.

In the last fifty years alone, residential design norms and housing typologies have changed significantly. Four major socio-political shifts have influenced housing form in South Africa:

- Firstly, the end of apartheid has informed the typical form of 'middle class white homes'. For example, 'servant quarters' aren't needed, and the kitchen is no longer necessarily 'back of house', although historical trends still prevail towards this segregation.
- Secondly, women's social change in status has reinforced the notion of the kitchen becoming part of the home.

- Thirdly, a growing recognition that household forms do not necessarily conform to the 1950s ideal of two parents with 2.3 children, has required the development and availability of varying alternative housing forms. Changing economic environments also result in changes to household form, as young adults may choose to live together to share the rent, or multiple families may share one house.
- Fourthly, last year it was announced that for the first time in history, more people live in cities than have ever before (World Watch Institute, 2007). As mentioned in the introductory chapter, Africa will be host to the majority of slums in the world which poses additional challenges. With increasing numbers of people living in urban areas such as Cape Town, a concomitant increase in density has placed a premium on space, and hence the response of more compaction with an emphasis on efficient spaces. A building that is going to stand the test of time must also allow for alteration, with the form of the alteration being appropriate to the typology and to the architectural logic.

While it could be argued that an architecture that requires continuous maintenance provides social value – as a modus for learning, exchange, and training in the craft of making buildings; as a means of generating non-consumptive economic growth through service provision; and as a creator of jobs that will ultimately address unemployment and poverty concerns – it must be recognised that high maintenance costs will also diminish the disposable income available to a residential household, either directly through materials and labour, or indirectly through time spent maintaining a building instead of bringing in additional revenue.

D.S. Macozoma of the Council for Scientific and Industrial Research (CSIR) has written extensively about the reclamation of building materials for recycling. At a 2001 conference, he urged designers to consider the end of a building's life in the design process, and argued for a cradle-to-cradle approach that considers disassembly for re-use, biodegradability and recyclability. In other words, the onus rests with architects to interpret and consider the destruction of their designs (Macozoma, 2001). While the necessity to do so is well-founded, it is counter-productive to design, because it changes the focus from longevity and durability – which ultimately lowers resource consumption in the first instance – to one of building recycling, which has the benefit of stimulating a non-consumptive economy. Both focuses clearly have their merits, but should be executed by different role-players.

In his 2006 dissertation, Macozoma significantly shifts the focus (Macozoma, 2006). A major barrier to the recycling of buildings has been the relatively high cost of labour compared with materials. In simple terms, the act of reclamation from demolition historically costs more than the resources it yields for sale in secondary markets. In recent years, the mechanisation of many of the processes has made reclamation more viable. Macozoma therefore argues for the stimulation of secondary markets. In this way, the barrier to building recycling (cost) is reduced through technical innovation and the responsibility is shifted from the designer to government incentive for secondary market stimulation. This shift should be coupled with an appeal to designers to consider designs as evolving buildings that will last for centuries. Architects should therefore design as if the buildings are going to last forever.

In some ways, Macozoma's historical shift in focus adds fuel to the argument for a sustainability that relies on intersubstitutability – the notion that human beings continually innovate and find new technology, and that technological advancement will ensure ultimate sustainability. This begs the question: At what cost? While the case for technological innovation is assisting in the creation of a secondary industry for recycling (for example), it must not be relied upon in the design of buildings, and hence the most sustainable solutions for the current time must be sought. For example, on the project of the Public Transport Shared Services Centre in Athlone, the engineers argued that it is sustainable to specify aluminium without limiting its application, because it is 100% recyclable. Currently aluminium is still the most energy-intensive metal available on the market. The world recycles 50% of all aluminium, but it is predicted that natural sources will run out in approximately a hundred and fifty years at current consumption trends (Aluminium Federation of South Africa, 2007). The argument that a building material is sustainable because it is recyclable is ill-founded when considered in terms of reduction of current ecological footprint and when weighed against a potential lifespan of, say for example, over four hundred years in relation to the sense of limits imposed by a hundred and fifty-year resource pool.

2.4 The Creative Leadership Vacuum and the role of emerging professions

Without a tangible presence of creative leadership, buildings are sacrificed on the altar of expediency.

The second half of the twentieth century was a highly dictatorial time, thus minimising the possibilities for creative leadership. The planning profession was used as an instrument of apartheid, contributing significantly to the legacy of apartheid planning and the 'apartheid city'. Spatial planning was an extremely effective instrument in entrenching separation and exclusion of specific races from economic opportunities. More than ten years after the official demise of apartheid, economists and planners continue to grapple with the question of how to undo the damage done by the apartheid regime.

In addition, the architectural profession undermined itself through the ill effects of the worst practices of modernism, embodied in a complete disregard for heritage and pursuit of an ideology of dominion over nature. As a result, both the planning and architectural professions find themselves in an apologetic position for the inherited legacies. In addition, the rise of the project and cost-management professions have elevated time and money considerations to a position of prominence within project leadership structures and undermined the leadership role which architects have historically taken.

In truth, architects are trained in design quality and aesthetic considerations that form the basis of sustainable design interventions. Many of the oft repeated recipes for sustainability are in fact constrained by case-project circumstances. For example, northern orientation (in the southern hemisphere) may be constrained by unsightly views, a tight site that is orientated differently, or a street-front that conflicts.

The GBCSA's rating system provides a reference point for all professionals on the team. It provides a check-list of considerations for interpretation that moves beyond the early postulations of principles for sustainable design. This will potentially stimulate a more detailed list of considerations that will ensure that buildings will perform in the way that they are supposed to. For example, the commissioning standard requires that a specialist engineer commission the air-conditioning system on a commercial building to ensure that it functions in the way that it is designed to (i.e. at maximum efficiency). Having said that, the accreditation process is one of documenting good sustainable design, and without the latter, the former is rendered impossible.

It remains to be seen how much of the slack the GBCSA will pick up, but by providing a rating system (initially for office buildings), they will give legitimacy and credibility to corporate initiatives that are successful in implementing strategies for sustainable development. This makes the goal of sustainable design more attractive to the client, as opposed to the historic situation where any attempt to 'be green' is often met with stark criticism, contestation, and debate. Among other benefits, the GBCSA will provide a system for ensuring that what is proposed finds continuity and results in on-site manifestation through a strict document tracking process that holds all parties accountable.

3.   Looking forward – Recommendations

The application of sustainability as a focus area calls for innovative approaches to the re-interpretation of existing professions. It becomes clear that for sustainable architecture to be realised in the South African context, certain adjustments in approach to the construction industry are required. We believe there are five specific preconditions that are required for the larger systemic context to support sustainable architecture:

1. *A shift in focus in the cost-control sector, so that insurance companies and banks provide incentives for innovation development.* This would be coupled to campaigns that create aspirations for innovation among end users, and the adoption of cost analysis over a building's whole-life as opposed to upfront budgets from quantity surveyors.
2. *The development of a culture of innovation generation through cross-disciplinary communication and product development.* Ideally, this would need to be initiated in partnerships with manufacturers and engineers
3. *Consideration of adaptability and long-term durability, and a design aesthetic that can be cherished by future generations in the design process,* so that buildings can become viable for use over longer time frames.
4. *Deceleration of construction time frames, so that there is time to slow down to allow for reflection and innovation development* during the design and construction phases of building.
5. *Upfront investment implementation capacity in terms of the right knowledge, skills, tools and methods to deliver green and sustainable solutions* is critical.

We will now illustrate the possibilities and challenges of sustainable architecture by considering two case studies.

4. Case Study 1: The Public Transport Shared Services Centre

Address: Corner Vanguard Drive and the N2, Athlone

Client: Western Cape Provincial Government (PGWC), Department of Public Transport

Project description: 6000 m2 shared services facility with customer interface, office space, public transport testing and licensing facility, and public administration headquarters.

The Public Transport Shared Services Centre (PTSSC) was one of the first administrative state buildings in South Africa to have a sustainability mandate. The project began with clear intentions to meet this mandate. The following strategies were intended:

4.1 Summary of Strategies

4.1.1 Narrow building modules (maximum 14m)

4.1.2 Orientated with long façade facing North

4.1.3 Passive ventilation (solar towers, atrium louvers)

4.1.4 Super-efficient hybrid heating, ventilation and air-conditioning (HVAC) system

4.1.5 Adjustable louver shading system

4.1.6 Light shelves to increase daylight

4.1.7 Intelligent lighting system

4.1.8 Planted sod roof

4.1.9 Vertical gardens

4.1.10 Solar water heating

4.1.11 Grey-water recycling system

4.1.12 Responsible material procurement –
- non-toxic
- locally manufactured
- low embodied energy
- durability
- natural/raw materials from a renewable source
- recycled content

4.1.13   Fly-ash in concrete

4.1.14   Insulated walls

4.1.15   Empowerment strategy for unskilled labour and the employment of women

4.2   Barriers to Sustainability as Experienced on PTSSC project

The interpretation of sustainability in the context of this building was applied in the realm of the technological interventions. Design aspects governing form were driven by the architectural team and found manifestation from inception to construction. At different stages in the project, different stakeholders/role players carry the responsibility of interpreting design intentions.

Relating this to the typical design process, one can highlight which role players come to the fore as follows:

Work stage	Description	Decision-makers
Site appraisal, briefing	Analysis of site opportunities and constraints. Client describes brief and accommodation schedule requirements	Client, legislation, site informants
Concept design	Interpretation of brief and site elements	Architect
Sketch design	Formalisation of and tuning of design	Architect, and engineer
Council Submission & Design Development	Detailed development and interpretation of the design	Technologist, engineer, quantity surveyor, specification specialists
Contract and site administration	Construction	Project manager, building contractor

This clearly shows that the significant decision-makers in the first stage of building design are completely different to those in the final stages of construction – a discrepancy that resulted in a range of impediments.

Firstly, the early stages were dominated by instructions from the client, based on a limited precedent study. One particular example that was used, was the precedent set by the proposed World Trade Centre in Dubai, which sports a giant wind-turbine between two aerodynamically shaped towers. Without digressing into a detailed description of the engineering requirements for a building like this, the key point is that the wind conditions on the site are not suitable for this intervention.

Secondly, there was a lack of intellectual continuity on the part of client representation, so that there was insufficient clarification of user requirements and/or developmental targets. Over the course of the project, there were three changes in engineers and two changes in the Project Architect responsible for project implementation. Each held a different set of work ethics, interpretation, and commitment to the sustainability mandate. This presented a challenge to the architectural team, who strove to maintain conceptual and design integrity throughout the process. So, for example, while the engineer in the earlier stages of the project argued for insulated wall cavities, these were not installed by the contractor when construction began. Exacerbating the problem, subsequent engineers refuted the necessity for such an intervention. Post-installation would have set the project back on its timeline by three to four months, and resulted in an extended cost to the client. The engineer appointed during this stage modelled the effects of the insulation and found that it would make a 1% impact on the overall heat efficiency of the building. This was not considered significant enough to warrant holding the project schedule back.

Thirdly, once on site, many of the specifications were overlooked by the cost-cutting, time-saving contractor, with the result that what was built barely complied with existing norms, let alone the higher required sustainability standards. This is due to lack of sustainability experience and know-how on the part of South African contractors, with a poor appetite for learning in an environment that is notoriously under-skilled and unprofessional in its undertakings in comparison to international markets.

Fourthly – particularly with respect to the specification of materials according to the strategy outlined above – many of the available options which met the criteria simply did not fit into the budget as proposed by the quantity surveyor. For example, the budget allocated for carpets only allowed for the cheapest carpets available on the market, in spite of the identification and specification of carpets which were made from recycled products and a system for minimising the use of glues which are known to be highly toxic.

The PTSSC case study outlines how the professional team structure and varying levels of engagement with, and commitment to, sustainability influenced the final outcome. With the advent of the Green Star rating system, a strict document tracking regimen is required to ensure that sustainability designs are followed through in order to establish a precedent for a more accessible coordination of objectives, thus effectively addressing some of the technical barriers associated with lack of continuity.

Currently, sustainability is treated as a supplementary service. In the case of the PTSSC, it was well-integrated into the initial phases of the project, but was not properly integrated into the final specifications, because the project manager usurped the traditional role of the architect as principal agent. This mandate was used to make strategic design decisions that favoured an expedited project delivery over the final performance of the building.[1] Without a tangible presence of creative leadership, buildings are sacrificed on the altar of expediency. It is not uncommon for contracting companies to dissolve shortly after project completion, which results in reduced accountability and the nullification of many of the legal instruments that can be pursued by clients against poor contractors.

 A further hurdle can be attributed to lack of experience, accessibility, and knowledge pools in the technical specifications for sustainability, and, more specifically, to budgets that do not allow for the additional funds needed to ensure sustainability. Xanita's *Green Forum* (2008) cites PriceWaterhouseCoopers UK:

"…the high prices associated with fair trade and organic products remained the main inhibitor to further growth. Over 48% of UK consumers stated that they were unwilling or unable to pay the premium for more sustainable products. Further studies concluded that they were correct in this assumption – sustainable goods are priced up to 45% higher than ordinary retail products."

Recent intelligent approaches to these supposed 'surplus costs' have demonstrated that there are substantial offsets in whole-life costing analysis. For a minimal premium from so-called 'additional' funds, long-term savings can be achieved. It is important that the impression that sustainable buildings are cheaper should be measured across time and ongoing maintenance.

Different considerations are revealed in our second case study in Cape Town.

5.   Case Study 2: The Oude Molen Eco Village

Address: Alexander Road, Mowbray

Client: Western Cape Provincial Government (PGWC), Department of Public Works

Project description: Urban design framework for an integrated eco-village.

At the time of writing, the Oude Molen Eco Village was in a stage of design development that precedes the technical resolution process exemplified by the PTSSC case study. There was political and community support for the vision of technical sustainability. The social agenda was strongly

---

1   The Principal Agent has veto powers of decision on a construction project. The Project Manager's traditional role is to ensure that a building is delivered on time according to budget. The Architect's role is to ensure that a quality design is implemented according to the client brief. The Architect's mandate is traditionally holistic, whereas the Project Manager's role is administrative and managerial.

addressed through inclusive planning, a robust social engagement process, and a planning scheme that would provide a range of affordable housing – from low-cost housing to middle income. The proposed model developed for this project was based on a public value approach. The land would remain the property of the state, while the right to use the super-structure would be tradable. Proceeds from the sales of these rights would be re-invested into the same site for cross-subsidisation of gap and low-income housing, and provide a float for maintenance. The project proposed to be a mixed-use development, well-connected to public transport. In this way, the project addressed the need to lower carbon footprint through planning schemes.

5.1 Barriers to Sustainability

In this case, the main barrier to the realisation of the sustainability of the project was presented by the complexity of the political environment. The neighbourhood itself is positioned between the Valkenburg Psychiatric Hospital, the Maitland Garden Village, and Pinelands.

The site itself is the former 'blacks only' wing of the Valkenburg Hospital. After it was decommissioned, squatters moved in and an Oude Molen Villagers Association was formed – ironically of a predominantly white constituency. The predominantly non-white Maitland Garden Villagers objected to whites being the beneficiaries of housing subsidies when the resident Maitland Village Backyard Dwellers Association had been on the housing waiting list for several years. They raised concerns over who would ultimately benefit from the project.

Similarly, the residents of Pinelands, a Garden City-paradigm planned suburb – formerly 'whites only' under apartheid regulations, and now a middle-class suburb – were concerned about who would move into the proposed development, and objected in terms of the NIMBY ('Not In My Back Yard') prejudice (discussed in Charlton, 2006). The primary motivation was the concern over a potential rise in crime due to the proximity of additional poor people.

The traffic impact assessment, conducted as part of the Environmental Impact Assessment process required for approval of the development plan, showed that the new development would require additional traffic access alongside the Valkenburg Hospital land, through an existing, but closed, road. The closure of this access road isolated Valkenburg from its surrounding areas entirely, as it was bounded by arterial roads on two sides and a river on a third. The land was owned by the Department of Health and its cooperation was required in order to see it re-opened. The access route was viewed as a security risk to patients.

5.2 Lessons Learned – Overcoming Barriers

As quoted in Charlton (2006), Mcarthey suggests that "the most effective impacts on the fortunes of the poor will most likely be achieved through leveraging market forces with earmarked public funds… yielded partly through facilitating… middle or higher-end developments".

The precedent is to sell off land to the private sector for profit, which is then invested in low-cost housing schemes elsewhere. The public value approach retains the land as an asset, but mobilises funds through the sale of the rights to middle- and upper-income interests, but ensures that provincially owned land is inalienable.

The Oude Molen project proposes to adopt a public value approach to this development. It was proposed that a special purpose vehicle for processing the public value approach be created. A housing special purpose vehicle would retain the ownership of the land for unlocking maximum opportunities and value for the public.

Historically there is a precedent for political parties utilising such vehicles to put land in the hands of the political elite. As a result of continuing political uncertainties, any unusual actions taken in the future will be viewed with suspicion and potentially met with resistance from political opposition.

The Oude Molen project exemplifies the more macro-scale planning issues for sustainable neighbourhoods. Planning is a political engine and cannot function in isolation of its political contexts. It has been used at different times, in different places, both for control and oppression, as well as for advocacy of the needs of the marginalised.

The ability to take cognisance of the political context, while continuing to search for innovative solutions to the barriers, is of paramount importance to the success of sustainable projects. Untangling the political challenges is the key to sustainable architecture becoming a practical reality.

6. Conclusion

There is light at the end of the tunnel. We predict that, as professionals grapple with sustainability, they will develop knowledge and experience in the most effective solutions, and fine-tune technical documentation and communication. Furthermore, increases in the knowledge base and a demand for sustainable products will make it more viable for suppliers and architects to engage with innovation more constructively. However, for the process to be properly integrated, there needs to be a dramatic shift in values from a tendency to focus narrowly on up-front costs to whole-life cost analysis. And last, but not least, architects will have to claim a stronger leadership role throughout the creative process in order to make sustainable buildings a reality.

References

AFSA (Aluminium Federation of South Africa). 2007. *Aluminium and the environment. AFSA Newsletter.* (Spring/Summer):5-7.

Beck, A. 2005. *Sustainability and Facility Management.* CRC for Construction Innovation Forum. Presentation in Sydney, Australia on 9 August. Available from http://www.construction-innovation.info/images/pdfs/PublicPresentations/Presentation_Beck.pdf (Accessed 31 July 2008).

Charlton, S. 2006. *Making urban land markets work for the poor: Synthesis paper.* Paper commissioned by Urban LandMark for Land Seminar, University of the Witwatersrand, Muldersdrift, South Africa. Available from http://www.urbanlandmark.org.za. (Accessed 30 July 2008).

Damanpour, F. & Wichnevsky, J.D. 2006. Research on Innovation in Organisations: Distinguishing innovation-generating from innovation-adopting organizations. *Journal of Engineering and Technology Management,* 23:269-291.

Design Indaba. 2009. *The Most Beautiful Object in SA 2009.* Available from http://www.designindaba.com/video/most-beautiful-object-sa-2009 (Accessed 31 March 2010).

Du Plessis, C. 2006. Agenda 21 for Sustainable Construction in Developing Countries – Project description. Available from http://www.buildnet.co.za/akani/2001/july/03.html. (Accessed 30 July 2008).

GBCA. 2008. *About – Green Building Council of Australia.* Available from http://www.gbca.org.au/about/ (Accessed 14 May 2010).

Green Globe. 2010. *Green Globe History.* Available from http://www.greenglobecertification.com/about.html (Accessed 14 May 2010).

Holden, M., Roseland, M., Ferguson, K. & Perl, A. 2008. Seeking urban sustainability on the world stage. *Habitat International,* 32:305-317.

Leonard, A. 2008. *The Story of Stuff.* Available from www.storyofstuff.com. (Accessed 29 July 2008).

Macozoma, D.S. 2001. *Secondary Construction Materials: An Alternative Resource Pool for Future Construction Needs.* Concrete for the 21st Century Conference, Midrand, South Africa.

Macozoma, D.S. 2006. *Developing a Self-Sustaining Secondary Materials Market in South Africa.* Dissertation. Johannesburg: University of the Witwatersrand

Napier, M. 2007. *Making Urban Land Markets Work Better in South African Cities and Towns: Arguing the Basis for Access by the Poor.* Paper presented at Fourth Urban Research Symposium: Washington DC.

NAO. 2004. *Getting Value for Money from Construction Projects – How Auditors Can Help.* Available from http://www.cabe.org.uk/files/getting-value-for-money-from-construction-projects-through-design.pdf (Accessed 14 May 2010).

PricewaterhouseCoopers. 2008. *Sustainability and Climate Change.* Available from www.pwc.com. (Accessed 29 July 2008).

SA Government. 1999. National Heritage Resource Act, 25 of 1999

The Slow Society. n.d. Slowness for Sustainability. Available from www.slowsociety.org. (Accessed 31 March 2010).

The Slow Society. nd. Welcome to the Slow Society. Available from www.slowsociety.org. (Accessed 31 March 2010).

The Worldwatch Institute. 2007. *The State of the World; Our Urban Future.* New York: W.W. Norton & Company.

UNEP 2002. GEO 3. Available from http://www.grida.no/publications/other/geo3/?src=/geo/geo3/index.htm. (Accessed 31 March 2010).

Worby, M. 2006. Lecture on Ecological Design presented at the Sustainability Institute, Stellenbosch, South Africa, 22 June 2006.

Xanita. 2008. July Newsletter. Email circular about the *Green Forum.* Available from www.xanita.co.za. (Accessed 2 July 2008).

14

Governance, Housing and Sustainable Neighbourhoods

Paul Hendler

1. Introduction

The primary aim of this chapter is to explore sustainable housing in Cape Town through the lens of the political economy of housing delivery, because this is where the major obstacles exist to broadening access to decent, affordable, and sustainable residential accommodation. Through identifying political and economic obstacles to the delivery of affordable and sustainable housing, the chapter intends to contribute to a debate about the political and economic practices that would arguably begin to address and overcome these obstacles.

Sixteen years after the advent of universal suffrage in South Africa, and the introduction of lofty policies proclaiming the intention to deliver decent housing and enable thriving communities with integrated living and work places, the majority of our people are still plagued by housing problems (informal structures, overcrowding, inaccessibility of housing finance, unaffordable housing prices, relatively high cost of well-located urban land and spatial *apartheid* distortions). One of these documents, the Western Cape Sustainable Human Settlement Strategy, Isidima, takes "as its point of departure the constitutional right to housing and the existence of a market economy that is regulated by a developmental state" (Western Cape Department of Local Government and Housing, 2007: 12). How is it that the state, which proclaims itself as a 'peoples government', and which is based on the popular will of the vast majority of the electorate, is unable – or unwilling – to facilitate the emergence of the sustainable human settlements frequently referred to in all its major policy documents?[1]

To answer this question we begin by first addressing the nature of the South African (post-apartheid) state. In doing this we take a different approach to the official notion that the state is a developmental state[2] regulating a market economy, and instead identify the state as a capitalist state, founded on the social relations of production and the social division of labour (Poulantzas 1980). Thereafter we define what sustainability means in the context of the economics of decent, affordable housing. We consider the financial and ecological implications of the current resource use approach in relation to affordable housing over the long term. Attention is given to a theoretical consideration of criteria for sustainability in the context of the political economy of affordable housing.

Using this framework, the chapter identifies the housing delivery mechanisms that have been enabled through the 'Breaking New Ground' (BNG) approach to housing, adopted by the Government during 2004.

Within the context of BNG and the variety of housing delivery mechanisms – or 'housing typologies', as these mechanisms are often referred to – the chapter goes on to explore and discuss obstacles to implementing sustainable housing on a large scale and the challenges to delivery in the specific context of Cape Town. The theoretical framework of the South African state provides us with a context within which to ask questions about the current provision of affordable housing. Has the increased policy-level flexibility made it possible for cities such as Cape Town to integrate both economic and ecological sustainability more fully into the delivery of housing? Have these reforms resulted in the intended range of housing delivery systems? This leads to a focus on the function of spatial planning as a precondition for sustainable housing.

The chapter argues that the political preconditions for sustainable housing in Cape Town do not yet exist, and that as and when these conditions start emerging, it will be necessary for municipalities to have partnerships with community-based organisations to change existing urban spatial and housing patterns. This is because the motive force for changing the *apartheid* housing landscape lies within

---

1   Isidima is replete with in-depth analyses and radical proposals for restructuring the Western Cape towns and cities but three years on there is little to show for the achievement of its strategic objectives; this conclusion is drawn by the author from his involvement in the human settlement planning aspect of the Built Environment Support Programme (BESP) a provincial initiative to capacitate Western Cape municipalities to complete credible Human Settlement Plans (HSPs) and Spatial Development Frameworks (SDFs).

2   Democratic South Africa has also been characterized by growing inequality, which reinforces the questioning of the notion of the developmental state (cf. Terreblanche, 2002:33).

independent community-based organisations of the largely black working classes, rather than within the state itself which is a condensation of these and other forces opposed to this objective.

The chapter concludes with thoughts on the make-up of the state administration and the role of the state bureaucracy and its officials in undermining or enabling the achievement of the laudable objectives of Isidima. If officials are not the main initiators of the required policy changes, what role is there for them as well as for professional practitioners who are sympathetic to and support the changes envisaged in Isidima and the development of grassroots organisations that can contest for state power, and thereby ensure or enforce the implementation of policies for sustainable housing? The conclusions consider some practical recommendations about how professionals working with government bodies, like the Cape Town Municipality, can effectively begin to plant foundational ideas for municipal delivery on the imperatives as stipulated in more innovative policies.

2.   The role of the state vis-à-vis the people

To answer the questions referred to earlier we need to dig deeper into the material conditions within which the state is rooted, namely the capitalist mode of production, and in particular the social relations of production and social division of labour. This goes back to Marx's analysis of the capitalist mode of production and the location of the state in that mode, a view that many might find surprising given the fall of the Soviet Union, the heartland of Marxist socialism, almost 21 years ago, and therefore the perceived failure of Marxist theory. Instead, we are led to believe in the 'end of history' (cf. Fukuyama, 1989) - modern capitalist economies are said to have solved the problem of crises, and democratic states under capitalism legitimated themselves through the defence of individual freedoms. However, the Asian crisis and the collapse of the dotcom bubble in the late-1990s, the near collapse of the world economy in 2008, the 2010 Greek sovereign debt crisis as well as intensified conflict in countries like Kyrgystan, Greece, and Thailand calls this into question. Marxist theory is a critique of an existing system, the usefulness of which lies in its explanatory value of social phenomena, like the apparent contradiction between the stated aims of the South African government's housing policies and the reality on the ground.

Poulantzas (1980) articulated a theory of the state under capitalism based on Marx's theories by developing the notion of the dominant and dominated classes further, noting the emergence of what he called a "new" petite bourgeoisie, namely professionals like engineers, lawyers, doctors, and also state bureaucrats, etc. He argued that power is inextricably bound up with and present in class relations, that the state under capitalism functions broadly to reproduce the social relations of production and social division of labour and that the state is the condensation of a relationship of class forces. This last point effectively broke with a Marxist-Leninist view of the state as an instrument which cannot be influenced by the dominated classes and therefore needs to be encircled, captured and smashed in order to erect a new, truly workers' state. This allowed the possibility of the struggle of the dominated classes to impact on the state and more importantly, to influence state policy, and therefore allowed the prospect of a political process of struggle through which the state could evolve into a more democratic state orientated to workers' needs.

Poulantzas (1980: 127) argued that under capitalism "the state represents and organises the long-term political interests of a power bloc, which is composed of several bourgeois class fractions" (like competitive capital, monopoly capital). One of the fractions of capital is hegemonic in this bloc – Poulantzas referred to the hegemonic status of monopoly capital in Europe in the 1980s. In the current South African case hegemonic status in the power bloc could be ascribed to what Mbeki (2009) refers to as the Mineral Energy Complex – comprised of the mining industry, and its associated chemical and engineering industries and finance. Manufacturing and agricultural capital could be seen to play a subordinate role, with manufacturing being effectively de-industrialised as South Africa opened up its markets to international competition; conflicts within the power bloc also forms the basis for internal divisions in the state. The black bourgeoisie is also a subordinate party in the power bloc – according to Mbeki it emerged through a deal between the Mineral Energy Complex (which he refers to as an economic oligarchy) and the African National Congress (ANC) leadership at the negotiations at Codesa: the deal struck was that there would be political control and Black Economic Empowerment (BEE) for the ANC and its associated networks, in return for no real transformation in the social relations of production, influence (for the economic oligarchy) over ANC economic policy, and positioning of this oligarchy for state contracts. Terreblanche (2002: 95-139), in examining the interactions between the ANC and corporate South Africa during the early 1990s, in the latter's search for a new political form to reproduce the established regime of capital accumulation, makes the same point.

The state also plays a role *vis-a-vis* the dominated classes: according to Poulantzas (1980: 140) the state brings "the power bloc and certain dominated classes into a (variable) game of provisional compromises. The state apparatuses organise-unify the power bloc by permanently disorganising-dividing the dominated classes, polarizing them towards the power bloc, and short-circuiting their own political organisations. " Poulantzas (1980: 141) stressed that popular struggles "are not really external to the state"; these struggles are inscribed in the state "because the State itself bathes in struggles that constantly submerge it." ".. the dominated classes exist in the state not by means of apparatuses concentrating a *power of their own,* but essentially in the form of centres of opposition to the power of the dominant classes" (Poulantzas, 1980: 142). The conclusion that Poulantzas draws from this view is that the dominated classes could not hold on to state power without a radical transformation of the state. This arises from the unity of the state power of the dominant classes as well as mechanisms within the state which reproduce the domination-subordination relationship, meaning that the dominated classes are retained within the state precisely as the *dominated* classes, something which is reflected in the division of intellectual and manual labour in the state administration (or civil service). Whether or not the dominated classes and their organisations are able to concentrate the power of opposition in the state depends on the strength and *autonomy* of their organisations from the state.

The point of this theoretical piece is that if the South African state is not delivering on its promises of housing and sustainable human settlements to the South African workers it is because as a class through their various organisations (i.e. trade unions, civic groups) South African workers do not exert sufficient force within the state to defend policies like Isidima and therefore to enable their implementation (or partial implementation). This lack of working class force is nowhere more evident than at the local government tier of the South African state, where maladministration and poor governance have led to a lack of service delivery and increasing dissatisfaction amongst the working class electorate whose grievances build up and then spill over into direct action (often violent), but these do not yet reflect a coherent political (yet alone anti-capitalist) character. Bond (2009) refers to some 5 800 protests held in South Africa's towns and cities during 2004 and 2005. Neocosmos (2006: 125) argues that the 2008 wave of xenophobia is a critical indicator of the extent to which the dominated classes are fractured and disorganised. Nevertheless signposts of independent movements and structures are there in the examples of landless peoples' struggles, like the Federation of the Urban Poor (cf. Community Organisation Resource Centre, 2009; 2009a), the Khutsong community's dramatic independent revolt against the ANC government (Sunday Times, 2009) and struggles of informal settlers in and around Durban.

The class nature of capitalism and the embeddedness of the capitalist state in class division and power struggles often gets lost in most contemporary debates about the limits to economic growth, resulting in a failure to get progressive policies and regulations implemented, which could make an impact.

3.   Sustainability in an affordable housing context

Before housing can be provided, it has to be produced through a given economic system, in this case the South Africa capitalist system. The production of housing comprises both the costs of production and the surplus generated from the sale of the housing units, which accrue to developers and real estate agents, while the financial sector also realises profits from the financing of housing, through the interest charged on mortgage-secured loans. Existing housing sold through secondary housing markets attains a value (price) determined primarily through location. Functioning secondary markets enable housing to be a store of added value (i.e. assets) in addition to use-value as shelter. Sustainable housing therefore presupposes a sustainable market for the production and provision of housing, where effective demand stands in a sustainable relationship to the supply prices in both primary and secondary markets. Housing that is priced *affordably* for the poor is highly dependent on the cost of the land component (the economic costs of a top structure being given), and the cost of land is a function of its location, which is subject to a spatial land allocation policy as well as market speculation. Given the relatively high price of well located land and the relatively limited incomes of most working people in South Africa, the latter have virtually no access to the private housing market. This is the material reality which public housing policies in both the *apartheid* and post-apartheid political regimes have addressed. Since affordable housing is nested in a set of economic production, market and state distribution and political (or policy) relations, an appropriate political economy of housing (based on a transformed state) is a precondition for sustainable affordable housing. Furthermore, the political economy of housing is embedded in a set of natural systems that are key to the provision of eco-services. In order to support broader eco-sustainability, it is crucial that the relationship between the political economy of housing and natural systems is sustainable.[3] Therefore a housing policy aligned with the needs of the majority will have to also reflect a sustainable usage of energy (by households) and recycling of so-called waste.

In addition to appropriate spatial planning, new methods of building, renewable energy usage, and waste recycling technologies are also required to contribute towards achieving a sustainable relationship between eco-systems and the political economy of housing. Using the above technologies, as well as enabling appropriately located residential accommodation, the political economy of housing could contribute to – and align with – transformation of the broader economy to one that is less carbon-intense through shifting to non-fossil fuel (renewable) sources of energy, and waste recycling.[4] Conversely, fitting new (and retrofitting existing) housing with the requisite technologies has an impact on the economy of housing – through the financing of the costs of installation – and ultimately the affordability and therefore effective demand for affordable housing. From our perspective on the South African state we see that the pre-condition for sustainable, affordable housing that addresses the needs of the majority of South Africa's people for proper shelter, is a *transforming state* – this transformation depends on the extent and level of autonomy of organisations of the dominated classes that are able to articulate the needs and required programmes through a set of demands and campaigns aimed at influencing and transforming public human settlements and housing policies and practices.

Developing the debate for these programmes requires that we should also interrogate whether ever-increasing housing prices, generated through the growing velocity of secondary market transactions, is affordable – and therefore sustainable – for a growing number of people in the below-R12,000.00 per month income bracket.[5] Housing booms, such as experienced until recently in the USA as well as in South Africa, reflect and contribute to an exponentially growing economy, which itself will ultimately run up against the limits of a finite eco-system. Jackson (2009) argues compellingly that a sustainable relationship between the global economy and the planet's eco-system implies a no-growth (or steady state) economy – in his words "prosperity that would be compatible with a no-growth economy".[6] It follows from the above that a sustainable relationship between the political economy of housing and natural systems ultimately requires a slowing down in housing price rises, and eventually a halt to price increases. This contradicts the conventional wisdom that what makes

---

3   'Sustainable' means "that which can be maintained over time." (cf. Heinberg, 2007).

4   Environmental economist Herman Daly has suggested conditions for sustainability, focusing on the resource base (cf. Meadows, Meadows & Randers, 2004).

5   This is generally viewed as the floor household income above which a household is assumed to be able to purchase adequate residential shelter on the private market.

6   See also Daly (1996) and William Rees (in Heinberg, 2007) who emphasise the same point.

secondary housing markets sustainable is the fact that prices in these markets should be rising over the long term. What is required is a profound re-thinking of the function, as well as the sustainability, of secondary housing markets.

This re-conceptualisation of the role of secondary housing markets should inform an emerging pro-poor view of sustainable housing practice, which calls for a balance between the valorisation (through increasing prices) of housing in the affordable segment of the market and a decrease in the velocity of valorisation at the middle and upper income segments, in order to both address the need to slow down the market (at the top end) and facilitate social equity for households in the affordable segment (through their accessing higher value assets in the short to medium term).

Achieving this balance in practice requires the relocation of affordable housing in areas favourably located in terms of work opportunities, social facilities (like schools, clinics, etc.), retail outlets and transportation, because secondary market prices are primarily driven by location. To understand the full picture of how housing could be integrated with these facilities, opportunities, and outlets presupposes the assumption that the housing is located within a space-economy, or what Giddens (1981: 29-33) refers to as the "created space of contemporary urban living", which is the integration of three sets of market relations – housing, labour, and product markets – and their interlinking transport arteries. The failure to disburse adequate residential shelter to a broad range of the population has persisted since the 1980s (Hendler, 1993), and represents an enormous failure of one of the three pivotal sets of market relations that constitute the dynamic of metropolitan South Africa.

This failure is in the first instance one of spatial planning policy and practice, the extent of which will be explored in the third section of this chapter, which analyses Cape Town's housing situation. That section will also apply the definitions of sustainable affordable housing in exploring the trends that have shaped the development of the housing market in Cape Town in recent years. Prior to analysing the sustainability of the various segments of Greater Cape Town's housing market, we need to understand the Government's new approach to housing and human settlements (BNG which was launched in 2004) and also identify the different housing typologies intended to facilitate the realisation of this approach.

4. Delivery mechanisms to generate sustainable affordable housing

During the 1960s and 1970s capitalist states all over the world – including South Africa – played a significant role in providing mass housing to employees of both parastatal organisations and private capital. The reason was to reproduce the work forces for the requirements of large-scale industries – in South Africa this led to the construction of hundreds of thousands of housing units in Soweto as well as other townships adjacent to all the major South African towns and cities, and also in the Bantustans near to the border industries which had been established through incentives under the *apartheid* regime (Hendler, 1986). The state provided what was referred to by Poulantzas (1980) and Castells (1977) as 'the collective means of consumption', and struggles by the dominated classes (especially in South Africa) were often focused around grievances about the inadequacies of the housing provided. Since the privatisation of public housing stock, beginning in the mid-1980s, a major focal point for resistance to state policies has been removed. After 1994 the ANC government continued the process of moving away from public housing to privatised housing, first with the RDP and currently with the BNG policy, which emphasises the need to create integrated human settlements by facilitating the geographical spatial integration of all residential communities and the accompanying, necessary social facilities and places of work. The method to achieve this is State intervention in property markets, by making land available at affordable prices and through providing subsidies to affect the structure and outputs of the housing markets.

The logic is that the State provides land and housing subsidies to accelerate housing delivery within sustainable, coherent settlements, which will turn quality housing and homes into assets and create a single efficient formal housing market, as well as overcome the effects of *apartheid* spatial planning on the housing market. Accordingly, the Government has formulated policies regarding the planning of human settlements and the implementation and delivery of housing in these settlements.

BNG's point of departure is that there is no single housing delivery programme that addresses comprehensively the problems of housing shortages, affordability, and sustainable settlement. Instead, a range of housing delivery mechanisms are required to address the full range of housing needs.

As has been mentioned very few sustainable human settlements have come to pass under the *post-apartheid* state, notwithstanding the fact that there are also at least six housing delivery mechanisms that are aligned with National and Provincial housing policies, which could be applied practically (these delivery mechanisms are depicted graphically in the appendix). The lack of sustainable human settlements could arise from intra-state contradictions between branches and apparatuses of the state relating to the determination of priorities and counter-priorities as well as a process which filters decision-making through a bureaucratic hierarchy. According to Poulantzas (1980: 135) the appearance of state policy at the micro-political level as phenomenally incoherent and chaotic, is because one is only viewing the policy at a particular moment of its movement through the state administration but that in the longer term a certain coherence emerges.[7]

We turn now to an explication of each of the housing delivery mechanisms before assessing their application and impact in the Greater Cape Town area. There is currently a flexible and open-ended public housing delivery framework, which through partnerships with the private and community sectors, could be exploited to maximise the provision of the most appropriate and affordable housing for the poor. In certain cases these offer opportunities for organisational development of the dominated classes, such as the idea that the management and building of own structures can be a community building and development process (cf. Turner [1972] in this regard).

4.1 Incremental formal housing

Incremental formal housing refers to a process whereby people build their own structures informally (i.e. from unconventional building materials, usually resulting in what is referred to as 'shacks') and then formalise these over time (i.e. alter the structure by including conventional building materials, like bricks and mortar). These structures are typically referred to negatively as 'shacks', and the response of the authorities (e.g. the National Department of Housing in respect of the N2 Gateway project) has often been to engage in shack clearance. However, there has also been resistance to this approach by shack dweller organisations such as the Federation of the Urban and Rural Poor (FEDUP), who have organised as shack dwellers to initiate and participate in the upgrading of their accommodation and residential space (FEDUP, 2009; Community Organisation Resource Centre, 2009).

In order to facilitate these grassroots initiatives, a different, incrementalist perspective could be taken with regard to informal structures. A reframed approach could acknowledge that the poor, currently settled illegally, have actually invested in housing and are therefore sitting on a potential asset. Such housing could be appreciated as a locator that forms a viable basis on which the building (as asset) could be expanded. What starts off as an informal structure can, through incremental building, gradually be turned into an acceptable formal house over time. The location of the shacks is critical in determining the success of this mechanism. If, for example, these structures are erected on flood plains, they would not be sustainable.

A critical starting point is the location of the informal housing within the space economy. Currently, many of these structures in South Africa are relatively far from amenities and access to transport points – although the current bus rapid transport (BRT) intervention in Cape Town might improve this. Where informal structures are not well located with respect to facilities and opportunities, they serve as little more than the dumping grounds of the aged and the unemployed (as they functioned under apartheid, to relocate the reserve army of labour and the structurally unemployed), rather

---

7   What precisely lies at the source of the extreme contradiction between the state's human settlement policies and social reality is beyond the scope of this paper, but could form part of the content of an agenda for further investigation.

than the locators for opportunity and housing consolidation. This point is especially important given the Western Cape Government's recent concerted move to emphasise site-and-service as its primary response to the housing backlog in and around Cape Town.

4.2   Subsidised housing – giveaways

Housing for the urban poor was (and still is, notwithstanding BNG's lofty vision) regarded as a welfare function, and as such directly facilitated by the State. Following the 1994 White Paper on Housing, State policy under Joe Slovo (then National Minister of Housing) and Billy Cobbett (then Director-General in the National Department of Housing) attempted to locate housing for the poor in appropriate parts of the metropolitan space-economy and channel substantial allocations of subsidies via mechanisms that enable access for the poor to established housing and property markets. Their approach was to provide subsidies for developers to deliver houses for the poor. Over one million so-called 'reconstruction and development programme' ('RDP') units were delivered in South Africa between 1994 and 2004, which is probably a world record for quantitative outputs. However RDP housing is perceived to have failed qualitatively: it has been criticised for being an inadequate form of shelter for people due to inadequate quality control over workmanship, lack of privacy (the standard RDP structure had no separate rooms), and location. More importantly it has also been criticised for reproducing the spatial marginalisation of the dominated classes: their strategy was at best only partially successful, because there was no coherent and integrated planning at local authority level.[8] This marginalisation reinforced a vicious cycle of poverty that disables poorer households from participating in the private accumulation of housing assets. Notwithstanding its growing unpopularity and the BNG move away from RDP as the major housing type, there are probably still many RDP subsidies in the pipeline for housing units yet to be built. Given the time that it takes to create alternative programmes, it might be prudent to use the RDP mechanism in a modified form to address some of its critical weaknesses.

A modified version of RDP design might be incorporated as a delivery mechanism within a sustainable settlement delivery strategy. Through the investment of public funds from the Municipal Infrastructure Grant (MIG) and other programmes of the Treasury,[9] infrastructure could be installed. Then – through the current capital subsidy programme – a defined number of units could be developed and integrated with existing settlements and Council-owned land. These RDP units could be laid out in denser formats, e.g. as cluster housing.[10] This would not only serve to minimise sprawl, but also enable more effective use of available land.

4.3   Social housing

Government policy facilitated the emergence of social housing as a delivery mechanism that enables a choice for alternative forms of tenure in addition to outright ownership – like rental, rent-to-buy, cooperative housing, and instalment sale. Government provides subsidies to accredited social housing institutions (SHIs) that also raise loans either to purchase and refurbish existing buildings, or to start new developments. These institutional subsidies are available for medium to high density social housing units. Recently, the social housing restructuring grant became available for SHIs operating in so-called 'restructuring zones' – this has significantly increased the quantum of public capital available for social housing units to about R170,000 per unit, and is an incentive to contribute to the integration of cities and neighbourhoods.

The SHIs' task is to effectively manage the properties to ensure decent and affordable housing for the sector of the population earning between R2,000 and R7,500 per month. Both public

---

8   By contrast, the delivery of housing to upper income housing classes has been left almost entirely to the for-profit developers, operating via traditional market mechanisms and mortgage-secured financing. The resources located in the spaces of the traditional housing and property markets made possible the significant annual price increases in housing assets in the richer suburbs of Cape Town.

9   From 2009 the Treasury will provide municipalities with funding from the central fuel fund (cf. Hendler, 2008a).

10 In a cluster arrangement, houses would not be laid out within planned grids, but rather clustered closer to each other.

and private institutions (including non-profit companies) can manage and maintain such stock. Since 1994, a social housing sector has emerged in the South African housing market. To date, 15 non-profit SHIs affiliated with the National Association of Social Housing Organisations (NASHO) manage approximately 20,000 accommodation units (http://www.nasho.org.za). In the inner cities of Johannesburg and Durban, where hundreds of thousands of working class residents located themselves since the 1980s, rented accommodation, most of it privately provided, has had a positive impact on the quality of accommodation for many tenants. Private sector participation in the subsidised rental market in Gauteng has been severely limited by provincial inefficiencies and red tape related to the disbursement of the institutional subsidy (specially crafted to address the refurbishment of inner city buildings for residential rental purposes).[11] Across the country the non-profit SHIs have suffered from a range of problems that have blunted their effectiveness (cf. Support programme for Social Housing, 2005; Hendler, 2009). At the same time that inner city accommodation stock has improved, there have been removals of the unemployed squatting in buildings to the urban peripheral townships - this raises questions about the interpolation of social housing into the total spectrum of housing markets and the inclusionary/exclusionary process based on access and affordability to various housing typologies.

4.4 Communal/transitional housing

Communal and transitional housing is a form of social housing targeted at households earning less than R2,000 per month, including the unemployed. As referred to earlier, a very high number of those households that make up the housing backlog also lack meaningful income-generating work. This mechanism is potentially very relevant to contributing to a sustainable housing process in Cape Town, given the relatively high levels of unemployment in the informal settlements of the South-East.

'Communal' refers to the fact that the cooking and cleaning (ablution) facilities are all shared within a single building or project, and 'transitional' refers to the fact that accommodation is only provided for a limited period (say one year), after which an individual or household is expected to assume responsibility for seeking their own accommodation on the housing market, either through private market or social housing delivery mechanisms. Communal/transitional housing has been implemented in larger cities, where transformed inner city buildings have provided accommodation for the poor or destitute, although the extent of this is probably limited. A particular housing institution in the Johannesburg inner city[12] has developed a unique service that includes occupational training and/or counselling, to assist residents to find employment. The service also monitors their employment history and encourages sustainable income generation. The outcomes are however quite limited numerically compared with the need for employment and job creation. Nevertheless communal/transitional housing remains a powerful instrument – as demonstrated in the Johannesburg case – for directly addressing the income generating as well as the shelter needs of the destitute, and therefore offers a form of empowerment to resist the marginalisation of this fraction of the working poor.

4.5 Formalised home ownership: mortgaged property and gap housing

This is the established delivery mechanism for private, outright ownership. Commercial banks can fund this effectively, and private developers manage the process of acquiring and servicing the land, constructing the houses, and transferring of title to new owners. The problem is that there are usually no houses in the price range that is affordable for people in the above subsidy income category (i.e. between R7,500 and R12,000 per month), for whom there is a supply-side gap – in recent years there has also been a rapid increase in the value of all housing properties, including those in previously segregated black townships (which also experienced the housing 'boom'). One way of addressing this gap is to bring down entry level prices for home ownership through the Cape

---

11 I had direct experience of this in applying for and sourcing institutional subsidies for a major private sector rental company in Johannesburg between 1998 and 2004.

12 Madulomoho Housing, which together with MES Aksie, a faith-based social service group, provides a package of services, including occupational training, job placing, and monitoring.

Town and other municipalities rapidly releasing municipal land at significantly discounted prices - in some instances at no charge. In this way, municipalities can facilitate affordable home ownership through in-fill sites and/or greenfields. Municipalities usually resist doing this because they see their land as an asset that they are expected to realise the greatest economic value from, an approach that ultimately leads to them facilitating the emergence of gated, upper income communities.

Nevertheless, if implemented the provision of cheaper municipal land would be relevant to address the needs of many households in overcrowded formal housing units – and probably also some who reside in informal structures due to a lack of formal units available in the housing market rather than simply their own affordability constraints.

However, until this happens on a significant scale the pressure of ongoing housing shortages will continue to fuel overcrowding and informal settlement.

4.6 Private rental market

Private high-, medium- and low-density units can be brought onto the market for rental under normal market conditions.

The providers of private rented accommodation are private landowners who rent their units to tenants. In many cases in existing townships, these landlords live on the same properties and rent out structures in their backyards. What is required here is a flexible application of municipal by-laws so that health and safety can be ensured without negatively impacting on the affordable private rental market by increasing the costs of compliance. At the same time backyard dwellers are reportedly being exploited by some through the charging of exorbitant rentals – increasing the overall supply of affordable housing in advantageous locations would go a long way to addressing this problem (Community Organisation Resource Centre, 2009a).

4.7 Employer housing

In line with international trends by businesses to focus on core functions and out-source peripheral functions, businesses in South Africa have tended to avoid involvement in securing housing for their employees. At best, businesses tend to provide financial support in the form of guarantees, and sometimes even loans, but the trend is to provide a 'clean wage', i.e. a wage that assumes the employee can look after the financing of housing themselves.

In the absence of sufficient trade union power and a local state to support greater involvement of business in employee housing (through a developmental agenda in sustainable housing) it is an open question why businesses should continue to play a significant role in employee housing.

5. Analysing Cape Town's housing situation from a sustainability perspective

Since 1994, the geographical area of the City of Cape Town (CCT) has expanded by over 40 per cent (Boshoff, 2008). During the period 1994 to 2004, the traditional for-profit housing and property markets in Cape Town thrived, with annual capital growth exceeding returns of 25 per cent (Haskins, 2008). By 2004, over 95 per cent of Cape Town's poor households had access to basic services, even though a much higher proportion – 350,000 out of 800,000 – did not live in formal housing. This is illustrated graphically in Figure 1.

There are clear indications that the Cape Town Municipality has consistently pursued a pro-poor 'services for all' policy, which has been cross-subsidised by richer rate payers and business (see Chapter 2 in this volume).[13] However, this was done at the expense of investment in maintenance,

---

13 Cape Town's Municipality appears to have been noticeably successful in extending infrastructure services to the urban poor, despite limited impact on economic growth – this is directly opposite to the Johannesburg Municipality, which enabled higher economic growth rates, but limited infrastructure upgrading for its urban poor (Schmidt, 2008).

operations, and upgrading. The chickens are now coming home to roost: infrastructures start to collapse under the strain, resulting in what could have been avoidable infrastructure expenditures.

During this period, the housing market in the City developed along – and therefore consolidated – historical spatial patterns, with poorer communities living in both informal and subsidised formal structures, south-east of the City, in what were once the dormitory suburbs of *apartheid*. As Turok and Sinclair-Smith (2009) point out, people in the poorest areas have to travel considerable distances to work (40 per cent of the opportunities of which are located in the city centre) and city centre facilities. Spatial planning has done little, if anything, to stop this trend.

Housing market values have been realised within this spatial context. While the expansion of middle class suburbs has been driven by developers and land speculators, the primary aim of local government has been to focus on extending basic services to the urban poor. Both these strategies – while successful on their own terms – served to reproduce a dualism in the housing market, where the upper income and rich continued to live close to the city centre and its amenities and opportunities, while the working classes and poor lived in poorly served peripheral areas.

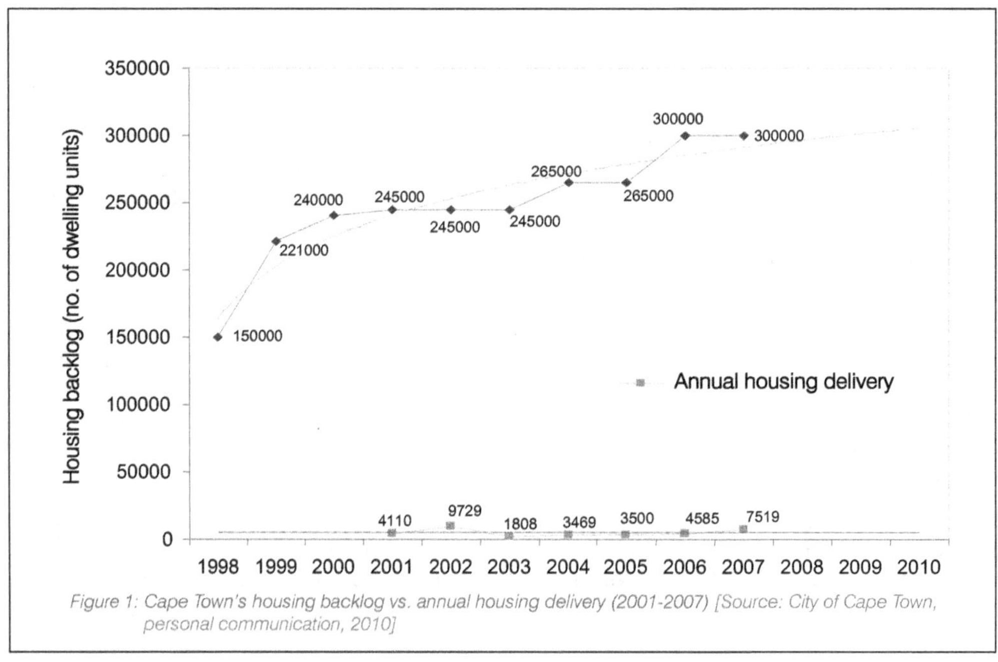

Figure 1: Cape Town's housing backlog vs. annual housing delivery (2001-2007) [Source: City of Cape Town, personal communication, 2010]

Generally, the subsidised and informal housing of the south-east does not trade in a functioning secondary market, meaning that many of these structures are 'dead' assets that offer no wealth creation opportunities to the poorer communities, but only owner-occupiers or tenants. Ecologically appropriate technologies have been installed only on a limited scale, mainly due to costs, but also due to ignorance and resistance to change. Since accessible subsidies for solar water heaters (SWHs) have become available (from 2008), there has been a great deal of work installing SWHs and energy efficient lighting in low cost housing developments.

In line with BNG, the Cape Town Integrated Development Plan (IDP) has identified seven strategic areas of focus, one of which is integrated human settlements. The core objectives for integrated human settlements include:

- Improvement and development of integrated human settlements by:
  - Transforming dormitory suburbs into areas which support a greater mix of land uses, offer a range of amenities, and have socially mixed facilities;
  - Putting in place policy and spatial planning frameworks that will facilitate the development of integrated human settlements; and
  - Developing and implementing an incremental housing programme.
- Delivery of housing opportunities by:
  - Developing new housing opportunities;
  - Increasing rental stock via social housing partnerships;
  - Redressing land ownership inequities by providing housing based on restitution claim settlements;
  - Facilitating gap housing programmes through partnerships with banks and private sector developers; and
  - Developing and maintaining zoned public open spaces, cemeteries, resorts, etc.

The Development Action Group (2006) has assessed the following lead and pilot projects of the Western Cape Department of Local Government and Housing (covering a wide range of issues, such as integrated approaches to development, mixed-use, informal settlement upgrading and sustainable housing, and categorised according to housing typology) in terms of whether or not they contribute towards sustainable human settlements:

- *Large-scale integrated projects:*
  - N2 Gateway Project – the largest of the pilot projects, involving the provision of 22,000 houses, a National pilot for the informal settlement upgrading programme.
  - Integrated Serviced Land Project (ISLP) – a major project involving the provision of 32,000 houses, which was initiated in 1994 and completed in 2005.
- *Mixed-use projects:*
  - Marconi Beam/Joe Slovo Park, Milnerton.
  - Westlake Village.
  - Ekupumleni, Philippi.
- *Informal settlement upgrading projects:*
  - Freedom Park, Tafelsig.
  - New Rest, which is part of the N2 Gateway Project (but the project has a long history that predates the N2 project).
  - Kosovo, Philippi.
  - Imizamo Yethu, Hout Bay.
- *Consolidation projects:*
  - Sinakho Ukuzenzela, Khayelitsha.
- *Hostel upgrading projects:*
  - Lukhanyo Co-op (formerly the Consol Hostel, Guguletu).
  - Ilinge Labahlali, Nyanga.
- *Gap market:*
  - Stock Road, Philippi (although technically not aimed at the gap market due to the institutional subsidy regulations of the time, the project has lessons for gap market housing), Royal Road, Maitland.

Most of the above projects are within the extensive informal settlements around Cape Town. This means that they should be assessed in terms of the extent to which they encouraged and progressed formalisation of housing units and the emergence of primary and secondary housing markets, as discussed in the second section of this paper. In addition, and linked to the criteria of formalisation and housing markets, they should also be assessed in terms of other BNG and Isidima objectives, namely upgrading of level of services and integration with the urban fabric.

There are many poor people in the Cape Town Municipal area who have constructed their own informal dwellings. Unemployment in the City rose from 13 per cent in 1997 to 18 per cent in 2007; the number of people in informal settlements rose from 23,000 families in 1993 to 115,000 families in 2005, and then increased to 117,000 families in 2007;[14] and households living below or marginally above household poverty line, rose from 25 per cent in 1996 to 38 per cent in 2005 (City of Cape Town, 2007; Haskins, 2008). The responses to this need in informal settlements have taken the form of attempts to upgrade their level of services and integrate them with facilities – here their location is critical with respect to existing urban facilities.

The ISLP and the N2 Gateway Project included the provision of facilities such as schools, clinics, and others. The availability of linked funding and cooperation between a range of stakeholders enabled integration of a range of programmes that covered a wide range of sectors, including health, child care, community safety, skills development, and support for small business. On a smaller scale, projects such as Freedom Park used a bottom-up approach to integration, based on an understanding of livelihoods within that specific community. The highest form of participation is a community-managed project, where decisions are taken by representatives of the community who are accountable to the community, such as characterised by Sinakho Ukuzenzela, Marconi Beam and Ekupumleni.[15]

Where the informal settlements were located is also an important factor influencing their formalisation and development into sustainable settlements for their inhabitants. The Development Action Group (2006) opined that in a small mono-centric town, the issue of location was fairly simple, but in a large polycentric metropolitan area such as Cape Town, location was a complex issue that depended on a wide range of factors (geophysical, proximity to social and economic opportunities, transport, etc.). It was clear, however, that some of the case study projects were better located than others in relation to existing facilities. The Marconi Beam/Joe Slovo Park project was assessed as the best located of the case study projects, being within walking distance of the Milnerton CBD and Montague Gardens industrial area, close to a major transport corridor (Koeberg Road), and a short journey by bus or taxi away from the Cape Town CBD. Both the Marconi Beam and Westlake projects showed that it was possible to have low income projects located adjacent to rapidly growing commercial/industrial areas with job opportunities. In the context of Cape Town, it could be argued that any project not in the poorly resourced south-east sector of the city could be regarded as relatively well-located. For example, Westlake (near Muizenberg) and Imizamo Yethu (in Hout Bay), although on the urban periphery, are located close to job opportunities. The Westlake case, however, showed that proximity to job opportunities did not necessarily mean that residents would be the beneficiaries of those job opportunities. Projects in or adjacent to old established townships, such as the Joe Slovo component of the N2 Gateway Project and New Rest, and the hostels upgrading projects in Guguletu and Nyanga (Lukhanyo and Ilinge Labahlali), could also be regarded as relatively well-located, as there were established social and economic networks, and reasonably adequate public transport linkages (compared to more recently developed areas, such as Delft South).

New Rest, Freedom Park, and Kosovo were assessed as examples of participatory integrated approaches to informal settlement upgrading. None of these projects were complete at the time of assessment, but they all involved intensive beneficiary participation processes and undertook major needs assessments (such as the participatory livelihoods assessment in Freedom Park). As a result, the upgrading processes, and the planned outcomes, responded to the real needs of beneficiaries and

---

14 However, there are over 300 000 households living in informal structures, which means that there are another 200 000 households living in informal structures but because these structures are on serviced sites largely owned by those who inhabit the structure, these households are excluded from the definition of informal settlement.

15 Freedom Park and Ilinge Labahlali are examples of very participatory projects that included participatory design processes, which involved workshops at which beneficiaries actively participated in the design process (for example, in the Freedom Park case, through formulating planning guidelines for the town planning consultant and then assessing the plans against the guidelines) (Development Action Group, 2006).

would help create a vibrant urban environment that was supportive of existing social and economic networks (*ibid.*).

According to the Development Action Group (*ibid.*), only three of the projects seriously challenged *apartheid* urban patterns, i.e. Marconi Beam/Joe Slovo Park, Westlake, and Imizamo Yethu. All three cases involved (or will involve) the upgrading of informal settlements that spontaneously developed in locations that were suitable for various reasons. The Marconi Beam project showed how 'not-in-my-backyard' (NIMBY) resistance could potentially be overcome through a (lengthy) process of negotiations in a negotiating forum involving beneficiaries, representatives of surrounding communities, and the Municipality. Two of the projects demonstrated that racial integration was possible – parts of the ISLP (e.g. Delft South) had both African and Coloured beneficiaries (as a result of a 50/50 breakdown between informal settlement residents of Guguletu/Nyanga and the waiting list of the City of Tygerberg), and the Cape Town Community Housing Company's Stock Road project also had both African and Coloured residents (as a result of beneficiaries being recruited from a city-wide waiting list based on savings records).

The interesting question would be regarding the extent of asset creation in these informal settlements. It seems safe to assume that while there is little, if any, buying and selling of structures, there probably is a rental market in the extra structures on the sites of these settlements. Therefore, these structures could be regarded as assets that yield future income streams to their owners.

5.2 Assessment of communal/transitional housing in Cape Town

It does not appear that anything along these lines has been applied in Cape Town yet.

5.3 Assessment of RDP housing in Cape Town

According to Haskins (2008), some 34,720 of these units were delivered in the City between 2001 and 2007, while the backlog expanded from 240,000 to 350,000 – indicating that the quantum provided was inadequate to meet a growing need. What would a sustainable approach, adapting the current RDP one, look like for Cape Town? Ultimately, we would need to know the extent to which innovative RDP projects – including modified design of structures – have been implemented appropriately in Cape Town. The impression from housing sites near Somerset West and Kraaifontein is that RDP housing in the Greater Cape Town area has been rolled out on the conventional scale and layout, rather than innovatively as suggested earlier, and that it is unlikely to have stimulated a secondary ownership market – although a backyard rental market is likely to have taken root, given the overall shortage of accommodation. This is significant from an asset creation point of view, because rental turns a structure into an asset that yields a cash stream. At the same time the conditions and rentals paid by the backyard tenants should be known to assess the impact on their lives.

5.4 Assessment of social housing in Cape Town

The first SHI in South Africa, the Citizens Housing League – now Communicare – is located in Cape Town and run as a non-profit company. The City has entered into partnerships with Communicare and also with the Social Housing Company (Sohco) (another non-profit company), in addition to the Cape Town Community Housing Company (CTCHC), in which the City used to be a 50 per cent shareholder with the National Housing Finance Corporation (NHFC) – the NHFC is currently the sole shareholder. The terms of these partnerships are that the City gives these three companies preferential access to Municipal land and institutional subsidies (approximately 2,000 for each entity) in return for their providing social housing units at scale.

To date, Communicare has delivered the most rental units, but only recently entered the market for black working class lessees. The company provides social rental housing for households earning between R1,500 and R7,500 per month, and presently holds 2,957 rental housing units, reportedly all in well located areas (www.communicare.co.za/pdfs/Tenant _Newsletter.pdf). Two thirds of Communicare's current rental stock is occupied by those with special needs (the elderly and/or people with disabilities). Besides developing housing, Communicare also provides management services and community development programmes for tenants and surrounding communities. Communicare's three area housing offices manage approximately 900 housing units each, home to some 7,000 persons. During 2010 the Drommedaris project will provide 239 units for families and those with special needs, and the Bothasig Gardens project will provide 120 social rental units. Communicare's Ruyterwacht neighbourhood regeneration project entails the upgrade of existing stock and building another 350 to 400 units.

Sohco recently entered the Cape Town market. Sohco's first social housing project site is in the Steenberg area of Cape Town. It reportedly has good transport links, and is close to social facilities and commercial nodes (http://www.sohco.co.za/capetown.asp). There is clear market demand in the area, and extensive civil construction and earthworks have already been completed on the site. It is anticipated that the site will yield approximately 400 to 450 units, which will be a mixture of one, two- and three-storey buildings and one, two- and three-bedroom flats. The other two sites that are currently being considered for allocation to Sohco are smaller infill sites in the City bowl area, one in the Bo-Kaap and one in Woodstock. Together, these sites will yield approximately 200 to 250 units. It is anticipated that the bulk of the units within these projects will be released on a rental basis, with small portions of instalment sale units.

CTCHC has been dogged by problems relating to rent defaulting on public housing stock and appears to have re-focused on the ownership gap market (www.ctchc.co.za/portfolio.asp).

There has been an innovative use of institutional subsidies in the informal settlement of Masiphumelele, where with the assistance of the City a social housing project[16] arose to address the needs of dwellers of informal structures, who were facing the ruin of their homes (and their lives) through fires that periodically ravaged their area. Through a grant from overseas donors the project was able to secure equity, and this was packaged with institutional subsidies to cover the required funding. The project did not however secure the required institutional subsidies without a protracted struggle against both the local ANC branch as well as the then Department of Local Government and Housing, which at its height involved a sit in at the Department's offices in Wale Street. This project articulated closely with a broader community initiative involving HIV-Aids support, and would be an interesting case study of the limits and possibilities for building autonomous community organisations through a housing struggle process.

The largest social rental project in Cape Town was the N2 Gateway project, which was initiated in 2005 with the aim of providing decent alternative accommodation to the thousands of people forced to live in what were seen as unsightly informal settlements alongside the N2 highway and into Crossroads. The project was dogged by controversy from its inception and caused a great deal of unhappiness among local small developers and builders, who felt that they were excluded from the process.[17] Later, the project aroused the ire of some surrounding communities in informal settlements, who feared that they would be moved to further locations and not get access to the new units. There were also protests and a rent boycott from the approximately 1,000 tenants to whom the units had been allocated, following significant problems with the quality of the structures – ostensibly the result of poor workmanship and negligent quality control.

The 2003 situational analysis, which formed the basis of the Cape Town Housing Plan referred to earlier, noted that of Cape Town's stock of dwellings, houses on separate stands, townhouses, and informal dwellings were in the main owned units. The analysis also noted that, while most of those residing in shacks owned their units, in general they did not enjoy security of tenure. The implication is that rental housing in the low income market is concentrated in public, municipal housing estates and backyard dwellings.

---

16 The Amakhaya ngoku housing project.

17 This viewpoint is based on the author's perceptions at the initial briefing of the project during January 2005.

The extent of the private rental market has not been documented, but it was noted that this predominates in informal backyard structures, which represent assets for the owners. Cape Town has large numbers of publicly-owned rental stock, which dates from the late-1960s and the 1970s, that is the era during which the state provided housing as part of the collective means of consumption. The reasons for the move away from state involvement and the attempts to privatise the housing should be sought in policies and programmes that were founded on the changing rhythms of economic accumulation, particularly in changes in the social relations of production and the social division of labour occasioned by new technologies and way of work. The record of public rental in Cape Town is not encouraging. There are many cases of inadequately maintained stock, rent defaulting, and the hijacking of units and entire blocks by gangs of criminals. Recent attempts to distance the City from these estates and outsource their management to the CTCHC have had mixed results, and political interference in the tenant management and rent collection functions of the business appears to have been a problem.[18] The precise role of the Municipality and the SHI partners would have to be looked at and thought through coherently if outsourcing of these critical property management functions is to succeed.

5.6 Assessment of the gap market in Cape Town

Communicare provides ownership housing for those with incomes between R5,500 and R15,000 (www.communicare.co.za/pdfs/Tenant_Newsletter.pdf.). A more recent CTCHC project, Royal Road in Maitland, has reportedly successfully targeted the gap market (www.ctchc.co.za/portfolio.asp). But overall, there appears to be a very limited gap market, suggesting the relatively small size of the class of people in the above income category.

5.7 Key challenges for affordable housing in Cape Town

It is apparent that much energy has been expended in addressing the need represented by the vast informal settlements that have sprung up mainly in the south-east, and some of these projects have made inroads into the core of urban facilities and work/life opportunities. Likewise, the social housing and private ownership projects have gained favourable access to facilities and opportunities for their residents, albeit on a limited scale when measured against total need. Efforts have been focused on upgrading and integration with facilities, and the development of housing assets has probably been only along the lines of informal rental yields, but there have been questions raised by some community groups about the exorbitant cost of this accommodation to the backyard tenants. Other typologies have contributed less output to the housing challenges of creating a diversified market. In general the following overall challenges have been noted by the CCT itself (2007):

- For the backlog to be effectively addressed, 20,000 housing units need to be delivered per year over the next 10 years;
- Available financial constructs will only deliver 7,500 housing units per year;
- The lack of qualified project managers and developers, and sufficient planning, as well as social and technical facilitators, has resulted in an actual delivery rate of 4,500 housing units per year;
- The City faces management and financial challenges in administering some 43,000 rental units; and
- Land invasions have taken place, which slow down effective housing delivery.

The above list of challenges suggests that the most pressing obstacles are limited financial resources and technical competence. However, underlying these obstacles is a fundamental determinant that has limited, and sometimes undermines, genuine attempts to address the backlog in affordable, decent, and well located accommodation for the majority of the City's citizens, namely a lack of concerted forward planning towards clear developmental goals for the City as a whole. The obvious manifestation of this lack of planning is that housing is delivered in locations that put their inhabitants at a severe disadvantage to capitalise on access to work and urban facilities.

---

18 This information was gleaned in 2005 during interviews by the author with the then Chief Executive Officer of the CTCHC.

Boshoff[19] (2010) points to the absence of a strong plan focused on the needs of ordinary people as the root of the problematic dualism in Cape Town's housing market. In support of this point, he refers to the success of the uncompromisingly driven *apartheid* spatial plans – success, that is, in its own terms, but with devastating impact on the majority of Cape Town's population. His point is that a strong plan, with a coherent logic, that links the interest of the ordinary citizenry to the possibilities of the space economy, will work for everyone. Instead, what has happened to planning in Cape Town – as well as other South African towns and cities[20] – is that plans have been based on assumptions about economic growth, where the conventional drivers of that growth are seen to be the private sector, rather than on the principle of what works for the poorest citizens. The result has been spatial frameworks that are the outcome of contestation by powerful economic and political interests in the motor transportation, property development and retail sectors, and which refract these interests rather than those of the poorer communities – many of whom are increasingly marginalised in the local space economy. The physical manifestation of this is the plethora of gated townhouse developments, golf course estates, and middle to up-market shopping malls that have arisen in the City and its environs between 1994 and 2010. Of course, this not how the process is represented by the groupings that benefit most in terms of accumulation opportunities, who prefer to see the latest developments in contemporary Cape Town as 'world class' (*ibid.*), and therefore as representing the best of urbanism and modernity to the citizenry.

5.8   Cape Town's unsustainable resource usage

From a sustainable resource use perspective, both past and present housing developments in Cape Town have been undertaken in an extremely unsustainable way. Both have been characterised by:

- Massive urban sprawl and the destruction of potentially productive land;
- Low numbers of housing units per kilometre of infrastructure line (energy, water, sanitation, storm water drainage, roads, and rail);[21]
- Rising levels of waste output; and
- Increasing levels of energy and material use.

Indeed, as the urban poor were located further and further outside of the city, so too did transport subsidies increase (Behrens & Wilkinson, 2003), thus increasing the impact of rapidly increasing oil prices (particularly during 2008) on the poor. Similarly, no provision was made for the fact that water and energy resources in the Western Cape are facing depletion.

Cape Town is facing the dual challenge of massively expanding the size of its formal housing stock to meet the needs of the poor, while simultaneously facing an increasing demand in the middle class markets.[22] However, Cape Town needs to both respond to these multiple demands and remain within the now fully accepted ecological limits with respect to energy, water, landfill space, sewage disposal, biodiversity, and food supplies which account for the great bulk of Cape Town's footprint (Gasson, 2002).

The challenge is to imagine a massive housing programme for Cape Town that would –

- comply with the recently adopted densification policy (which may be too moderate), as well as with the spatial configurations of an appropriate space economy;

---

19 Stephen Boshoff was Executive Director of Planning of the City of Cape Town during the period referred to in this article.

20 Similar trends in the actual role and function of town planning within six other Western Cape municipalities have emerged in the course of the Western Cape Government's Built Environment Support Programme (BESP), which assesses the Spatial Development Frameworks (SDFs) and Human Settlement Plans (HSPs) of these municipalities, and in which the author is a participating service provider.

21 Turok and Sinclair-Smith (n.d.) interestingly point out that residential densities rise from the city centre towards the south-east to peak in Khayelitsha.

22 This demand is reflected in property prices which, despite leveling off at the top end, have remained buoyant in the R300,000 to R700,000 categories, where returns remain at the 20 to 30 per cent level.

- utilise a range of new technologies and design features (enforced through bylaws) that massively reduce the average amount of energy, water, materials, and sinks (landfill space, air space for pollution, emissions, etc.) that each house needs to get built and operated over the life cycle; and
- boldly intervene in the agricultural food chain to localise food supplies, and thus create markets for urban agriculturalists and peri-urban small-scale farmers.

To meet these sustainability criteria this massive housing programme will have to be characterised by an overall slowdown in the velocity of the upper and middle income housing markets, as part of a generally 'cooling' macro-economy. This presupposes a radical change in macro-economic policy at national level and this will only happen under the impetus of a social force that has a political impact.

6. Transforming the political economy of housing

The persistence of spatial segregation and marginalisation, sixteen years after the advent of democracy in South Africa, should give pause for thought.

This paper has argued that the reason for the exclusion of the majority lies in the structure and power distribution of the South African state, through its organising the general interests of the dominant classes and its disorganising effect on the dominated classes. The impact of the state on the spatial configuration of housing is mediated through the various state administrations (at national, provincial and local level), as well as through intra-state departmental power conflicts and the power of struggles by the dominated classes themselves. To concretise these processes at a local level requires understanding the way in which powerful economic classes in the motor transport, property development, and retail sectors shape the framework of local spaces within which local economies function. The housing market forms an important and significant part of these local economies, providing a potential source of wealth creation for housing classes as well as forming the real and potential rates base for local governments. Through work with colleagues and clients, in a process separate from the Sustainability Institute's service to the City of Cape Town, and in framing municipal SDFs and HSPs over the past twenty-four months, I have crystallised the following question, which I think takes us close to the root of the inability – or unwillingness – of local municipalities to discharge their stated – and legislated – duties to the urban poor:

How does a local authority/municipality help to reproduce conditions that both enable housing markets and regulate these same markets in order to achieve socially equitable distribution of housing that over the medium term consumes less resources and produces less waste output than the rate at which these (resources) can be renewed and (waste) recycled?

Municipal officials dealing with housing grapple with the pressure from financiers and developers for new accumulation opportunities on the one hand, and the demand by communities for improved shelter and service delivery on the other hand. They have to try to get better housing and residential locations for the dominated classes, while simultaneously facilitating middle and up-market developments that shut out the poor, but at the same time increase the value of the municipal rates base.

Conventional wisdom is that the virtue of market-driven housing is that privately-owned housing – through exposing owners to both incentives and penalties – has shaped homeowner behaviour patterns that have maintained housing assets and added value to them, thereby laying the basis for the financing of local governments (via the tax base). Successful housing markets have however developed highly inequitable patterns of human settlement, the rectification of which requires transformation of the local space economies within which housing is embedded. In addition these markets account for an unsustainable carbon footprint – housing asset values are the security against which consumers in unsustainable growth economies borrow in order to fund economic activities. The growth is unsustainable, because there are limits to debt funding on a global scale, as the collapse of the global credit system during 2008 demonstrated.

The economic transformation required for sustaining the majority of South Africa's (and the world's) population in decent and affordable settlements is resisted by suburban housing classes, precisely because it threatens their tried and tested asset growth model: falling house prices are anathema to them, but this is precisely what is implied in locating decent, affordable, and integrated housing for the urban and rural poor. In addition, extensive retrofitting of existing houses and installation of appropriate technologies into newly-built ones in order to align them with sustainable carbon foot-printing will add further costs and act as an additional tax on already devaluing assets. The above theme could usefully contribute to explaining housing practice in Cape Town since 1994.

Given the contradiction referred to above, where is the motive social force likely to come from to drive the required transformation? The reactionary tendency of established housing classes, and the conquest of local public institutions and authorities by powerful private financial and development interests lead one to the conclusion that a new 'politics of the grassroots' is required, where urban social movements emerge to address the obstacles to transformation posed by a developer-local municipality-established homeowner classes complex. If urban social movements can emerge as independent organisations and grow grassroots, democratically-controlled structures, they represent a potential vehicle to fight for the formulation and implementation of the policies required to effect urban transformation and begin restructuring existing space economies.

The urban social movement in South Africa is at a nadir now compared to the heady days of the late-1970's and throughout the 1980's, a time when there was real potential for a transformative urban agenda. Partly this reflects the sucking off of the best leadership into the ANC governments and state administrative departments. At the same time, as noted earlier, there is widespread popular dissatisfaction with the state of existing and new dormitory townships and the lack of services there, that has led to the emergence of social protest across many towns and cities – from these, grassroots structures are emerging but the electorate generally still seems to view the ANC-as-government as the vehicle to deliver on their aspirations. This is manifested in the consistently high percentages with which they have re-elected successive ANC governments. There is therefore not yet a significantly powerful and widespread urban social movement and mass consciousness that can have the effects required on the government and state programmes.

Where does this leave us as practitioners of spatial frameworks and IHSPs? The well-known theorist of urban meaning, Manuel Castells, argued in 1983 that urban social movements were the drivers of changes in urban meaning and primary agents in creating new social effects (Castells, 1983). He argued that in this process, movements related to (and to some extent relied on) technically competent professionals, the media, and political parties. In post-*apartheid* SA, local municipalities have become a crucial actor that urban social movements and progressive operators should seek to influence and have an impact on their SDFs and HSPs, because they have the potential to refract the interests of the dominated classes and poorer communities through organised movements participating in – or impacting on – democratic elections. As Poulantzas (1980) emphasises, popular participation and impact will remain defensive until they start transforming the state, but that is another story.

Like Poulantzas, Castells stressed that urban social movements needed to remain independent in order to avoid becoming appendages of these operators' specific interests and agendas which are presented as the 'public' agenda. This point applies as much to a mass political party like the ANC as it does to teams of or individual professional operators. I am particularly concerned with our function and impact as professionals schooled in the notion that we are somehow above the social processes being played out in the municipalities and local space economies. Objectively we serve contradictory ends to the extent that class and other social power relationships are condensed and refracted in the state. I think that many of us would agree that much of our work over the past decade and a half has reinforced and reproduced the nice-sounding policies but that there is little to show on the ground for the sustainable human settlements that we have participated in the conceptualisation and planning of. Can we commit ourselves to a genuine class-based (or pro-poor) approach, which requires a political commitment to striving towards a City that works for all its citizens and not just the wealthier minority? If so, how do we make an impact given that we work to a brief from governmental clients, and are therefore restricted in influencing intra-state relations?

There are no easy answers to these questions. Two minimum guidelines for our practices could be usefully considered: first, that we do everything to enhance the emergence of and support of independent urban social movements; and, second, that we could advocate new ways of planning involving strong community *participation* (and not just *consultation*). One way to give effect to the latter would be to develop and transfer the *occupational skills* required to take up community demands in a holistic plan. In this regard we also need to be aware of the class position of the personnel in the social division of labour in the state.

7. Strategic funding options

In looking forward to a new politics of the urban, we can also discern spaces for interventions by social movements to begin having an impact on planning and housing, and through this, on urban meaning. In post-*apartheid* South Africa, these spaces appear in the notion of funding for neighbourhoods and projects through partnerships with the local authority, which have hitherto been taken up mainly by private sector developers and financiers.

7.1 Funding interventions

Linked to and underpinning the sustainability of housing delivery mechanisms is the crucial factor of housing finance – the state's emphasis on public-private-partnerships (PPP) has laid the basis for a strong private sector presence in the funding and also the implementation of projects in the fields of housing and also those of infrastructure and social services. In the housing field this was the basis for a strong private developer-driven RDP housing process, with all the inadequacies that we referred to earlier, and the effect of further marginalising and disorganising the dominated classes. (There have also been significant PPPs in the fields of toll roads and tourism). In short, the funding and financing of projects, many of which are supposed to benefit the poor, is not a mere technical process, effectively neutral to the outcome and its impact on the balance of class forces; it is woven into the power imbalances that exist between the contracting (private) parties and the beneficiaries. Nevertheless, for the foreseeable future many of the necessary financial and other resources to implement socially useful projects will come from the dominant classes, who will be looking for a return on their capital investment above all else. The challenge for the emerging organisations of the dominated classes is to understand the underlying class interests that drive the involvement of the parties as well as the specific implementing mechanisms, in order to wring out the best compromises from deal-making – these outcomes will also reflect the balance of power between the negotiating parties at each point in time. A short overview of the funds and finance available for various forms of housing delivery, as well as the public private relationships involved in implementing housing projects, is given. The following sections identify each of these mechanisms and how they might address the Cape Town housing shortage. They form the basis for thinking about new ways to deliver affordable and social housing.

The Municipality will need to attract a range of role players as key stakeholders in viable and sustainable housing initiatives:

- institutional investors (e.g. Sanlam or Old Mutual);
- banks;
- international donor and aid organisations;
- development finance institutions;
- local industry players;
- municipal guarantee instruments;
- packaged and prioritised infrastructure financing from organisations like the National Treasury, the Development Bank of Southern Africa (DBSA), and the municipal infrastructure grant (MIG); and
- National subsidy funds specifically targeted at housing.

The PPP approach is relatively novel in South Africa, where municipalities have traditionally regulated residential (and non-residential) development by the private sector in terms of planning bylaws and building regulations. Typically, they have not taken the initiative to shape the type of developments that would be appropriate and sustainable for the majority of the population living in their jurisdictions. The conventional approach is for each municipal department to vet a development application, in a linear way, which can take up to twenty-four months. Conventionally, departments operate as silos, often isolated from each other and often in competition. Because of these organisational realities, municipal silo organisations are themselves serious obstacles to sustainable development. Transforming silo organisational structures into flat hierarchies with a different emphasis is an example of transforming the state and the way the local state carries out its functions. This is unlikely to happen until the emergence of autonomous grassroots-based organisations with sufficient power to make this impact. There is therefore a time period that it will take for this to happen, depending on the nature of the issues and on the conditions in specific localities and amongst specific fractions of the dominated classes. For instance there already appears to be an autonomously organised shack dwellers coalition with national as well as international linkages.

Ideally, the municipalities should be driving development through a dedicated structure that includes officials from all key departments and relevant councillors, as well as community representatives and outside specialists. This range of stakeholders would write the broad development criteria that would be advertised through Requests for Proposals (RfPs) to the private and community sectors. The details of the development would form part of the private and community sector bids, which they would submit in order to compete for participation in the development.

Six funding and/or contracting mechanisms are in place for a municipality to use where appropriate to attract and enlist the services of the successful private sector bidders to fast track delivery of housing: services contracts, management contracts, leasing contracts, investment-linked contracts (or concessions), joint ventures (or corporatisation), and private ownership and operation (also known as build-operate-transfer [BoT]). A special contract could make provision for community-based organisations (CBOs) to participate in PPPs – this would be a unique contribution to the PPP practice as it has been carried out until now in South Africa, by creating a stake in development for authentic, grassroots-based organisations that can demonstrate that they represent a significant constituency.

A 'community ownership' option, which will form a critical component of a development coalition, rests on collective savings to generate individual and collective equity that could leverage further public and private funds. A pertinent example of this is the savings clubs approach of FEDUP. This option does not stand alone: it can interface with most of the other public and private relationships referred to above. 'Community ownership' means that participants contribute equity, which means they also take risk and therefore contribute to the protection of the asset and its continuous improvement. Furthermore, 'community ownership' builds bonds of solidarity and reciprocity in the community, resulting in a greater sense of community, which, in turn, often translates into more reliable repayments, higher levels of ongoing investment in individual and collective assets, and reduced levels of conflict, violence, and crime. The institutional arrangements can range from simple savings groups to mobilise community equity and investment, through to housing associations and even housing cooperatives.

Community partnerships, however, are likely to happen only on the basis of strong community advocacy, which presupposes the organisation of strong urban social movements.

8.   Conclusion

This paper has argued that Cape Town's housing crisis will not be resolved if existing housing and property markets are left intact. Although the BNG policy framework has created a range of flexible opportunities for intervention that are responsive to the wide range of different housing needs,

Cape Town's housing and property markets operate within a space economy that continues to concentrate housing for the urban poor within particular locales. This reflects the nature of the political economy of housing: in the city the municipal-ratepayer-developer axis has a definite interest in maintaining the status quo, which broadly aligns with the role of the state to reinforce the domination/subordination relationships between classes in favour of the bourgeoisie. Cape Town's Municipal Government is clearly the key role player when it comes to envisioning and planning a pro-poor alternative, but without a strong social movement to support such a vision, it is highly likely that property development interests will make sure that the *status quo* remains intact. These vested interests will continue to criticise Government for failing to resolve the housing crisis, but they will resist any attempts to mix low income housing into middle income suburbs or secure creative financing flows to supplement Government subsidies.

It is therefore proposed that progressive professionals interested in pro-poor reforms consider ways of debating and developing ideas about a different, transformed way of living in houses and being located in space with the leadership and grassroots of emerging urban social and protest movements, as well as with local municipal and provincial officials grappling with the contradictions of the political economy of housing. This conversation cannot however happen in a voluntarist way: professionals individually and collectively reflect the broader social division of labour of our society, and many of us will continue to perform functions for public sector clients that objectively serve the interests of the dominant classes without being perturbed. However, those of us who are conscious of and live the contradiction between our ideal of a pro-poor city and our objective functions in professional work, neither individually or collectively have the power to effect the significant political changes required to make sustainable human settlements not only possible but also likely to happen. Ours is a role of encouraging debate, planting ideas and being available to support the birth as well as the autonomy of emerging social movements. An empowered citizenry that is able to start thinking about a new way of being in the urban areas, formulating appropriate sustainable settlement programmes, and acting politically to achieve programmatic demands, is the precondition for institutional reforms that could over time change the exclusionary logic of the space economy, re-orient housing and property markets, support grassroots empowerment, and promote more sustainable use of resources.

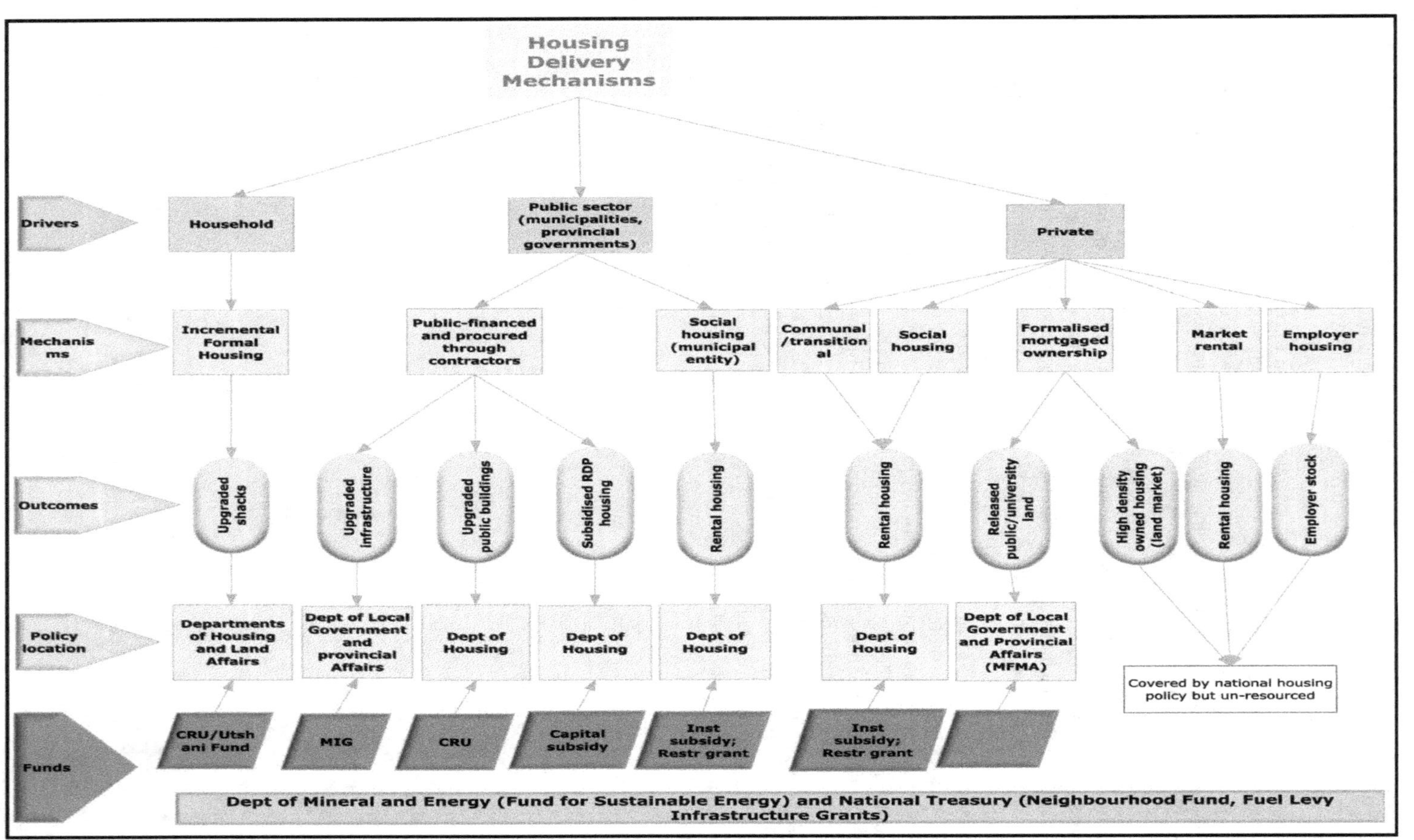

Housing Delivery Mechanisms
Drivers
Household
Public sector (municipalities, provincial governments)
Private
Mechanisms
Incremental Formal Housing
Public-financed and procured through contractors
Social housing (municipal entity)
Communal /transitional
Social housing
Formalised mortgaged ownership
Market rental
Employer housing
Outcomes
Upgraded shacks
Upgraded infrastructure
Upgraded public buildings
Subsidised RDP housing
Rental housing
Rental housing
Released public/university land
High density owned housing (land market)
Rental housing
Employer stock
Policy location
Departments of Housing and Land Affairs
Dept of Local Government and provincial Affairs
Dept of Housing
Dept of Housing
Dept of Housing
Dept of Housing
Dept of Local Government and Provincial Affairs (MFMA)
Covered by national housing policy but un-resourced
Funds
CRU/Utshani Fund
MIG
CRU
Capital subsidy
Inst subsidy; Restr grant
Inst subsidy; Restr grant
Dept of Mineral and Energy (Fund for Sustainable Energy) and National Treasury (Neighbourhood Fund, Fuel Levy Infrastructure Grants)

References

Behrens, R. & Wilkinson, P. 2003. *Metropolitan Transport Planning in Cape Town, South Africa: A Critical Assessment of Key Difficulties*. Working Paper No. 5. Faculty of Engineering and the Built Environment. Cape Town: University of Cape Town.

Bond, Patrick. 2009. 'Are those Planact's fingerprints?', in *Making Towns and Cities Work for People – Planact in South Africa 1985 to 2005*. Johannesburg: Planact.

Boshoff, S. 2008. *Cape Town 2030*. Lecture delivered on the Sustainable Cities Module, Sustainability Institute, May 2008, Stellenbosch.

Boshoff, S. (2010). The 'Cape Town 2030' Initiative: A Contested Vision For The Future. In Pieterse, E. (Ed.). *Counter Currents: Experiments in Sustainability in the Cape Town*. Auckland Park: Jacana-Media.

Castells, M. 1977. *The Urban Question – A Marxist Approach*. London: Edward Arnold.

Castells, M. 1983. *The City and the Grassroots*. Berkeley: University of California Press.

City of Cape Town. 2007. *5-Year Plan for Cape Town*. IDP document delivered to Council, 30 May 2007, Cape Town.

Community Organization Resource Centre, the Joe Slovo Blocking Project – A Diary of Events, March 2009 (electronic memo).

Community Organisation Resource Centre, 2009. *Backyard Fax - KZN Chapter of a National Shack Dwellers' Coalition Demands Control of Provincial Development Process by The Poor*.

Community Organisation Resource Centre, 2009a. *Cape Town's Backyard Dwellers*, unpublished memorandum.

Daly, H.E. 1996. *Beyond Growth: The Economics of Sustainable Development*. Boston: Beacon Press.

Development Action Group. 2006. *Western Cape Strategy for the Development of Sustainable Human Settlements*. Commissioned report for the Western Cape Department of Local Government and Housing. Cape Town: Development Action Group.

De Villiers Leach, S. Interview between Sarah De Villiers Leach and the Sustainability Institute, 9 September 2008, Cape Town.

FEDUP. 2009. Cape Towns Backyard Dwellers (electronic memo).

Fukuyama, F. 1989. 'The End of History?' Article published in *The National Interest*, and located on web address: http://www.wesjones.com/eoh.htm.

Gasson, B. 2002. *The Ecological Footprint of Cape Town: Unsustainable Resource Use and Planning Implications*. National Conference of the South African Planning Institution, 18-20 September 2002, Durban.

Giddens, A. 1981. *A Contemporary Critique of Historical Materialism*. Basingstoke and London: Macmillan Press.

Haskins, C. 2008. *State of Cape Town*. Presentation to SACN Renewable Energy Workshop with Mayoral Sub-Committee, June 2008, Cape Town.

Heinberg, R. 2007. Five Axioms of Sustainability. *Museletter*, No. 178, February 2007.

Hendler, P. 1986. *Capital Accumulation, the State and the Housing Question: The Private Allocation of Residences in African Townships on the Witwatersrand 1980-1985*, unpublished MA Dissertation. Johannesburg: University of the Witwatersrand.

Hendler, P. 1993. *Privatised Housing Delivery, Housing Markets and Housing Policy: Residential Land Development for Africans in the Pretoria/Witwatersrand/Vereeniging Region between 1975 and 1991*. Unpublished PhD Thesis. Johannesburg: University of the Witwatersrand.

Hendler, P. 2008a. Personal communication with Bernard Pick, National Treasury RSA, SA Cities Network Renewable Energy Conference, 13 May 2008, Spier Estate, Stellenbosch.

Hendler, P. 2008b. Personal communication with Jacus Pienaar regarding the Community Residential Units (CRU) subsidy, 2 June 2008.

Hendler, P. 2009. 'The relevance of NGOs to housing delivery', in Making Towns and Cities Work for People – Planact in South Africa 1985-2005, Johannesburg: Planact.

Hendler, P. 2010. Personal Communication with Gary Fahy of Project 90X2030.[23]

http://www.nasho.org.za, web address of the National Association of Social Housing Organisations (NASHO).

http://www.sohco.co.za/capetown.asp, web address of the Cape Town division of the Social Housing Company (SOHCO).

Irurah, D. 2004. Ecological Design Module at the Sustainability Institute, 31 May-5 June 2004, Cape Town.

Jackson, T. 2009. Prosperity without growth? Sustainable Development Commission, March. Located on web address: www.sd-commission.org.uk.

Khan, F. 2007. Towards 2010: Mumbo Jumbo, 2010 and the masses. *Cape Argus*, 25 April 2007. (Quoted in Boshoff [2010].)

Marx, K. 1977. *Capital: A Critique of Political Economy Vol 1*. London: Lawrence and Wishart.

Mbeki, M. 2009. *Architects of Poverty – Why African Capitalism Needs Changing*. Johannesburg: Picador Africa.

Meadows, D.H., Meadows, D.L. & Randers, J. 2004. *Limits to Growth: The 30-Year Update*. London: Earthscan.

Neocosmos, M. *From 'Foreign Natives' to 'Native Foreigners' – Explaining Xenophobia in Post-Apartheid South Africa*. Dakar: Codesria monograph Series.

Poulantzas, N. 1980. *State, Power, Socialism*. London: Verso.

SHF. 2006. *Community Residential Units (CRU) Policy Framework and Programme Guidelines*. November 2006 (unpublished memorandum).

Support Programme for Social Housing, 2005. Synopsis of the findings of the curative organisational diagnosis of social housing institutions, Programme Management Unit, Johannesburg (internal memorandum).

---

23 Project 90X2030 grew out of discussions during 2006 within the Goedgedacht Forum for Social Reflection. The project's main purpose is to challenge South Africans to change the way they live and the way they relate to the environment. Its vision is that South Africans from all sectors of society must do their bit to preserve the environment, committed to changing the way they live by 90% by the year 2030.

Sunday Times. 2009. The Khutsong Rebellion, 15 March.

Swilling, M. & De Wit, M. 2008. *Sustainable Urban Development in Cape Town: Planning for a Natural Resource-based Service Provision with a Systems Dynamics Model*. Paper presented at CNRS Colloquium on North-South Urban Planning, 24-25 January 2008, Paris.

Swilling, M. (forthcoming). 'Greening Public Value: The Sustainability Challenge', in Benington, J. and Moore, M. In *Search of Public Value: Beyond Private Choice*, London, Palgrave (forthcoming).

Terreblanche, S. 2002. *A History of Inequality in South Africa 1652 -2002*. Pietermaritzburg and Sandton: University of Natal Press and KMM Review Publishing Company.

Turner, J. 1972. 'Housing as a Verb', in Turner, J and Fichter, R *Freedom to Build: Dweller Control of the Housing Process*, New York: The Macmillan Company.

Turok, I. & Sinclair-Smith, K. 2009. Article in *The Cape Times*, 26 October 2009.

Western Cape Consortium. 2003. City Housing Plan. *Business Plan*, Volume 3, November 2003.

Western Cape Consortium. 2003. City Housing Plan, *Draft Situational Analysis*, Volume 1, May, 2003.

Western Cape Department of Local Government and Housing. 2007. *Western Cape Sustainable Human Settlement Strategy – The Road to Dignified Communities*. Cape Town: Department of Local Government and Housing.

Winkler, H. 2006. *Energy Policies for Sustainable Development in South Africa*. Cape Town: Energy Research Centre, University of Cape Town.

www.communicare.co.za/pdfs/Tenant_Newsletter.pdf, page on the web address of the Communicare Social Housing Company.

www.ctchc.co.za/portfolio.asp, page on web address of the Cape Town Community Housing Company (CTCHC).

In the Introductory chapter I argued that our imagination of Cape Town's futures will inevitably be shaped by the twin dynamics of the global polycrisis and the second urbanisation wave. We need to connect our 'sense of the city' with the kind of 'sustainability thinking' that will make it possible to see the potential for innovations in the challenges we face every day, from the smallest decisions to recycle our waste or replace a high energy light bulb with a low energy one, to the big decisions that will have to get made by those elected to govern regarding the bulk supplies of energy, water and food. After reading the chapters of this book it is my hope that you will never again see the city – and Cape Town in particular – in the same way. Although these chapters were written by people from different disciplines using different conceptual languages, they were all trying in one way or another to make this connection between a 'sense of the city' and 'sustainability'. To this extent this is a book that is more than just about Cape Town. It demonstrates the rich dimensions of the discussion about the complexities of city dynamics and takes a small step towards an analysis that makes it possible to go beyond the rather narrow confines of environmental impact assessments.

We need to be far more impatient than we are with the self-satisfying banalities about how bad things really are, especially if the underlying suggestion is that we can do nothing to change things. We need to be particularly frustrated with ecologists who deliver long talks about how the environment is being destroyed before concluding that if more people were aware of this destruction, things would change. How would things change, we must ask? What specifically must be done when? Believing in the mystical possibility that somehow we will figure it out if everyone was just a bit more aware is not good enough. Many people *are* aware, but what can they do if they are locked into urban systems that condition environmentally destructive behaviour? So what practical examples are there of different ways of living? The problem is that most ecologists cannot answer these questions. Nor, for that matter, do many social scientists have answers, not least because most ignore the ecological context of the socio-economic challenges.

Maybe what we really need are *patterns of thinking* informed by what Gibbons *et al.* have usefully called "socially robust" knowledge about future orientations at different scales (Gibbons *et al.* 1994). This, rather than prescriptions for a specific alternative is what the chapters of this book have offered. Although academic writing is often dismissed because it seems overly concerned with its own internal (often arcane) debates, the advantage of the academic discipline is that it maps out the complex overlapping trajectories of thought that shape the assumptions that lie at the centre of the (often partially understood) common sense ideas that condition everyday life. It is not possible to understand our world without understanding the language we use and the origins of the concepts that are embedded in the common sense ideas that get mobilised in everyday conversations, in the media, in politics and the markets all around us at all times. Academic writing should be the 'global positioning system' of a conceptually cluttered city.

Many University-based researchers complain about the large gap between their research and the knowledge sets that are used by decision-makers. Decision-makers, in turn, complain that researchers do not generate the kind of applied research that they need to tackle practical challenges. These are familiar refrains which tend to reinforce the stand-off. In my view, what gets missed in this stand-off is the more creative discussion about the kind of research that needs to get done. I strongly agree with a recent article by the ecological economist Robert Costanza who argued that we need to restore the "balance between synthesis and analysis". In his words:

"Science, as an activity, requires a balance between two quite dissimilar activities. One is analysis – the ability to break down a problem into its component parts and understand how they function. The second is synthesis – the ability to put the pieces back together in a creative way in order to solve the problems. In most of our current university research and education, these capabilities are not developed in a balanced, integrated way."

(Costanza 2009: 359)

Because the chapters of this book originated in a set of papers jointly commissioned by the CCT and the Sustainability Institute, they needed to be more than just analysis. Each chapter in different ways provides analysis *and a synthesis* that contributes to the kind of socially robust knowledge about future orientations that will be needed to inspire and support what will be a prolonged and difficult transition to a more sustainable city.

It may be useful to conclude by drawing out three cross-cutting themes from the chapters of this book. These are the way innovations can be spurred by resource limits, the significance of seeing flows rather than structures, and the role that networks play at different levels.[1]

1.   Resource limits and innovation

Resource limits are often misunderstood. For some this suggests that we are 'running out of resources', a view that is often articulated as part of a generalised critique of everything to reinforce the notion that we face cataclysmic and catastrophic collapses in the near future. The collapses are inflated to such proportions that all space for thinking what happens beyond them is obliterated. For others, resource constraints will not be a major problem because they will be reflected in price rises that, in turn, will trigger market responses that will bring on line a range of substitutions. An alternative approach is to see existing or looming resource limits as a set of necessary, but not sufficient, material conditions that can spur a set of innovations. If there are a set of networks in place that connect an appropriate set of capacities for innovation, then resource limits can give rise to innovations that could over time create the technologies, processes and systems that become the basis for new modes of production and consumption.

Innovations, however, do not necessarily make things better, they can also make things worse. For example, not so long ago biofuels were seen as the solution to problems caused by the emissions generated by fossil fuels until it was realised that the energy return on energy invested did not make sense and that land for food crops could be negatively affected.

All the chapters related in one way or another the challenge of resource limits and innovation. The photo essay in Chapter 7 captures most vividly the remarkable impact we have on the landscape as we assemble spatial plans, infrastructures, houses, lives, cultures and flows into urban settlements over very short spaces of time. In Chapter 2 the visual sense of rapid change was expressed in financial terms as resource constraints affect municipal budgets forcing policy makers to adopt policy frameworks that suggest that more sustainable ways of doing things are required. This basic argument is reinforced in Chapters 3, 4 and 5: if low density urban form and unsustainable consumption continues into the future, many potential social and ecological benefits will not be achieved. Chapters 3, 4 and 5 suggest alternative spatial and socio-economic arrangements that could be incorporated into policy frameworks and the way developers and their professional consultants start thinking about urban development in the future.

Chapters 6, 7, 9 and 10 address the most significant networked urban infrastructures that make it possible for the city to function and for people to live, work and play. The large bulk of any city's budget is spent on these infrastructures. If we see these networked infrastructures as conduits for a set of flows (as suggested in the next section), then it becomes possible to figure out ways of modifying these infrastructures in order to redirect the directionality (linear or circular), distribution (affordably for all users) and efficiency (more for less) of these flows. When these infrastructures were designed, there was no need to think about directionality and efficiency, and only after 1994 was it necessary to think about distribution. But resource constraints are already triggering technology and system innovations that are affecting directionality and efficiency.

Although Chapters 11, 12, 13 and 14 refer to changes that will depend on socio-spatial innovations discussed in Chapters 3-5 and flow innovation as suggested in Chapters 6-10, Chapters 11-14 touch on a set of cultural innovations that will be needed to redefine a set of rather intimate relationships that we have with nature, food production, built spaces and the class character of our neighbourhoods.

---

1   The discussion that follows draws extensively on Swilling, M. & Annecke, E. *Just Transitions: Explorations of Sustainability in an Unfair World*. Cape Town: Juta, forthcoming.

All these relationships are highly contested in South African cities: natural space is under pressure as demand for land escalates; local food production as both employer and provider of affordable food competes with the supermarketisation of food sales; buildings that can be built more sustainably will more than likely not get a certificate of approval from the National Home Building Registration Council (NHBRC) which means they will not be bonded by the banks; and there is now a strong multi-racial middle class that resists the incorporation of low-income housing into their neighbourhoods. Addressing these challenges will be primarily about system innovations coupled to regulatory interventions and market-based incentives.

2.   Flows and structures

When searching for connections between 'urban systems' and 'eco-systems' we need to complement the traditional concerns of urbanists with structure, levels, physical spaces, built forms, boundary lines, technical constraints, spatial functionality, the architectures of built forms and even the image of inter-locking systems and look in relational ways for *processes* which connect urban living, working and playing with the resources that flow into, through, and then out urban systems – what is referred to as the 'socioecological metabolism' of the city (Heynen, Kaika & Swyngedouw 2006, Guy, Marvin & Moss 2001). Cities are not fixed physical artefacts or historical subjects, nor are they simply spaces within which other things happen. Cities from this perspective are, pre-eminently, emergent outcomes of complex interactions between overlapping socio-political, cultural, institutional and technical networks that are, in turn, in a constant state of flux as vast sociometabolic flows of material resources, bodies, energy, cultural practices and information work their way through urban systems in ways that are simultaneously routinized, crisis-ridden, unpredictable and transformative.

Seeing cities this way means paying attention to the role that urban infrastructure networks play in the metabolism of the city – a theme addressed in various ways by most of the chapters. Here we could think of almost anything that conveys a flow of some sort: the roads, pavements, footpaths, electrical cables, sewage pipes, water lines, waste sites, fibre optic cables, servers, vehicles and trucks, telephone lines, food supply lines, canals, airways, shipping lines, railways, quarries, dams, aquifers, slopes and soils that the users of the city as a whole depend on for bringing resources, bodies, water, goods, nutrients, information, wastes and energy into the city, and getting them out transformed and re-combined as they are after use. But none of these are fixed artefacts that exist outside of everyday life: we take them for granted as we use them to go about our daily business, and they have been imagined, designed, negotiated, funded, constructed and managed by an incredible range of actors along the way, all of whom do what they do with particular preferred outcomes in mind that cannot be divorced from their socio-cultural backgrounds, value-preferences and interests they are aligned with. We are, as Latour argues, "*many participants* ... gathered in a *thing* to make it exist and to maintain its existence." (Latour 2004: 246)

Once we see flows and infrastructures as the embodiment of designs and actions, then we can begin to imagine what it will take to change these flows and infrastructures with a more sustainable future in mind.

3.   Networks and innovation

None of the changes that are suggested by all the chapters have any chance of materialising without innovations. Innovations, however, do not emerge simply because there is a material or logical need for them. Much will depend on what we mean by innovation and how seriously investments are made in the human, relational and institutional preconditions for really transformative innovations. The problem with the national innovations systems that have been promoted by many governments around the world over the past two decades is that they have been aimed almost exclusively at promoting economic growth with very little attention paid to the various dimensions of decoupling (cleaner production being an obvious major exception, plus sector-specific investments in innovation such as renewable energy in Germany). Running in parallel to this has been the critique of growth by ecological economists who have argued that growth is responsible for the environmental destruction that is now undermining traditional growth models. In other words, innovation is not in and of itself a good thing from a sustainable resource management perspective. Maybe the word itself is irredeemably tied to creating more rather than less stuff and should possibly be replaced by a new word such as, for example, *exnovation* so that the intention is clear from the start. What is obvious is that a new conception of innovation is required. Carlos Montalvo, a researcher at TNO in The Netherlands, has suggested that innovations for sustainability deserves to be further explored (Montalvo 2008).

Innovations for sustainability are specific interventions (at the technological, institutional and relational levels) that result in the dematerialisation of economies by increasing resource productivity and reversing environmental degradation. Such innovations will take to scale what is already emerging out of life-cycle analysis, material flow analysis, ecological design, environmental impact and fair trade thinking: long-life design instead of rapid obsolescence, zero-waste, cleaner production, product stewardship, re-manufacturing, factor 4/5 improvements in productivity, fair trade agreements, green taxation, and so on. As Montalvo puts it: "The challenge lies in replacing a large proportion of our current technological stock with new technologies underpinned by new science and applied knowledge that do not violate but accommodate the first and second laws of thermodynamics." (Montalvo 2008: 3)

Four key insights can be drawn from the innovation literature that are relevant for sustainable resource management (Lundvall 2007), namely:

- Innovations are different to inventions – an invention is when a new idea emerges for a new product or process, while an innovation is the synthesis of the idea with a complex set of financial and institutional arrangements to implement the new idea on a broader scale;
- Innovations are not random events, but are rather the function of specific incentives and investments;
- Innovations do not arise from single individuals or single firms, but rather from well networked economic agents working collaboratively with knowledge institutions (such as Universities) and in ways that are open, creative, problem-driven and connected to learning from practice;
- Innovations are not about building up stocks of knowledge capital (patented ideas) created for trade in the so-called "knowledge economy", but rather innovations are continuous learning processes that are responsive to the fact that in a highly complex globalized world, fixed bits of knowledge rapidly become obsolete – we live, therefore, in a learning economy, not a knowledge economy.

Innovation, however, is not simply about technological solutions (the so-called 'techno-fix' approach). We need to think of innovation as a process that manifests at three different levels, namely:

- technological innovations – specific techniques for managing/processing materials and energy (e.g. the steam engine, hydrogen fuel cell, micro-chip, or any process that achieves more with less);
- institutional innovations – for managing on a society-wide basis – or even globally – incentives, transaction costs, rents, benefit distribution, dispersal, contractual obligations, precautions, and individual obligations;
- relational innovations – for managing cooperation, social cohesion, solidarity, social learning and benefit sharing.

As already suggested, innovations emerge when a set of networked capacities exist that can both compete and cooperate to generate new ways of doing things. Cape Town may have a set of unique conditions for enhancing such networked capacities (Republic of South Africa. Western Cape Provincial Government 2007, Western Cape Provincial Government 2005), namely unlike Gauteng's cities Cape Town's economy is not dependent on resource extraction and energy intensive industries; it's economy is dominated by small- and medium-size enterprises; the average education levels are higher and unemployment levels lower than elsewhere in South Africa; it has the highest internet penetration per capita compared to other cities; it has four major universities all of which have substantial research and development capabilities (with two of them being the two leading research Universities in South Africa); and Cape Town is well known for being the home base of a large number of NGOs, consulting firms and businesses that are in one way or another engaged in work related to sustainability. This mix of conditions is arguably conducive for the kinds of networks that are needed for translating innovations for dealing with resource limits into major system and infrastructure investments that transform the relationship between the way the city works and the resource flows it depends on.

To conclude, the challenges Cape Town faces are not substantially different to those that cities face in most parts of the world, in particular in the developing world. All face the challenge of building more inclusive equitable cities that can continue to access, use, dispose and re-use the key resource flows that have been discussed in this book. Each city, however, can choose to face these challenges and invest in the innovations required, or a choice can be made to either deny the problem or delay the process of dealing with the challenge. There is sufficient evidence that across a range of fronts various public, private and civil society stakeholders in Cape Town are responding to the sustainability challenge by initiating innovations that could over time become the basis for how the city will work in future.

References

Costanza, R. 2009. Science and Ecological Economics: Integrating of the Study of Humans and the Rest of Nature. *Bulletin of Science, Technology and Society*, 29(5): 358-373.

Gibbons, M., Limoges, C., Nowotny, H., Schwartzman, S., Scott, P. & Trow, M. 1994. *The New Production of Knowledge: The Dynamics of Science and Research in Contemporary Societies*. London: Sage.

Guy, S., Marvin, S. & Moss, T. (Eds.). 2001. *Urban Infrastructure in Transition*. London: Earthscan.

Heynen, N., Kaika, M. & Swyngedouw, E. (Eds.). 2006. *In the Nature of Cities: Urban Political Ecology and the Politics of Urban Metabolism*. London and New York: Routledge.

Latour, B. 2004. "Why Has Critique Run out of Steam? From Matters of Fact to Matters of Concern". *Critical Inquiry*, 30: 225-248.

Lundvall, B.A. 2007. *National Innovation System: Analytical Focusing Device and Policy Learning Tool*. Osterund: Swedish Institute for Growth Policy Studies.

Montalvo, C. 2008. "Sustainable systems of innovation: How level is the playing field for developing and developing economies?", Paper presented at Globelics Conference, 23-24 September, Mexico City.

Republic of South Africa. Western Cape Provincial Government. 2007. *iKapa Provincial Growth and Develoment Strategy*. Cape Town: Department of the Premier, Western Cape Provincial Government.

Western Cape Provincial Government. 2005. *Micro-Economic Development Strategy Synthesis Report: Version 1*. Cape Town: Western Cape Provincial Government, Department of Economic Development.

Contributors

Roger Behrens is an Associate Professor in the UCT Department of Civil Engineering. He is Director of the Centre for Transport Studies, and of the African Centre of Excellence for Studies in Public and Non-motorised Transport. His current research activities relate to analysis of the dynamics and pace of changing travel bsehaviour, the regulation and integration of paratransit systems, and analysis of the use of transport systems by pedestrians.

Peta Brom completed her B. Consumer Sciences (Housing) in 2004 at the University of Stellenbosch. Thereafter she secured a position at Eco-Design Architects and in 2006 she joined MDL as their internal sustainability researcher and advised on sustainability issues for key projects such as the Public Transport Shared Services Centre and Cape Town Station 2010. In 2008 she completed her B. Phil Sustainable Development (Planning and Management) achieving cum laude from the Sustainability Institute. Since 2009 she has been an Environmental and Sustainable Design (ESD) Consultant at PJCarew Consulting.

Matthew Cullinan has a Master's in City and Regional Planning and a BA in Economics and Geographical and Environmental Science from the University of Cape Town. With over 15 years of experience, Matthew has worked in metropolitan spatial planning, urban renewal, eco-housing and sustainability strategies. More recently he has extended his work areas to include organisational and resource management systems.

Amy Davison studied Environmental Science and Social Anthropology at the University of Cape Town and holds an honours degree in Environmental Management. She works as the State of Environment and Sustainability Co-ordinator in the City of Cape Town's Environmental Resource Management Department. She is currently working on environmental policy implementation, with a focus on the targets that make up integrated metropolitan environmental policy as well as the City's Environmental Agenda 2009-2014.

Prof **Martin de Wit** is an economist focused on environmental and developmental problems. He is Director of his own consulting firm De Wit Sustainable Options (Pty) Ltd, an Associate to the University of Stellenbosch and Director of ASSET research, a section 21 company focused on building student capacity on the interface between economics and ecology.

Sally-Anne Engledow is an Associate of Jeffares & Green Engineering and Environmental Consulting Firm based in the Cape Town Office. She is the Manager of the Environmental Division with a particular focus on integrated waste management plans, environmental impact assessments (EIAs), cleaner production for manufacturing industries, environmental footprinting (including carbon, water and waste), environmental auditing and a variety of other environmental consulting areas. Her main focus and passion is to assist clients in terms of á systems approach to all projects to ultimately move to a more cyclical economy.

Mazibuko Jara has worked for more than 8 years in the non-governmental sector in South Africa in the areas of human rights advocacy, HIV/AIDS, media liaison, communications, public and community education, community development, local government transformation, local economic development, promotion of co-operatives, alternative economic transformation, financial sector transformation, social and economic justice, and land and agrarian reform. He currently works as a researcher, trainer and monitoring and evaluation specialist.

Kathryn Ewing is an urban designer and architect. **Nisa Mammon** is an urban and regional planner and Director of NM & Associates Planners and Designers. Their planning and design work focuses mainly on making the public spatial realm accessible to the majority of people to help create an inclusive city. Both Nisa and Kathryn are involved in research directly related to urban planning and design practice.

Gareth Haysom is programmes coordinator at the Sustainability Institute, involved in a number of the research and academic programmes. For the past three years Gareth has played an active role in the Agriculture programme, initiating the programme and then playing a role in various research components within the programme. Some of the core research areas are Urban Agriculture and Food Security.

Paul Hendler received his PhD from the University of the Witwatersrand (1993), on the privatisation of housing in the PWV region between 1975 and 1991. He is a specialist in planning and implementing sustainable human settlements, with a focus on the political economy of housing, the financing and implementation of social housing projects and capacity building through training for these functions. He manages Probitas, a company that provides these services to the public and private sectors.

Mokena Makeka is principal and founder of Makeka Design Lab. He sits on the World Economic Forum's Global Agenda Council for Design, is an external examiner at the Columbia University School of Architecture and lectures at the University of Cape Town.

Luke Metelerkamp holds a BA in Applied Design and has worked as a commercial photographer for the last four years. He is also in the process completing his MPhil in Sustainable Development Planning and Management at the University of Stellenbosch.

Barbara Nussbaum is a thought leader, visionary and a coach based in the Cape. She has recently co-authored *Personal Growth African Style* (Penguin 2010): a bold new book celebrating the promise of Africa's gift towards creating a more caring and sustainable world – communally expressed humanity.

Frank Spencer is the CEO of Emergent Energy, a Sustainable Energy solutions provider in Cape Town. He has a Masters degree in Engineering and Bachelor of Philosophy in Sustainable Development. Projects he has worked on have ranged from government policy, to designing and commissioning solar and wind renewable energy systems, to optimising energy efficiency in the commercial and agricultural sectors.

Professor **Mark Swilling** is Programme Director of the Sustainable Development Programme in the School of Public Leadership, Stellenbosch University, and Academic Director of the Sustainability Institute. He is responsible for the coordination of a Masters and Doctoral programme in sustainable development and conducts research on ways to make urban systems more sustainable.

Lisa Thompson-Smeddle is the Programme Coordinator for the Sustainable Neighbourhoods Programme at the Sustainability Institute. She manages applied research, sustainable human settlements and infrastructural projects and works as a consultant in the field of sustainable development.

Associate Professor **Peter Wilkinson** teaches in the City and Regional Planning and Transport Studies programmes at the University of Cape Town. His research interests, undertaken through the Centre for Transport Studies at UCT, focus on the governance and transformation of urban public transport systems.

Dr **Kevin Winter** is an academic in the Environmental and Geographical Science Department, University of Cape Town, and a lead researcher in the interdisciplinary Urban Water Management Research group at UCT. His research activities focus on the management of grey water in informal settlements, sustainable urban drainage systems, and impacts of pollution on freshwater systems and water resources at a catchment scale.